The Biology of Coral Reefs

THE BIOLOGY OF HABITATS SERIES

This attractive series of concise, affordable texts provides an integrated overview of the design, physiology, and ecology of the biota in a given habitat, set in the context of the physical environment. Each book describes practical aspects of working within the habitat, detailing the sorts of studies which are possible. Management and conservation issues are also included. The series is intended for naturalists, students studying biological or environmental science, those beginning independent research, and professional biologists embarking on research in a new habitat.

The Biology of Coral Reefs

SECOND EDITION

Charles R. C. Sheppard
School of Life Sciences
University of Warwick
UK

Simon K. Davy
School of Biological Sciences
Victoria University of Wellington
New Zealand

Graham M. Pilling
SPC
Oceanic Fisheries Programme
New Caledonia

Nicholas A. J. Graham
Lancaster Environment Centre
Lancaster University
UK

OXFORD
UNIVERSITY PRESS

OXFORD
UNIVERSITY PRESS

Great Clarendon Street, Oxford, OX2 6DP,
United Kingdom

Oxford University Press is a department of the University of Oxford.
It furthers the University's objective of excellence in research, scholarship,
and education by publishing worldwide. Oxford is a registered trade mark of
Oxford University Press in the UK and in certain other countries

© Charles Sheppard, Simon Davy, Graham Pilling, and Nicholas Graham 2018

The moral rights of the authors have been asserted

Second Edition published in 2018

Published in the United States of America by Oxford University Press
198 Madison Avenue, New York, NY 10016, United States of America

British Library Cataloguing in Publication Data
Data available

Library of Congress Control Number: 2017959221

ISBN 978–0–19–878734–1(hbk.)
ISBN 978–0–19–878735–8(pbk.)

DOI 10.1093/oso/9780198787341.001.0001

Printed and bound by
CPI Group (UK) Ltd, Croydon, CR0 4YY

Links to third party websites are provided by Oxford in good faith and
for information only. Oxford disclaims any responsibility for the materials
contained in any third party website referenced in this work.

Preface to second edition

In the 8 years since this book first appeared, tropical coral reefs have deteriorated markedly in almost all parts of the world. At the same time, there has been a huge increase in scientists' understanding of coral reefs, of how they function and of the myriad reasons for their decline. It was felt that, given both our increase in understanding, coupled with the remorseless decline of this habitat, a revision and update of this book would be valuable in informing readers about both the new findings and the increasingly precarious state of this major ecosystem.

Differences captured in this edition are essentially:

- An update specifically of the molecular details of the symbiotic relationship between coral host and their captive algae. This symbiosis is one of the keys to a healthy and functioning coral reef. Much work has been done on this in the past several years. New techniques are identifying the ways in which the crucial symbioses work, and what the more important aspects are of their vulnerability and susceptibility to warming water due to climate change.
- A review of some of the huge volume of work done on reef fish in the past few years. For this there is an additional author, and we focus particularly on trophic structure on reefs.
- An update of environmental issues affecting reefs. These are not different in essence, more a difference in magnitude. Reefs are declining at rates estimated variously between 1.0% and 2.5% per year, and in some areas at rates considerably greater. This has mean the near-elimination of this richest of marine habitats in several areas, as well as their decline across most of the tropical world.

A pertinent example of this last point: The first edition of this book had an image on the cover of a reef at 10 m depth in Salomon Atoll in the central Indian Ocean. It is as remote a reef, and one as free from pollution, fishing and other disturbances, as one can find, and the photo was taken in 2006. This is reproduced as Plate 1 in this edition. The figures below are (left) the same image, accompanied by (right) an image of exactly the same site taken in 2017, showing its corals to be about 95% dead. The coral mortality event of 2016–17 was widespread, profound and of extreme importance to the few dozen countries and millions of people living alongside them.

Illustrated by about 150 figures, this volume summarizes both the present knowledge of reef functioning and the risks they now face. These risks endanger their continued ability to support their diverse ecosystem as well as reducing their direct and substantial contribution to human needs and welfare. Reefs have been called the 'canary in the cage'—the habitat which is the first to suffer under environmental and climate change, and they provide a warning to us of the importance of these changes. This is certainly becoming clear today, and the question has perhaps shifted to one of what we can do or will do about arresting their decline.

C.R.C.S.

The same site on the ocean-facing reefs of Salomon Atoll, Indian Ocean, at 10 m depth. Left: 2006, with very high cover of diverse corals, especially table corals. Right: 2017, showing less than 5% live coral with much broken and tumbled dead coral skeletons. *Photo:* Anne Sheppard.

Contents

Abbreviations

Ω_{arag}	Aragonite saturation state
ABH	Adaptive bleaching hypothesis
CCA	Crustose coralline algae
CFB	Cytophaga–Flavobacteria–Bacteroidetes
CMT	Customary marine tenure
COO	Compatible organic osmolyte
DGGE	Denaturing gradient gel electrophoresis
DIN	Dissolved inorganic nitrogen
DIP	Dissolved inorganic phosphorus
DOC	Dissolved organic carbon
DOM	Dissolved organic matter
DON	Dissolved organic nitrogen
EIA	Environmental impact assessment
HRF	Host release factor
Ma	Million years ago
MPA	Marine protected area
NTZ	No take zone
PAR	Photosynthetically active radiation
PICT	Pacific Island countries and territories
PON	Particulate organic nitrogen
PSU	Photosynthetic unit
ROS	Reactive oxygen species
RuBisCO	ribulose-1,5-bisphosphate carboxylase
SRP	Soluble reactive phosphorus
SSC	Suspended sediment concentration
VLP	Virus-like particle

List of boxes

1 Coral reefs

Biodiverse and productive tropical ecosystems

1.1 Introduction

Coral reefs are the iconic ecosystems of tropical seas (Figure 1.1). Several entire nations are made from them, and many others contain extensive reef systems that support large portions of their populations. Rich in species and highly productive, coral reefs are the main source of protein for many millions of people in tropical countries, as well as providing shoreline protection for even more. From the more wealthy countries which do not have reefs of their own, thousands of visitors descend on countries that do, and images of species-rich coral reefs are contained in countless travel and natural history magazines, and beam out on many television channels. For the most part, this is because of their beauty, and because of the attraction felt for an unfamiliar environment. For some, the allegedly dangerous forms of life on reefs, like sharks, offer a frisson of excitement too.

Reefs are economically important to most of the countries that contain them. Many countries depend on reef-dependent tourism for substantial proportions of their total revenue or foreign earnings, and this can be one-third or even two-thirds of foreign earnings in countries like the Seychelles and Maldives, for example. Much of this comes from tourists wishing to dive on the reefs, but it also comes from those seaside holidays that are based around white beaches made from the coral sand derived from natural processes of growth and erosion that takes place just offshore.

Tropical coral reefs are one of Earth's critically important life support systems, and the way they work is the subject of this book. Reefs are one of those few ecosystems that make their own substrate. They are made of limestone

The Biology of Coral Reefs. Second Edition. Charles Sheppard, Simon Davy, Graham Pilling, and Nicholas Graham, Oxford University Press (2018). © Charles Sheppard, Simon Davy, Graham Pilling, and Nicholas Graham (2018). DOI 10.1093/oso/9780198787341.001.0001

Figure 1.1   The richest part of any coral reef lies below the depths where waves break, where light is abundant for photosynthesis, sedimentation is low, the salinity is near that of the open ocean (about 31–34 ppt) and the temperature is about 20–29 °C.

derived from several different but interconnected sources. Much comes directly from the skeletons of corals themselves and from other marine groups; when these organisms die, a myriad of other species erode these skeletons away, producing rubble and sand. At the same time, several other, less well understood, processes make limestone 'cement', whose deposition consolidates the skeletal rubble and sand, binding it into a more durable reef substrate. The result is growth of the reef, its extension upwards to the low-tide level and to seaward, in constant opposition to various forms of erosion that tend to reduce that growth. Reef rock, therefore, is biogenic, meaning it is made by biological processes. A reef makes its own substrate in a way which, in a healthy reef, is constantly balanced but which is, overall, growing and accreting more than it is eroding. But limestone is a soft rock and, were it not for its continual growth and replacement, it otherwise would not last long in the shallow seas.

Coral reefs, which are the focus of this book, are ecosystems restricted to the tropics, for reasons explained later. There are many coral species that do grow in deep, cold waters, but these are, for the most part, small, slow-growing forms; there are a very few that do form hugely important reefs in deep waters, and these are the subject of 'Cold-water corals' in Section 1.7.

The energy input that shallow, tropical reefs receive from the sun is high. More species of life occur on them than in any other marine habitat. In addition, more kinds of life at the higher taxonomic levels of order and phylum occur on them than in any other terrestrial habitat. More kinds of feeding, more different methods of reproduction, of growth, of predation, of symbiosis and of locomotion are seen on a small area of reef than in any other ecosystem on land or sea. Their fascination for biologists, therefore, has been immense and this has increased exponentially since the intro-duction about 70 years ago of scuba equipment, which permits the close examination of reefs.

Before much direct observation was possible, those parts of reefs that lay beneath the shallowest expanses and were accessible by foot were known as the *Mare Incogitum*. Reefs were known more for the hazards they presented to ships' captains than for their biological wealth. Yet, some hints of the rich-ness of their natural history gradually became available:

> I have already had occasion to speak, in the course of my travels, of the astonish-ing mats of works formed by marine infects; namely the immense banks of coral bordering, and almost filling up [the sea]. The reader may therefore conceive with himself what a variety of madrepores and millepores are to be met within these seas.   (Niebuhr 1792)

The inability to explore beneath the surface was still evident 50 years later:

> There are secrets on the surface as well as within the bosom of the ocean, which lie shrouded from human observation and research ... Where there is mys-tery there will always be interest, and the greater the one, the more intense the other.   (Wellstead 1840)

Then, Darwin opened up the subject of geology and development of reefs. Nowhere is biology and geology so intertwined as is the case with reefs, and he ends:

> I may be permitted to hope, that the conclusions derived from the study of coral-formations, originally attempted merely to explain their peculiar forms, may be thought worthy of the attention of geologists.   (Darwin 1842)

Darwin not only addressed reef development but, by implication, challenged the contemporary view of the unchanging nature of huge geological struc-tures which seem immovable and constant in terms of human lifespans. Spe-cies evolve, as he explained at great length in a later book, but he showed that these massive geological features upon which entire human communities lived were not permanent after all, but also evolved, grew and were reduced again by weather and by subsidence of the very seabed itself. Reefs had existed for aeons, and some were clearly built by other, now extinct groups of species (see 'Ancient reefs'). The obvious impermanence of these apparently solid features was unsettling to many people.

Ancient reefs

Over geological time, several different groups of organisms have developed reefs, which, while widely separated taxonomically, have shared an ability to extract calcium carbonate from the water and deposit it in solid, skeletal form. Sometimes this has been in the

Figure 1. Neoproterzoic reef built by stromatolites, in Nama Basin, Namibia. **Figure 2.** Late Devonian reef complex, Canning Basin, Western Australia. Stromatoporoids were important ancient reef builders in this region.

Source: Photos by Dr Rachel Wood.

Continued

Ancient reefs *(Continued)*

crystalline form called aragonite (as is the case with modern corals), or as calcite or a form called high-magnesium calcite.

Stromatolites formed abundant reefs in pre-Cambrian times, and living stromatolites can still be seen, notably in Western Australia where high salinities and temperatures in bays have prevented grazing of the biofilms and especially cyanobacteria that are responsible for them. These organisms trap and deposit sedimentary grains which are now seen in several places as exposed, fossil reefs (Figure 1). Some forms of stony algae were also amongst the first reef builders, dating back to Cambrian times and continuing today, where they develop highly wave-resistant ridges along seaward sides of modern coral reefs. In the late Cambrian, archaeocyathids, sponge-like animals, formed reefs for a few tens of millions of years before becoming extinct. Brachiopods were next to build reefs; these were shelled animals which still exist today in much more modest densities.

During the long period between the Ordovician and Permian periods, from about 500 to 225 Ma, bryozoans, stromatoporoids, tabulate corals, rugose corals and sponges all formed reefs. The two corals mentioned were cnidarians, although not related to today's corals. Bryozoans were, and still are, primitive colonial animals of tiny size, and stromatoporoids were a kind of primitive sponge. Most of these older reef-building groups disappeared in the great extinction at the end of the Permian period. These were followed in the Triassic and then Jurassic periods by reefs of rudists, a group of giant bivalve molluscs, now also extinct, and also by today's stony corals.

All of these reefs built by extinct groups can be seen in various geological structures, usually well inland, commonly in the form of low hills, cliffs and other large outcrops of limestone. Reasons for the successive extinctions are not well known, although several of the major reef-building groups became extinct during the five well-documented major global extinctions. Postulated causes include meteorite impact and intense volcanic activity, the latter causing the pH of ocean water to fall. This issue of lowered alkalinity is undergoing considerable research now in connection with the present, measured drop in oceanic pH caused by solution of rising amounts of CO_2 in the atmosphere. In the past, when reef formation ceased at these extinction boundaries, it has taken hundreds of thousands, or even millions of years, for reef building to begin again (Veron 2007). In the same manner as today's reefs, each ancient reef type supported a high biodiversity of other forms of life.

Prof. Charles Sheppard, University of Warwick, UK.

Still, the biology of reefs and the mechanisms by which life built them remained very poorly known. For a start, people could not really see much of them. Coral colonies dredged up had limestone skeletons that were similar to the material making fossil reefs and similar also to the sand on beaches behind modern reefs. But the mechanisms of construction were unclear. People saw and marvelled at huge schools of fish and at numbers of turtles scarcely imaginable today. They could see vague shapes below calm waters but, as late as the nineteenth century, it was still argued what exactly corals were. A number of now famous naturalists found them fascinating:

> As humming-birds sport around the plants of the tropics, so also small fishes, scarcely an inch in length and never growing larger, but resplendent with gold, silver, purple and azure, sport around the flower-like corals. (Ehrenberg 1834)

But these biologists were still limited by accessibility, and some found it necessary to explain their undignified ways of working, at a time when the dress code for tropical fieldwork still insisted on the wearing of respectable clothing, as explained in this jubilant note by a Red Sea naturalist of great renown:

> Nowhere can one contemplate the life of the corals, and what belongs to it, more quietly and comfortably than here, although he has to lie on his belly—a trifling matter for the naturalist—and hold his magnifying-glass at the point of his nose above a coral bush. (Klunzinger 1878)

Yet, although naturalists were still able to look only at shallow reef life at low tide, progress in the reefs' natural history was clearly being made. Biology was heavily taxonomic during that period of early reef science, when, in fact, naming the various products of God's creation was seen as a very noble calling. Describing and illustrating was paramount and highly skilled in the nineteenth century (Figure 1.2). So, there might have been some fun being poked in the following quote:

Figure 1.2   Early drawings of corals.

Source: From Goldsmiths Animated Nature of 1852, based on material from the expedition of Baron Cuvier.

There are below the waters charming genii who are eager to marry human beings, though to be sure, only when the latter have mortified themselves for months previously with unsalted bread and water, so as to give their flesh and blood a half-ethereal character. The naturalist, however, cannot allow himself to be allured.   (Klunzinger 1878)

Geology, meanwhile, was progressing faster, and several theories were developing around the formation of reefs, some competing with and some adding to that of Darwin. Geology is more important to the study of reefs than it is to the study of many other marine systems because of the very fact that the corals make their own substrate: the biology makes the geology, and that geology then supports the next generation of biology, as it were. At about this time, the fact that sea level had fluctuated substantially over historical time was becoming understood, and it was being concluded that this was enormously influential with respect to reefs too. Growth and erosion of reefs therefore took place at different elevations at different times. Daly's (1910) 'glacial control' theory of reefs was influential and added to (or conflicted with, according to view) that of Darwin. With this theory, the huge volumes of water that periodically became removed from the global ocean to become locked up in ice caps would cause successive deep immersion and then exposure of the limestone mountains that were being made by corals wherever shallow waters existed. Measurement of past sea levels was difficult, but it was increasingly realized that understanding past sea levels was critical to any explanation of coral reefs (see 'Coral microatolls and sea level').

Wave erosion at the contemporary sea level was one thing but, added to that, it was also known that rain with its mildly acidic pH would also etch limestone that was exposed to the atmosphere. Accordingly, various modifications to ideas of reef formation were derived, not least by Purdy (1974) later on, observing that even depressions the size of atoll lagoons could be formed from such solution by slightly acidic rain falling on exposed limestone.

Progress in understanding came in steps, with the next step sometimes being delayed while scientific corners were fought, lost or held, or sometimes after apparently competing explanations were reconciled and seen to be complementary sides of the same coin. Reefs, and especially the life on them, were more complicated than first thought. The same result or the same reef shape, for example, could arise from different causes, or at least they could derive from different variants of the same cause. Both growth and erosion took place, but at different rates in different places according to geography, latitude, exposure to cyclones, rainfall, and so on. Darwin's underlying scheme of reef development was shown, by boreholes drilled approximately a century later, to be absolutely correct, but more recent and local changes greatly added to or influenced the result, and affected the present shapes of coral reefs seen today.

Coral microatolls and sea level

Coral microatolls are individual colonies of coral which grow mainly on tropical reef flats. They are known as microatolls because of their small size and because their shape resembles an atoll in that it has a circular rim with a depressed central section. Many species form microatolls but massive corals such as *Porites* are the most common. The colony grows upwards with normal morphology until its growth is inhibited by the air–water interface around the low-tide level. After the coral colony reaches this elevation, the surface of the coral may die and erode, but the microatoll continues to grow outwards, because the polyps on the sides of the colony remain permanently submerged. This creates their characteristic disc shape (Figure), which can reach up to several metres in diameter. This contrasts with the more usual domed shape of the species, which occurs when the colony's upward growth is not constrained by tidal exposure.

Coral microatolls are widely used to reconstruct Holocene sea level changes in tropical locations because of the close correlation between their upper surfaces when alive with the air–sea interface around the lowest tide level. This makes them precise fixed indicators of past sea level changes on coral reef systems, especially in areas of low-tidal range. However, care is needed when interpreting the low-tide level: although microatolls may occur in settings where there is a free connection to the open sea at low tide, they also grow in protected moated settings where water levels are ponded back and held above

Figure. A fossil coral microatoll on the reef flat at Middle Island, Central Great Barrier Reef coastline, North Queensland, Australia. Professor David Hopley standing on it gives scale. This particular specimen dates from the mid-Holocene sea level highstand and shows the characteristic disc shape formed by the limiting level of tidal exposure during the reef's growth. A specimen of this size will have lived for many hundreds of years.

Source: Photo by Dr Roland Gehrels.

Continued

Coral microatolls and sea level (*Continued*)

the open-sea low-tide level by geomorphological features such as rubble ridges. In the Cocos Keeling Islands, Indian Ocean, a detailed study of over 280 individual living coral microatolls revealed that open-ocean microatolls typically form midway between Mean Low Water Spring Tides and Mean Low Water Neap Tides, while moated colonies may grow above Mean Low Water Neap Tides (Smithers and Woodroffe 2000). The height of living coral in unmoated colonies varied in the Cocos Keeling Islands by a maximum of 6 cm (Smithers and Woodroffe 2000). This study shows that fossil unmoated microatolls preserved on reef flats can be very precise proxy sea level indicators.

Sea level change reconstructions

In many tropical locations, the relative sea level fell slightly through the mid-late Holocene, and fossil coral microatolls are often preserved on reef flats above their present growth position. Where a series of fossil microatolls are preserved in a vertical sequence in a single location, tight constraints can be placed on the position of former sea levels through time using their levelled altitudes and radiogenic methods for age determination. Relative sea level fall over the past 6,000 years on the Great Barrier Reef coastline of North Queensland, Australia, is revealed by a large body of coral microatolls in the region, now stranded above their present growth position (Chappell 1983).

At a much smaller scale, individual long-lived microatolls also preserve a filtered record of recent water-level changes within the annual growth bands preserved in their upper-surface morphology. In tectonically stable mid-ocean locations, they are ideal indicators of sea level changes over the last century, especially important in areas where there are no long tide-gauge records. If the height of each annual growth band, seen under UV light, is measured and the age of the band calculated, high-resolution sea level reconstructions from individual microatolls can be made (e.g. see Smithers and Woodroffe 2001). However, at the individual microatoll scale, these organisms only record a dampened sea level record, because coral growth rates limit their response to rapid sea level rise and therefore they do not record small magnitude and frequency fluctuations.

Potential problems with using coral microatolls as sea level indicators

Although coral microatolls are precise proxy sea level indicators in tropical locations, they have some problems and potential limitations. Using them to create continuous Holocene sea level histories depends on the presence of a staircase of preserved specimens in a local area. When using a fossil microatoll, it is also important to ascertain whether it was freely connected to the open ocean when it was alive. It is clearly important here to assess the geomorphological setting within which the coral grew before using fossil specimens to reconstruct sea level changes. Similarly, the upper surface of the specimen may have eroded, or the whole specimen may have moved in relation to its substrate after death. The most robust sea level reconstructions in tropical locations utilize a suite of evidence from different proxy and direct indicators, including fixed intertidal shells, buried sediments, tide-gauge records and coral microatolls to produce a full picture of the changing sea level through time.

Dr Sarah Woodroffe, University of Durham, UK.

As for other branches of science, the past few years have seen enormous increases in the study and understanding of reefs, and this has come just at the time that reefs are changing most markedly in response to human-caused stresses. The possibility of increased study, of course, followed on from the development of scuba equipment, which permits scientists to actually see reefs and their species directly and in close-up. This was combined with improvements in taxonomy and in the wide range of instrumentation available today, which can view optical properties of reefs from orbiting sensors and permit measurements of numerous key chemical, physical and biological processes. Methods were developed to study not just each species one by one, but the important interactions between different species, and between those species and their environment.

The heart of a coral reef's biology lies in the well-lit waters just below the breaking waves. The area which so enthralled Klunzinger and other naturalists of past centuries is indeed rich in life, but it is now recognized as being just the shallowest, sun-baked or rain-soaked outpost of a much more bewildering kaleidoscope of life, merely one that most diving scientists pass by on their way to examine the richest habitats on Earth.

This book focuses on several key groups, particularly those that are most important in the functioning of the reef, and on processes. It cannot describe in detail each of the many groups of animals and plants that inhabit the reef, and has to include many only briefly. Indeed, many species, perhaps even most, have yet to be properly identified and studied, and every year new insights come from unexpected directions or from research on species that may have been previously neglected. The reef is enormously complex, and scientists try to distil from the whole structure the essential features, in order to understand the working of the reef and, ultimately, how to identify those key aspects which are now under greatest threat, in order to inform those responsible for their management.

1.2 Areas and distributions of reefs

The total area of coral reefs and their substantial contributions to global biodiversity and productivity make them important to the world and to human society. Mapping of reefs and exact calculations of total area have produced differing estimates, partly because of differing methods of measurement, but rough estimates are available (Table 1.1). The boundaries of reefs are defined by the kind of survey method used, such as satellite imagery (see 'Remote sensing on coral reefs' in Chapter 9) or depth sounding, and may, in any case, not relate very closely to the full ecological extent of coral communities on the raised limestone structure. Backreef areas, which are ecologically integral to the hard substrate of the forereef slope and reef crests, may be mainly sand and thus included or not, as the investigator requires. Spalding et al. (2001) estimate that coral reefs cover 284,300 km^2, with 91% of this area in the Indo-Pacific region. Costanza et al. (1997) used a much higher value (920,000 km^2)

Table 1.1 Areas of coral reefs in different parts of the world.

Rank	Country	Area (km²)	Per Cent of World Total
1	Indonesia	51,020	17.95
2	Australia	48,960	17.22
3	Philippines	25,060	8.81
4	France, including territories	14,280	5.02
5	Papua New Guinea	13,840	4.87
6	Fiji	10,020	3.52
7	Maldives	8,920	3.14
8	Saudi Arabia	6,660	2.34
9	Marshall Islands	6,110	2.15
10	India	5,790	2.04
11	Solomon Islands	5,750	2.02
12	United Kingdom and territories	5,510	1.94
13	Micronesia, Federated States	4,340	1.53
14	Vanuatu	4,110	1.45
15	Egypt	3,800	1.34
16	USA and territories	3,770	1.33
17	Malaysia	3,600	1.27
18	Tanzania	3,580	1.26
19	Eritrea	3,260	1.15
20	Bahamas	3,150	1.11
21	Cuba	3,020	1.06
22	Kiribati	2,940	1.03
23	Japan	2,900	1.02
24	Sudan	2,720	0.96

Note: Estimates are difficult to carry out and depend on several factors, not least of which is the definition of how much of the integral habitat around and behind a reef is included. Thus, other estimates for individual countries exist, usually showing more reef area, but the list shows all in a consistent manner, and the ranking is probably unchanged whatever measure is used. A total of 80 countries have reefs; this lists only those 24 with approximately 1% or more of the world total.

Source: This list of areas is from <http://coral.unep.ch/atlaspr.htm>, which accompanies Spalding, M.D., Ravilious, C. and Green, E.P. (2001). World Atlas of Coral Reefs. University of California Press.

when they produced estimates of dollar values of reefs, and many estimates in limited areas likewise suggest that somewhere between half a million and nearly a million square kilometres is a reasonable estimate of ecological extent. The differences arise not only because of different methods of measurement, but also because of different ideas of how much associated habitat constitutes part of a 'coral reef'.

Reefs are tropical, bounded very roughly by the tropics of Cancer and Capricorn, and, because they need light, they are limited to areas where there is shallow water (Figure 1.3). Controls on their distribution are described later, but are such that they have strict requirements of temperature, salinity

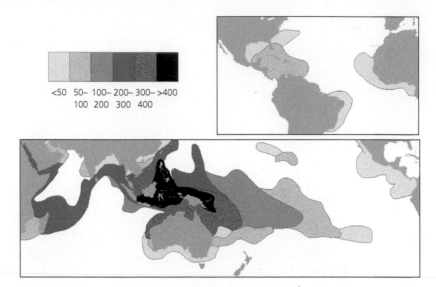

Figure 1.3   Map of the distribution of coral reefs, including species diversity of reef-building corals.

and suitably stable and illuminated substrate. Where warm currents flow poleward, reefs extend further outside the topics. Likewise, cold-water incursions into tropical areas restrict reef development; such incursions may be from currents or from upwelling systems. The most northerly reefs of the Indian Ocean and its peripheral seas are those near Suez, at the north of the Red Sea, at nearly 30° N. In the Pacific, the northern extent lies in the southern Japanese islands at almost 31° N. In the Atlantic, the most northerly reefs are those of Bermuda, at nearly 32°30′ N. These northerly extremes are habitable due to local conditions. In the case of the Red Sea, the water is warmed by both the currents within the Red Sea and its partial enclosure; in the other two cases, warm northerly flowing currents extend the area of warm water northward. In the southern hemisphere, southward-flowing warm currents allow reefs to grow as far as South Africa, at 28°30′ S, while, around Australia, they extend as far as 31° S at Lord Howe Island, whose 6 km of reefs are the southernmost reefs in the world. In Western Australia, good reefs are found at about 29° S at the Houtman Abrolhos Islands, although some fairly extensive coral communities extend further south, including to Rottnest Island near Perth. In the Atlantic, Brazil supports reefs to little more than 18° S, some extensive ones being very newly discovered, while, on the African side of the Atlantic, true reefs have not been found at all to date, although some coral communities occur on rocky substrates. The suppression of coral and reef growth on this eastern Atlantic is caused by low salinity from several major rivers draining this wet part of Africa, and by heavy sedimentation in some areas also. Outside the region of heavy rainfall, cold currents and upwellings suppress coral and reef growth.

1.3 Biodiversity on coral reefs

At the level of phyla, the diversity of coral reefs exceeds that of any other habitat on Earth by a large margin (Porter and Tougas 2001). There are currently (by some estimates) 34 animal phyla, of which 32 occur on reefs (the other 2 are found in moist forest and deep sea, respectively). As well as having a high biodiversity, reefs are very productive, supplying protein in tropical countries for many millions of people, many of whom survive on less than US$1 per day, or its equivalent. Many of the poorer countries on Earth tend to have some of the richest reefs.

Estimates of the number of species of animals and plants on reefs range widely from 600,000 to over 9 million worldwide (Knowlton 2001). The true diversity of reefs remains unknown and perhaps only 10% of species have been described. Some patterns across broad biogeographic scales are clear (Table 1.2). For instance, it is widely recognized that the greatest diversity of species and genera occurs in the Indo-West Pacific, which extends from East Africa to Hawaii and Easter Island in the East-Central Pacific. This vast area encompasses half of the tropics and the majority of the world's coral reefs, and it is home to at least ten times more coral and fish species than are found around Eastern Pacific sites such as the Galapagos Islands (Briggs 1999). Within the Indo-West Pacific, the highest species diversity is centred

Table 1.2 Regional patterns of species diversity on coral reefs and in related ecosystems (i.e. seagrass meadows and mangrove forests), highlighting the biodiversity 'hotspot' of the Indo-West Pacific region.

Taxonomic group	Indo-West Pacific	Eastern Pacific	Western Atlantic	Eastern Atlantic
Stony corals	719	34	62	
Soft corals	690+	0	6	
Sponges	244		117	
Gastropods				
Cowries	178	24	6	9
Cone shells	316	30	57	22
Bivalve molluscs	2,000	564	378	427
Crustaceans				
Mantis shrimps	249	50	77	30
True shrimps	91	28	41*	
Echinoderms	1,200	208	148	
Fish	4,000	650	1,400	450
Seagrasses†	34	7	9	2
Mangroves	59	13	11	7

Note: The diversity of sponges is for genera rather than species. No value indicates data unavailable rather than no presence.
*Shrimp species for the whole Atlantic.
†Seagrass values include species with warm-temperate distributions.
Source: Data from Spalding,M.D., Ravilious, C. and Green, E.P. (2001). *World Atlas of Coral Reefs*. University of California Press; *and references therein.*

on the Indonesian–Philippine region (Gaston 2003) in an area known as the 'Coral Triangle'. By comparison with the Indo-Pacific, the number of species on reefs of the entire Atlantic, the heart of which (in terms of reefs) is the Wider Caribbean Region, is just 10%–20% of that in the most species-rich regions of the Pacific (Karlson and Cornell 1998).

Longitudinally, reef distribution in all three tropical oceans is concentrated along their western sides. In part this is because of the distribution of islands, but reef distribution is also controlled by salinity, dissolved nutrients and temperature. Several coasts experience upwellings, for all or part of each year. Upwelling water is characterized by lower temperature and rich nutrients, both of which generate some of the world's richest fisheries areas, but which inhibit coral growth. The western coasts of North and South America are examples of areas where rich pelagic fisheries exist but where coral reef development is restricted to only a relatively few islands. This is the case for western Africa also. In the Indian Ocean, a very strong seasonal upwelling occurs off the coasts of Arabia and North Africa (see Figure 3.1 and Chapter 3) which inhibits reef development along those shores, giving rise instead to communities of macroalgae, amongst which are scattered coral colonies. The western side of Australia differs somewhat from this pattern, because of warm currents which flow southward along that coast from Indonesia.

The marked difference in diversity between Indo-Pacific and Atlantic reefs has arisen as a result of major tectonic, geological and climatic events. First, as tectonic forces caused the continents to move and as sea levels lowered, the Tethys Sea—the tropical sea that once connected the Indian and Atlantic Oceans—eventually closed in the late Miocene (>5.2 million years ago (Ma)). This led to the waters of the Indo-Pacific becoming separated from those of the Eastern Pacific and the Western Atlantic. In consequence, the reef biota in these two regions diverged and developed their own distinct characteristics. However, about 3.0–3.5 Ma, the Isthmus of Panama also closed, thus separating the Eastern Pacific reefs from those of the Western Atlantic. It is thought that the biota of both the Eastern Pacific and the Western Atlantic regions was more diverse initially, but that glaciation during the Pliocene/Pleistocene removed most of the coral reefs from the Pacific coast of the Americas. The Eastern Pacific has subsequently been partly recolonized by reef organisms from the Indo-Pacific, but the huge expanse of open ocean between the Indo-Pacific coasts and the Eastern Pacific coasts (the 'East Pacific Barrier'), along with largely westward-flowing ocean currents in that region, has meant that diversity in the Eastern Pacific continues to be low, even though a number of endemic species have evolved there. The low diversity of Atlantic reefs also came about as a result of the Pliocene/Pleistocene glaciation as well as subsequent glacial activity. The barrier imposed by the Isthmus of Panama has meant that this region's biota is now very different from that of the Pacific, at the species level at least. Today, the Atlantic and Indo-Pacific have no reef-building coral species and almost no soft corals in common. A much more detailed account of the tectonic, geological and climatic events that have led to the distinct diversity patterns of the Pacific and Atlantic reefs is given by Veron (1995).

The pattern of high diversity in the Coral Triangle of South East Asia, and the decline in diversity with increasing distance from it, has long been recognized (Stehli and Wells 1971). However, the patterns of currents, salinity and nutrients in the oceans greatly modify the distributions of reefs and of the species on them both latitudinally and longitudinally. The general pattern of declining diversity is seen on land and in the sea, across almost all groups of living organisms. Foremost hypotheses to explain the pattern in the sea are (1) high diversity results from the South East Asian region being a centre of overlap or accumulation, acting as a 'catch bag' for the westward-flowing currents of the Pacific; (2) high diversity in the South East Asian region results from it being a centre of survival that has provided refuges over geological time from the extinction seen in more peripheral locations and hence more opportunity for speciation; and (3) reduced diversity decreases away from the South East Asian region as a result of distance and dispersal patterns from this diversity 'hotspot' (Veron 1995; Palumbi 1997). However, all of these are ultimately dependent on the huge area of shallow-water habitat available for colonization by coral reefs in the South East Asian region. In their study of fish and coral distributions across the Indo-Pacific, Bellwood et al. (2012) calculated that large-scale patterns in habitat area and, in particular, the area of suitable habitat within 600 km of a particular study site constituted by far the major determinant of diversity. Longitude alone (i.e. distance from the diversity 'hotspot') was far less important, and reef type (i.e. offshore versus continental) had little influence either. Of note, despite the loss of diversity away from the South East Asian region, there was a degree of constancy in the composition of key fish and coral taxa across the Indo-Pacific. For example, 6%–22% of fish species were always damselfish (pomacentrids), and 4%–28% were groupers (serranids); 14%–43% of coral species were always acroporids, and 7%–16% were always poritids. All of the ranges are wide but, together, they hint at an underlying pattern.

There is also a marked decline in the diversity of most, but not all, taxonomic groups along a latitudinal gradient from the equator towards higher latitude reefs. This is seen very clearly in the decline in diversity of coral species along the Great Barrier Reef (GBR), from 324 and 343 in the northern and central regions, respectively, to 244 in the southern region, while just 87 coral species occur at Lord Howe Island, at 32° S (Veron 1993; Harriott and Banks 2002; also see Table 1.2). Similarly, just 21 species of corals are known from Bermuda (32° N) in the North Atlantic, compared with two or three times more in the main parts of the Caribbean.

There are a number of possible reasons why species diversity is greatest in the tropics and falls off towards the poles. These include (1) the relatively great age of tropical biotas, providing more time for species diversification; (2) more rapid rates of species diversification in the tropics than at higher latitudes; (3) greater environmental stability in the tropics than at higher latitudes, providing more opportunity for niche specialization; and (4) more area within each isothermal belt in the tropics than occurs at higher latitudes, and hence more potential habitat for species. Added to this is the simple fact that energy input (from the sun) per unit area of Earth's surface is greatest in the tropics and declines substantially towards the poles (because the Earth's surface is increasingly 'slanted' in relation to the sun's incident rays towards to poles). A 'surplus' of incoming radiation over outgoing or reflected radiation occurs between the tropics and about 35° north or south of them, poleward of which there is a 'deficit', in which outgoing radiation exceeds that striking the surface. The greater energy input in the tropics may underlie a greater speciation, along with greater productivity in general. Thus, the species diversity of a particular taxonomic group declines as the environment becomes more 'marginal' at high latitudes.

In the case of reef corals, species diversity drops off towards the latitudinal margins of coral reef development as a result of several abiotic factors, not only reduced temperature. Under marginal conditions, coral species present must not only be able to tolerate the prevailing abiotic conditions but also be able to grow fast enough to withstand competition, in particular overgrowth by macroalgae.

Set within these broadscale latitudinal and longitudinal variations, there are also more localized patterns resulting from smaller-scale abiotic environmental characteristics. Thus, even within this tropical zone, there are many areas where particular coral species struggle to grow and where reefs develop poorly or not at all. For example, a relatively low diversity is seen on reefs at locations made suboptimal by factors such as increased dissolved nutrients derived from upwelling systems such as those of the Arabian Sea (see Figure 3.1). Salinity and temperature extremes also constrain corals; while over 400 species of corals have been recorded around the diversity hotspot of South East Asia (Veron 1993) only about 60 species are known from the high-latitude but relatively hot and saline Persian Gulf. There, low temperatures in the winter are as limiting for corals as are high temperatures in the summer months. A location's remoteness in the sense of its position relative to the prevailing currents which flow towards it can also influence local diversity, because of the extent to which larvae and other propagules can successfully be transported to that location from source populations. For example, just 19 coral species are known from the coast of Brazil, where oceanographic isolation from the Caribbean has resulted both from the west- and northward flow of the South Equatorial Current, and from the massive barrier caused by freshwater and sediment discharges of the Amazon and Orinoco rivers. These barriers have led to low diversity but a high degree of endemism in the Brazilian region's coral fauna (Spalding et al. 2001). Along the West African coast too, sedimentation as well as upwelling means that there is very little coral growth and almost no reef development at all (see Chapter 3 for more details).

1.4 Coral communities and reef growth

There are, thus, major geographical patterns in the distribution of both corals and reefs. But the diversity of corals, and the associated invertebrates and fish on those reefs, have a fairly weak relationship with the actual size of a reef—with its area or thickness. Some of the largest and most solid reef structures of all, for example many Pacific atolls and all Atlantic reefs, are constructed by a relatively small number of species (Figure 1.4). In contrast, in several parts of the Indo-Pacific where diversity is very high, reef development may remain extremely weak. Thus, high diversity does not necessarily equate with good or strong reef construction. In some parts of the highly diverse South East Asian region, corals may be unable to create biogenic reef at all. There, typically, reefal communities exist on rocks of other, perhaps igneous origin, from which the corals readily detach after death, with no build-up at all of limestone reef structure (Figure 1.5). The distinction between corals which develop into coral communities without reef growth, and true reef growth, was first made long ago (Wainwright 1965; Hopley 1982). Coral communities which are not building reefs range from some which closely resemble those seen on true reefs, to being scattered colonies coexisting amongst brown or green algae. The substrate in such cases may be nearly obscured by corals so that the whole may have the appearance of a

Figure 1.4   Caribbean reefs have a relatively low number of reef-building coral species, yet the reefs and limestone banks they have produced are enormous (British Virgin Islands) (See Plate 2).

Figure 1.5   Corals growing on igneous rock in a very high diversity coral area (Malaysia). Top: *Acropora*. Bottom: *Platygyra*. Despite the profuse and diverse corals, no reef development occurs here for unknown reasons (possibly high dissolved nutrient loads) and, when colonies die, they readily detach from the underlying rock.

true coral reef. The associated benthic and fish fauna will also be those of a coral reef, but the surface corals are attached to a wide range of older rocks and may not develop any true reef at all. Reef development is not simply a case of 'corals growing on old corals'.

Another exception to the usual kind of reef growth, and an important one in many locations, is the 'mono-specific' reef, or one which is nearly mono-specific. In these, just one coral species occurs (or at least is heavily dominant). Large areas of many Indo-Pacific lagoonal areas may be dominated by just one species of *Acropora*, for example, while branching *Porites* or *Madracis* species may do the same in some Caribbean reef areas. In turn, where there is upwelling in Oman, many hectares may be covered by one species of *Montipora*, or by *Pocillopora* (Figure 1.6).

1.4.1 The fate of coral limestone

Most coral skeleton is eroded and reduced to sand and silt after the death of the living coral. Much of this is exported from the area completely, perhaps ending up in very deep water or deposited onto land, forming white coral sand beaches. But, on a reef where true reef growth takes place, much

Figure 1.6 *Pocillopora verrucosa* forms mono-specific reefs in parts of the Daymaniyat Islands off the coast of Muscat, Oman. The reef which is packed with this coral extends from near the surface to just 8 m deep, at which point the unconsolidated edge is visible.

becomes trapped in crevices in the rock, and subsequently becomes stabilized by overgrowths of algae or other encrusting life. Chemical changes also take place due to bacterial metabolism, so that the result is a reconsolidation of the sediment into hard limestone rock. Failure of any of these processes means that, while corals may grow profusely, reef growth may not necessarily occur. In such cases, a 'decoupling' of reef growth from coral growth takes place (Sheppard 1988). This is most commonly seen in high latitudes, usually because of a combination of increased nutrient levels in the water, lower temperatures and lowered alkalinity, so that the mechanisms of reef growth are inhibited, even though potentially reef-building corals continue further poleward for a few hundred kilometres. In all reef areas, a considerable amount of the limestone is worked and reworked by small organisms, including microbial life, and eventually becomes redissolved.

1.5 Types of reefs

The three classical types of reef are the fringing reef, the barrier reef and the atoll. The way in which one evolves to the next has been described and illustrated in detail and colour several hundred times since Darwin's original description (Figure 1.7), so the following is only a very brief summary. When corals fringe an island, they develop a fringing reef; when the island slowly subsides, coral reef growth keeps up with the surface of the sea. The line of reefs becomes increasingly distant from the shore and forms a barrier reef, commonly named as such when a navigable channel develops between the reef and the shoreline. Finally, when island subsidence is complete, the remaining ring of reefs enclosing a central lagoon is called an atoll. These are the three classical forms, which intergrade according to the stage of development.

But there are many variants on this theme. Added to this, during ice ages, when water was locked up in ice caps so that sea levels were substantially lower than at present, the location of where rain eroded the exposed limestone also differed. Many different shapes of lagoonal depressions and channels resulted from the differential erosion and etching by rain. In addition, local tectonic processes can govern reef development. Where uplifting of land occurs, essentially reversing the classical subsidence, a series of fringing reefs develop seaward, one after the other. Where land is stable relative to the sea level for long geological periods, the developing reef may become broken up into series of patch reefs. The name patch reef is itself also applied to almost any size of reef which is not one of the classical forms. Patch reefs may not reach the present sea level, or may themselves support isolated islands, and may or may not constitute an isolated part of a barrier reef.

In many cases, atolls do not support any islands at all, although they may have a rim that is partly awash at low tide, or the entire rim may even lie a few

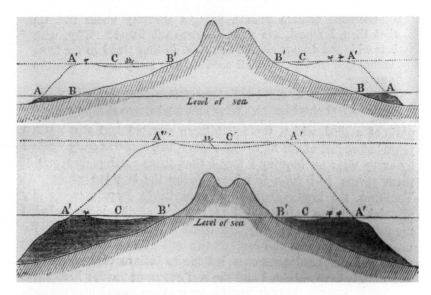

Figure 1.7   Darwin's original illustrations of the progression from fringing reefs, through barrier reefs, to atoll. Top: The first stage in the progression occurs when the sea level is at the line from A to A. A fringing reef develops, hugging the volcano (darkly shaded). Then, when the island has subsided further, the sea level is at the line from A' to A', and the developing barrier reef is shown by the dotted line, such that B' is the shoreline and reef flat, C' is a navigable lagoon (note the small ship beside the letter C) and A' marks the outer edge of the reef, which may have islands, and the reef crest. Bottom: Continuing further, the line from A' to A' is the same as before, with the shaded blocks representing the barrier reef. Then, with the island completely submerged, A" through C to A" marks the relative sea level. There may or may not be islands at A" but, either way, it marks the rim of the atoll.

Source: From Darwin, C. (1842). *On the Structure and Distribution of Coral Reefs.* Ward, Lock & Bowden Ltd.

metres deep at all states of the tide. The name 'atoll' comes from the Maldivian word for the abundant rings of islands which make up that nation. Other kinds of atolls in Western Australia have developed on the edge of a sinking continental slope; these do not fit Darwin's definition of having developed around a volcanic island, so some have argued about using the term 'atoll', but the principle of a sinking substrate matched by upward growth of coral reef still holds true, and they have all the appearance of an atoll.

Many other forms of reefs have been named. Guilcher (1988) described the 'ridge reef' in the Red Sea as 'a heretofore neglected type of coral reef' which should join the classical Darwinian forms of fringing reef, barrier reef and atoll. These structures are longitudinal ridges lying along the axis of the Red Sea, and probably result from a combination of faulting from the progressive opening of the Red Sea and from being pushed upwards by underlying salt deposits. The only atoll described from the Red Sea, Sanganeb Atoll off Port Sudan, rests on a ridge reef, so in this case one reef could fit both definitions. Other kinds of reefs defined over the years include crescentic, ribbon, planar and reticulated reefs, as well as the faro reef, which is a ring structure found within lagoons of Maldivian atolls, and several others. There may be a descriptive advantage to defining new kinds of reefs, but the important points are only that reefs grow slowly and usually steadily when oceanic conditions permit, that they are limited in their upward extent by the contemporary sea level and that erosion of the exposed limestone of which it is made is relatively swift. Reefs, therefore, adopt shapes which reflect present and historical substrate and shorelines. As the substrates subside, become raised by uplift or sedimentary deposits or erode during millennia of rain, for example, reefs develop in shallow water, where limits of salinity, temperature and turbidity permit them to do so. It is not surprising that, while they occur in a huge number of shapes, in them one may discern a much smaller number of basic designs.

A major complication comes from the fact that successive ice ages have caused the sea to rise and fall massively, in cycles. At the start of the present era, the Holocene, the sea level was about 140 m below its current level, and it rose rapidly (in geological terms) to near its present level (Figure 1.8). Reef growth when the sea level was much lower continued at those lower sea levels, and reefs were, like present reefs, shaped by similar processes of erosion. When the sea level rose, those reefs became too deep for growth and so formed the foundation for more recent reef growth. Thus, the shape of a reef is also dependent on its antecedent platform. During and following the last set of ice ages, reefs developed across a vertical span of 150 m or more, giving great scope to the development of a wide range of reef types and depths. Of key importance throughout all this development and growth is that reef limestone is a relatively soft rock, easily eroded by rain, waves, storms and

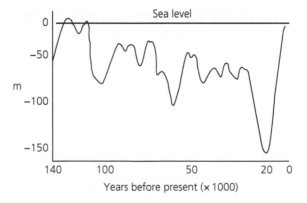

Figure 1.8   Changes in the sea level during the last 140,000 years, through the period of the most recent ice ages, into the Holocene period.

Source: From Sheppard, C.R.C. (2000). Coral reefs of the Western Indian Ocean: An overview. In: T.R. McClanahan, C.R.C. Sheppard and D.O. Obura (eds), *Coral Reefs of the Western Indian Ocean: Their Ecology and Conservation.* Oxford University Press, pp. 3–38.

Figure 1.9   Top: Height above sea level of 14 elevated Red Sea reefs, ordered left to right, youngest to oldest. Vertical bars are dates of six of them. Bottom: Living corals grow on the contemporary reef flat which extends out to sea, to the side of which is the youngest of the series of fossil reefs, uplifted by 3–5 m above the present level.

Source: Sheppard, C.R.C., Price, A.R.G. and Roberts, C.J. (1992). *Marine Ecology of the Arabian Area: Patterns and Processes in Extreme Tropical Environments*. Academic Press.

burrowing organisms (of which much more later), and it is also a rock which is deposited by corals fairly rapidly.

Sometimes reefs emerge completely. Examples of this phenomenon, which is caused perhaps by tectonic uplift, are common, for example the fringing reefs of the Red Sea, which is technically an ocean, not a sea, because sea floor spreading is widening the Red Sea by about 2 cm y^{-1} and causing a series of tectonically uplifted reefs to appear on each side. The top part of Figure 1.9

shows the ages and elevations of series of successively elevated reefs from different parts of the Red Sea over the last 200,000 years. The bottom part of the figure shows the present, living reef, which plunges deeply into the Red Sea, behind which is the reef which was most recently uplifted in the series, raised 3–4 m above the present sea level.

These processes of growth, erosion and land shifts result in the infinite variety of reef shapes that we see today.

1.6 Profiles and zones of reefs

Despite their varied nomenclature and variety, reefs do tend to have a general cross-sectional profile. Some are illustrated in Figure 1.10, which shows the slight differences that occur.

1.6.1 The reef flat

Almost all reefs which reach the surface today have a reef flat, which may extend just 10 m from the beach to the plunging slope, as in some popularly visited Red Sea reefs, but more often extends 100–1,000 m from the shore (Figure 1.11). Some may extend for a kilometre or more. Coral growth and reef growth continue but, because they cannot grow above water, the reef can grow only outwards. They continue to grow out to sea, steepening the angle of the reef slope, so that sections of reef may then slump downwards, leaving visible scars as huge blocks slide down to find a more stable resting place. Then, on the shallow reef flat, growth continues outwards again.

The broad horizontal expanse of most reef flats commonly dries at very low tides and is rarely deeper than about 1–2 m. Parts may be deeper in areas where wave or current scouring occurs. It is by far the largest expanse of most reefs in terms of planar area. It receives intense sunlight, so that water in its various pools at low tide may heat to over 40 °C, which is lethal to corals and to most (but not all) reef life. Further, about a third less oxygen can remain dissolved in water at 40 °C than is the case at 20 °C, adding to the biological stress. Evaporation in the hot sun may raise the water salinity to levels that are lethal to marine life while, in many parts of the world, the surface of the reef flat may be deluged with fresh rainwater on occasion, lowering the salinity to below lethal tolerances for most marine species. Therefore, conditions on the reef flat are relatively hostile, and biodiversity is correspondingly low. In the past, this was the part of the reef most studied for reasons of easy access, to the extent that many of the earlier accounts of 'reefs' really referred to the reef flat only.

The kind of life that lives there is mainly of two general kinds: species which live there either exclusively or nearly so—those with very high tolerance to

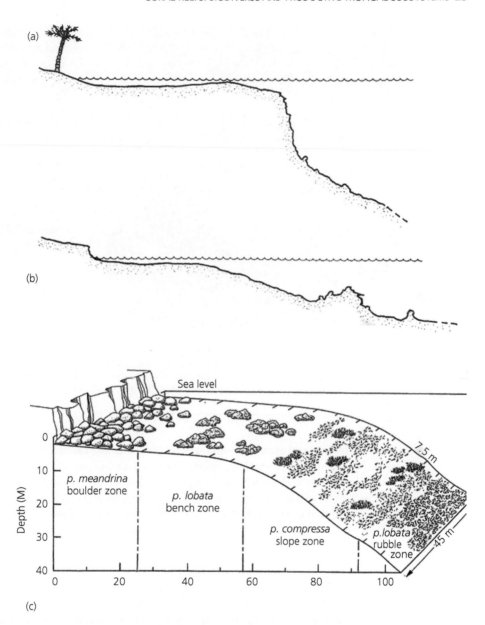

Figure 1.10 Sketches of a cross section of a typical reef. (a) A typical fringing reef of the kind seen in the Red Sea, surrounding most atolls and extending from continental shores. The length of the reef flat may vary from a few metres to over a kilometre. There is commonly a fairly sharp break between the reef flat and the reef slope, marked in high energy locations by massive constructions of red algae. (b) Many reef flats in sheltered locations have a less pronounced break at the seaward edge of the reef flat. (c) Profile of reef in Hawaii.

Source: (a, b) Sheppard, C.R.C., Price, A.R.G. and Roberts, C.J. (1992). Marine Ecology of the Arabian Area: Patterns and Processes in Extreme Tropical Environments. Academic Press. (c) Courtesy Dr Steve Dollar.

Figure 1.11   A reef flat extending about 300 m seaward of the beach. The line of breakers marks the edge of the reef flat; seaward of that line, the reef slopes downwards steeply. Darker patches in the water of the reef flat are patches of coral; the pale parts are sand. The scene is at mid-tide, giving just enough clearance for the small boat (Praslin Island, Granitic Seychelles).

these environmental extremes—and species for which the reef flat lies at the extreme edge of their normal or preferred range, the main part of their range being the deeper reef slope. The first group is low in numbers of species, although some of them may be abundant. Those species that can live here are well adapted to the extreme stress of heat, UV radiation, extremes of salinity, and so on, and they have the advantage that they do not face competition from many other species.

Reef flats which are located slightly deeper and which are rarely exposed above low tide are richer in abundance and variety of life. Sometimes, lagoonal reef flats of atolls are slightly deeper than those on the ocean-facing sides of the same islands. Many such flats support thickets of corals. Dense stands of staghorn corals reach the surface on reef flats which face sheltered locations, such as those which are protected by islands. Many massive or dome-shaped corals also grow, but the growth of their usually hemispherical colonies is truncated by low-tide levels into the annular rings of 'microatolls', which are constrained by low tide from growing on their upper surface.

As the seaward edge of the reef flat is approached, wave action is more vigorous, and this has at least two effects on the life which grows there: first, conditions are rougher, especially at periods of high tide, so more robust forms are generally seen—the staghorn forms seen in sheltered areas would simply be smashed in this environment. But, second, that same greater water agitation means that water exchange is greater, so this region does not heat nearly so much, nor does it suffer from depleted oxygen levels. Furthermore, the greater mixing ensures that salinity does not drop so dangerously during heavy rain. Therefore, different reef flats show very different patterns of zonation, according to their particular degrees of exposure to waves, their total width, the tidal range and the height of the reef substrate relative to the low-water level.

In all cases, however, the biota on the reef flat is limited to that which has adapted to survival in conditions that become periodically extreme. While this means that overall diversity is relatively low, it also means that the species which have adapted to living there have several physiological characteristics and tolerances which have been used by scientists who are specifically examining questions of living under stressful conditions and of climate change.

1.6.2 The reef crest

At the seaward edge of the reef flat is the reef crest. There, the reef grows actively, and this marks the point where the reef plunges steeply. It is sometimes outlined by boulders of coral rock and coral colonies tossed up onto the reef flat by storms. The angle formed between the horizontal reef flat and the slope may be sharp. This is the most wave-exposed part of the reef, and very few coral species can grow here. Instead, wherever wave exposure is high, the dominant forms of life are the calcareous red algae which are extremely wave resistant and which deposit a form of limestone known as high-magnesium calcite. These algae grow as a slightly elevated ridge and, in very exposed locations, a series of spurs made of the same algae extend seaward of the ridge, developing into a 'spur and groove system' (Figure 1.12). Spurs dip gradually into deeper water, and the grooves have steep sides, usually with a scoured appearance, while the widths of the spurs and grooves in any one series are similar both with each other and throughout their length. At their seaward ends, spurs fade out at a depth where average wave turbulence is greatly diminished.

The series of structures—the spurs—project into the prevailing waves, and the channels between them—the grooves—carry water at rapid velocity. The water oscillates in the grooves such that outgoing swashes collide with incoming waves, resulting in dissipation of energy. It is clear that this mitigation of energy is one of the ways in which reefs resist destruction (see Chapter 3).

The coralline algal genus *Porolithon* is the main agent of this growth in the Indo-Pacific, joined by *Lithophyllum* in the Atlantic (Adey 1975). The series of spurs obtain their shape from biotic construction of the algae rather than

Figure 1.12　Calcareous red algae thrive in the surf zone of many reefs, building ridges to just above the low-water level. They also build spurs which extend seaward. Harder than concrete, they play a major role in wave energy reduction.

from erosion in grooves; the latter have a floor usually at or about the elevation of the main base of the reef slope (Sheppard 1981). Strong water movement and high aeration (Doty 1974; Littler and Doty 1975) are required for these algae to grow to the required amount. It is not known how such structures and their regular spacing are initiated, but it has been observed that the spacing of the spurs may be some harmonic of the waves' mean wavelength (Munk and Sargent 1954). Thus, the mechanism regulating the spacing is a correlate of wave energy; the greater the energy, the larger and more widely spaced is the series of calcareous algal spurs. At all sizes, the structures greatly ameliorate the impact of breaking waves, in a self-maintaining arrangement. Even a few lagoonal reefs can show small, rudimentary series of algal spurs if the water is rough enough (Sheppard 1981), while the largest in the Pacific and in the Indian Ocean have developed into huge buttresses (Odum and Odum 1955; Pichon 1978; Sheppard 1981). However, some large structures appear to owe their shape partly to older, underlying erosional features too, and may have been cut into pre-existing limestone platforms (Hopley 1982).

While the stony red algae are the most conspicuous sources of limestone deposition, this turbulent area is also the site where wave energy forces sediment particles into each and every crevice and pore in the limestone rock. These then become re-cemented, and the result is a very solid rock matrix. The general principle is simple; namely, particles are stabilized and held in position, either by compaction or by being grown over and sealed off by more growth of the red algae, or are bound by other mechanisms of aragonite cementation. The exact way it occurs at the particle level is unclear; microbial action which perhaps changes the pH on the particles' surfaces, or whose metabolic actions temporarily redissolve the surface limestone, have both been suggested. The result is the creation of the most solid limestone found on the surface of a growing coral reef, in that part of the coral reef where robust construction is most needed.

1.6.3 Reef slopes

At the seaward edge of the reef flat, the reef plunges, usually steeply. The slope may be continuous or very irregular and, on atolls, there may be a second 'edge' or 'drop-off' below which the slope is steeper still (Figure 1.13). In clear water, such as in most of the Caribbean, the Red Sea and most atolls, for example, this steep slope may support corals to depths of 50 m or greater. Much of the rest of this book describes processes that take place on the reef slope, because this is the heart of a reef, the region of highest diversity and the most growth and activity. Therefore, its description here is brief.

Several gradients define the zones on a reef slope: a gradient of wave energy, which declines with increasing depth; another gradient of light, which also declines; a gradient of sedimentation, which generally increases with increasing depth; and a gradient of temperature, which may decrease gradually or, equally commonly, abruptly across a thermocline. These parameters control the broad distribution of species—the 'what species can live where' problem.

Figure 1.13   A sudden steepening of the reef slope occurs on many atolls at depths variously between 4 m and 20 m. On the steeper and deeper part, light is reduced and there can be a marked change in the coral community.

Cold-water corals

Although corals are usually associated with shallow, tropical seas, over half the coral species known occur in waters >50 m deep (Cairns 2007) and several of these form substantial structural habitats in deep, cold waters (Roberts et al. 2009) (Figure 1). These cold-water corals include anthozoan species of hard corals (order Scleractinia), zoanthids or 'gold corals' (order Zoanthidae), black corals (order Antipatharia), gorgonians or 'soft corals' (subclass Octocorallia) and hydrozoan species of stylasterids or 'lace corals' (family Stylasteridae).

Since the 1990s, work on cold-water corals has grown exponentially, with exploration of their biogenic habitats in fjords, on the continental shelf, on slopes, on offshore banks, on seamounts and on mid-ocean ridges. There has been considerable focus on the scleractinian reef frameworks formed by species including *Lophelia pertusa, Madrepora oculata, Solenosmilia variabilis, Goniocorella dumosa* and *Enallopsammia profunda*. These species each develop complex branching skeletons that, over time, trap mobile sediments, creating seabed mounds, often with morphologies sculpted by the prevailing near-bed currents.

Cold-water corals do not host photosynthetic microalgal symbionts and instead rely on food supplied from primary production in the surface ocean. Detailed studies of the hydrographic regimes surrounding coral habitats have revealed a variety of complex food supply mechanisms often related to internal wave dynamics (e.g. Davies et al. 2009). It was even recently proposed that some deep coral mounds have grown so large that, compared to areas beside the coral mounds, they preferentially draw down food from surface waters (Soetart et al. 2016), a phenomenon with great implications for understanding carbon flux along the continental shelf, where deep coral mounds are abundant.

These scleractinian coral reef frameworks support communities of animals often characterized by suspension and filter-feeding corals, sponges, bryozoans and hydroids that can be several times as diverse as nearby off-reef habitats (Henry and Roberts 2007). The dense biomass and relatively high food flux through deep coral mounds makes them enhanced sites of carbon turnover, compared with equivalent off-mound sites (Cathalot et al. 2015; Rovelli et al. 2015). Although substantially less diverse than their tropical counterparts, cold-water corals also support a characteristic deep-water fish fauna (Ross and Quattrini 2007; Milligan et al. 2016) (Figure 2) and play an important role as spawning sites for some species (Baillon et al. 2012; Henry et al. 2013).

However, over the last 20 years, as surveys of deep coral habitats have increased, the number of coral habitats found damaged or destroyed by bottom trawling has also increased (e.g. Fosså et al. 2002). This led to calls for their protection and the creation of closed areas not just in national waters but also on the high seas. At the time of writing, this discussion had reached an important point at the United Nations to negotiate a new legally binding instrument on the conservation and sustainable use of marine biodiversity in areas beyond national jurisdiction (Long and Rodríguez Chaves 2015).

Continued

Cold-water corals *(Continued)*

Figure 1. Top: A complex mixture of species building a deep-water reef. **Figure 2.** The deep reefs support a specialist fauna.

Continued

Cold-water corals (*Continued*)

Such protection measures are now even more important as we move further into a high-CO_2 world where the implications of global change on deep-sea ecosystems are becoming apparent (Sweetman et al. 2017). At colder temperatures and at greater ambient pressures, cold-water corals are already growing closer to the calcium carbonate saturation state than tropical corals. For the scleractinian reef frameworks, built from the more soluble aragonitic form of calcium carbonate, the future looks particularly bleak. Many present-day reef complexes are projected to be in undersaturated waters within the twenty-first century as the global aragonite saturation horizon shallows (Guinotte et al. 2006). Laboratory studies suggest that ocean acidification causes species like *Lophelia pertusa* to produce more chaotic calcification patterns and become structurally weaker (Hennige et al. 2015), although no studies have yet examined the adaptive capacity of deep corals to such changes.

Prof. J. Murray Roberts, School of GeoSciences, University of Edinburgh, UK.

In shallow water, light is abundant, and wave energy prevents sediment from settling on corals and reducing their efficiency or even smothering them, but too much wave energy readily smashes corals, so only the toughest forms survive in the shallowest parts of the reef slope. At the deepest extreme on the dimmest parts of the reef slope, leafy corals are mostly found, where they grow outwards to trap the diminishing amount of light. But trapping the most light would mean extending horizontally, which would also mean that they trap the rain of sediment. Therefore, leafy corals commonly find an optimum angle of about 45°, sufficient for light trapping yet sufficiently angled to shed sediment.

1.7 Biodiversity on reef slopes

The greatest diversity is found in intermediate depths. Most reef species can live there, but this also means that competition for space is most intense, and it is here where most different colony growth forms occur, where most different methods of predation and hiding are found and where the greatest range of reproductive strategies are found, all factors being used by different species in different ways to secure their space on the reef.

The most visual elements on a coral reef, to perhaps most people, are the reef fish. To a degree not seen in any other marine or freshwater habitat, most reef fish are extraordinarily colourful. A healthy reef is teeming with fish, both solitary individuals and large schools, some living within single colonies of a branching coral, and others living in the open or browsing in the sandy patches. Coral reef fish exercise considerable control over all aspects of reef ecology, feeding on plankton, feeding on other fish or grazing the algae which otherwise would smother most of the reef. Most are carnivorous and exert a dominant role in the trophic structure of the reef. Their roles are described in Chapter 6, while fisheries are covered in Chapter 7.

Coral reefs contain the richest biodiversity of macrofauna in the seas, and most occurs in this zone. A large proportion of reef life is cryptic, that is, occupying crevices and, in many cases, making their own crevices. When it is remembered that about two-thirds or more of the fish that swarm over a coral reef are carnivorous, with a varied food preference in many cases, a cryptic way of life becomes easy to understand. Evolutionary pressures— the survival of those which are hidden, and a rather short life for those which are not—have ensured that a hidden way of life is immensely popular. Such a way of life, however, has an obvious drawback, namely the inability to catch food other than that which drifts towards you. Hence, many cryptic forms of life are filter-feeders, trapping drifting food in tentacles or creating currents which pass water through the body, thereby trapping particles of food internally. The many kinds of polychaete worms are conspicuous members of the first group, and sponges are a good example of the second. Other forms hunt and trap larger animals in the maze of subterranean caverns and passages, while others carry protective armour around with them, molluscs and crustaceans being prime examples. But, for every development that evolves, countermeasures are developed by potential predators: some crustaceans can crack open or even peel mollusc shells, and some molluscs drill holes in other shells to get to the flesh beneath. The 'Red Queen' hypothesis, named after *Alice in Wonderland*'s Red Queen, who continually had to run just to keep up but got nowhere, holds good for life on the reef; there, an arms war takes place, where one defence is countered by another method of breaching the defence, and one way of feeding is countered by new defences, developed by successful species over evolutionary time. If a species cannot evolve continuously or keep itself immune in some way from predation or elimination, it becomes extinct. When it is realized that 99% of all species that have ever lived have become extinct over evolutionary time, it is easier to appreciate the need for continuous evolution by species on the reef.

This is a form of co-evolution, an important principle of the development of life over a long time. One species will evolve in conjunction with another, sometimes, as noted, with responses to forms of predation or development of better forms of predation. But, in many cases, co-evolution leads to symbioses of various kinds. Some species are wholly dependent upon others and, in this group, there is one of the most important symbioses of all, that of algal cells with the corals (Chapter 4). The extra boost that this symbiosis gives to coral growth is the reason why the stony corals and their relatives can create a reef at all, and why they can create a three-dimensional structure which is home to such a high diversity of life.

1.8 Values of reefs to people

The value of coral reefs lies in several main areas, some more important than others, depending on the region and current needs. They provide a variety of 'ecosystem services', including food production, coastal protection from wave energy, tourism and a general value often named high biodiversity. The last of these terms is rather nebulous but may be interpreted as the resources obtained from biopharmaceutical prospecting, as well as extend to the more general belief that the biological and aesthetic (and unknown) riches of biodiversity are too valuable to lose.

Food production is clearly key in many tropical countries. Reef-based fisheries encompass a wide range of species, both vertebrate and invertebrate. These are explored in Chapter 7. Wave energy reduction is important to many locations also. A large proportion of coral reefs, but far from all, grow upwards to the surface, where they form effective, natural breakwaters. Fringing reefs especially provide substantial protection to island shorelines, as do seaward-facing reefs of islands on all coral atolls. These natural breakwaters renew themselves at a rate which, in a healthy reef, slightly exceeds the rate of their erosion both by wave energy and by bioerosion caused by the many burrowing and rasping organisms. Unfortunately, wave erosion and bioerosion continue even when accretion is impaired, and this can have serious consequences to effective shoreline protection. Chapter 9 describes some examples of decreased shore protection once reefs are removed or damaged.

Tourism, focused sometimes around recreational scuba diving, is a major revenue earner for many countries where in some cases it is their principal source of foreign currency. It has been shown repeatedly that maintaining a reef in healthy condition pays. Tourism is a renewable resource when managed well, whereas a one-time use like, for example, extractive exploitation has too often lead to ecosystem deterioration or collapse, which then benefits nobody. Biodiversity is difficult to quantify as a value. It overlaps with tourism values and food production values. But it has separate value over and above these. The large numbers of species suggest a large number of interesting and useful biochemicals, of which several are increasingly being investigated. Important also, however, is an aesthetic sense that this is a remarkable ecosystem, one that needs to be better understood and which is simply much too valuable to lose. This book shows why this is the case.

2 The main reef builders and space occupiers

The dominant groups of biota on coral reefs are corals, soft corals, sponges and algae. The ratios of each depend on which ocean is being considered, on water quality and on the exposure and depth of the site. The term 'coral reef' itself implies strongly that corals are the main builders; this is certainly the case in most places, but the other components are integral components too.

The organisms that build reefs have a biomass which is a tiny fraction of the mass of the reef that they have created. As Hatcher (1997) put it:

> Coral reefs are gigantic structures of limestone with a thin veneer of living organic material—but what a veneer! Everything that is useful about reefs (to humans and to the rest of nature) is produced by this organic film, which is approximately equivalent (in terms of biomass or carbon) to a large jar of peanut butter (or vegemite) spread over each square metre of reef.

The groups of animals and plants that are mainly responsible for the occupancy on each of those square metres are the subject of this chapter. These benthic components do not act alone, of course, so fish, plankton and other key interacting organisms and processes are described in later chapters.

There are three major groupings of cnidarian animals which are both important and conspicuous on reefs: the hexacorals, which include stony corals, which deposit limestone skeletons and whose polyps have six or multiples of six tentacles; the octocorals, which include both the classical soft corals and the sea fans, whose polyps always have eight tentacles; and a third, smaller group which includes the black corals and sea whips. Some of the latter may show an outward resemblance to the sea fans in that the colonies of many consist of finely branched, tree-like structures.

Stony corals, in the order Scleractinia, occur abundantly on all reefs, and are the animals most responsible for production of calcium carbonate, on which reef growth depends. Octocorals occur in very different forms in the Caribbean and Indo-Pacific; in the Caribbean, they are primarily the tall,

The Biology of Coral Reefs. Second Edition. Charles Sheppard, Simon Davy, Graham Pilling, and Nicholas Graham, Oxford University Press (2018). © Charles Sheppard, Simon Davy, Graham Pilling, and Nicholas Graham (2018). DOI 10.1093/oso/9780198787341.001.0001

branching kinds, including sea fans but, in the Indo-Pacific, the dominant forms are the encrusting, lobed and low-branched forms. Branching sea fans may be locally abundant in all regions.

Sponges are major space occupiers on Caribbean reefs, where they are considerably more abundant than they are on Indo-Pacific reefs. Finally in this overview, the stony red algae are crucial in the shallowest and most wave-exposed regions of reefs. These algae occur throughout the world, but their most substantial structures are found on exposed Indo-Pacific reefs. These were described in Chapter 1 in the context of their position and construction of the reef and are not considered further here. Other macroalgae, however, may be major space occupiers (increasingly so in polluted areas) and these are included.

Amongst these major reef builders and components are several other smaller groups, such as fire corals, blue coral, organ pipe coral and others, which may be locally very important (see 'Related reef builders: *Heliopora, Millepora, Tubipora*'). Finally, two groups of angiosperms, the seagrasses and mangroves, are intimately connected with reefs in most areas, and their function with respect to reefs is briefly described.

Related reef builders: *Heliopora, Millepora, Tubipora*

Most hard corals on coral reefs are Scleractinia or hexacorals, but a few are octocorals or hydrozoans. Like Scleractinia, they produce skeletons that add to the building of coral reefs. Scleractinia are anthozoans with multiples of six tentacles, while octocorals are anthozoans with exactly eight tentacles, which have tiny side branches called 'pinnae'. Two species build solid skeletons: the blue coral *Heliopora coerulea* and the organ pipe coral *Turbipora musica*.

Blue coral

The name 'blue coral' refers to the aragonite skeleton, which is coloured blue due to iron salts (bilichromes) deposited in the skeleton. Blue coral ranges from the Red Sea to American Samoa, and can be common in places such as the Marshall and Gilbert Islands and Tuvalu, and even dominate reefs such as one reef at Ishigaki Island, Japan (Zahn and Bolton 1985). Polyps are quite small and appear as white fuzz on the light blue to brown living colony, and colonies may be large, with vertical branches (Figure 1, top right). Closely related species extend back 100 million years in the fossil record.

Organ pipe coral

Organ pipe coral produces a bright red skeleton composed of vertical tubes joined by thin horizontal layers. Each polyp grows within the upper end of a tube, and secretes tiny

Continued

Related reef builders: Heliopora, Millepora, Tubipora **(*Continued*)**

Figure 1. Top left: *Stylaster roseus* from the Caribbean. Top right: *Heliopora coerulea* from the Indo-Pacific. **Figure 2.** *Millepora complanata* from the Caribbean.

sclerites or spines of calcite, which it then cements onto the tube. Polyps are grey and about 5–10 mm in diameter; they cover the living colonies, obscuring the red skeleton. Colonies are rounded lumps. The geographic range of this species is similar to that of blue coral.

Octocorals

Most octocorals are referred to as 'soft corals' because their body is soft and they do not secrete a rigid skeleton (gorgonians secrete a flexible skeleton). However, most do secrete sclerites made of calcite. Some species concentrate those sclerites near the base of the colony's soft tissue. In some species, the sclerites are cemented together in a solid base which is located under the colony and called spiculite. In most of those species, the base

Continued

> ## Related reef builders: Heliopora, Millepora, Tubipora (*Continued*)

is a smooth, rounded, boulder-like surface, which can be seen if the soft coral dies. But, in a few species in the genus *Sinularia*, the solid base of cemented spicules is formed into a large branching structure that can be 2 m or more tall (Schuhmacher 1997). In a few places, soft corals have been so common for so long that the dominant component of the reef rock is the spiculite formed by soft corals. These are widespread in the Indo-Pacific (Fabricius and Alderslade 2001).

Fire corals

Fire corals (Figure 2) in the genus *Millepora* are hydrozoans, and they are common on both Caribbean and Indo-Pacific reefs (Lewis 1989). There are around 16 known species, which are major contributors to reefs in places, and they contain zooxanthellae. They earn their name because, like many hydroids, they have powerful nematocysts which can penetrate thin human skin and cause a burning sensation, although fire coral stings are not usually dangerous. The polyps are tiny and may be extended during the day, when they appear like white hairs. There are two kinds of polyps: short, feeding polyps with mouths, called gastrozooids, and longer, thin polyps with nematocyst batteries but no mouths, called dactylozooids. Each is set in a tiny pore in the smooth aragonite skeleton. The skeleton is porous, but the spaces are closed, and oxygen produced by photosynthesis by the zooxanthellae can build up in the skeleton.

The Western Atlantic has different species to the Indo-Pacific and Brazil has species that are not present in the Caribbean. Species are distinguished by colony shape and by the size and pattern of the pores that the polyps are seated in. Species range from encrusting to intersecting flat thin vertical plates, to thick paddles or rounded branches, to fans of branches and lumpy masses. One Caribbean species (*Millepora alcicornis*) commonly overgrows gorgonians, much as a vine may grow on a tree. Most species are yellow to tan coloured, but some can have various shades of pink, red or burgundy. Fire corals are very similar in their biology and ecology to Scleractinia in most ways, but there are a few differences. They appear not to be eaten by crown-of-thorns starfish, and diseases have not been reported in them; however, they are amongst the most sensitive corals to mass coral bleaching. They reproduce by producing tiny medusae, which are released from blisters and which, in turn, release the gametes. Sexes are separate and, as with several scleractinians, the eggs contain zooxanthellae.

Others

There are two genera related to *Millepora* that do not have zooxanthellae and which form small lacy hard skeletons. *Stylaster* (Figure 1, top left) has particularly thin lacy branches, while *Distichopora* has thicker, smoother branches. Both are found in shaded locations such as overhangs, and have bright colours like purple, pink, red or yellow. They are minor contributors to reefs. *Stylaster* has one shallow species in the Caribbean and several species in the Indo-Pacific. *Distichopora* has several shallow species in the Indo-Pacific (Veron 2000).

Dr Douglas Fenner, Pago Pago, American Samoa.

2.1 Corals

The unit of the coral is the polyp, which resembles a small sea anemone, to which it is related. It has one or more rings of tentacles surrounding a mouth, leading to the main body cavity. The polyp's tissues deposit calcium carbonate around themselves in the form of aragonite, such that a typical polyp occupies a cup of limestone, or corallite, which grows upwards or outwards. The fleshy polyps are generally emerged during the night (Figure 2.1), while in the day they are retracted into the corallites.

The polyps may be separated individually (Figure 2.2), in which case they grow upwards and outwards, depositing limestone as they grow, and at intervals each divides into two or more daughter polyps which continue to grow until they, too, divide. In many forms, the polyps are connected to adjacent polyps via a thin tissue called the coenosarc, which is also capable of depositing limestone beneath it, with the result that, instead of there being 'gaps' between each polyp, the spaces between the polyps also become filled with limestone (Figure 2.3, top left). In the species shown, the Caribbean *Dichocoenia*, the polyps elongate into oval shapes before constricting in the middle to pinch off two roughly equal-sized polyps. All the while, they grow upwards, with limestone being deposited by each polyp and by the connecting coenosarc, thus expanding the solid colony. This process of pinching off polyps is called intratentacular budding; specifically, the budding results from division within the ring of tentacles. Other species show extratentacular budding, in which the daughter polyp emerges outside the ring of tentacles, forming a new, small polyp, or it may bud from the stem of the parent polyp (see Figure 2.8).

Figure 2.1   Left: Tentacles of the Caribbean stony coral *Montastraea*, emerged at night. Right: Tentacles of an Indo-Pacific soft coral, showing the eight feathery tentacles.

Figure 2.2   The Caribbean coral *Eusmilia*. Each cup houses one polyp, and each growing tube shows division into two, sometimes dividing more than once.

Many species take this process of intratentacular budding to extremes, such that they continue to divide with very little or no 'pinching off'. The result is long, meandering ribbons of polyps, forming the brain corals; *Meandrina* from the Caribbean is a good example (Figure 2.3, bottom left). Each valley in such a brain coral is enclosed within one continuous ring of tentacles, and the valley may contain dozens or even hundreds of mouths with associated gastric cavities—they could equally be considered as one multimouthed polyp or many polyps, referred to as 'centres'—which are only partly separated from their neighbours. In the case illustrated, the series of polyps are closely adjoined, so there is no coenosarc and no area between adjacent valley walls.

There are many possible combinations of these budding patterns. The meandroid form may develop such that each series or valley becomes disconnected from others (Figure 2.3, right). Valleys may be short—two or three centres only. Some species have very small polyps and small rings of tentacles, and relatively large expanses of tissue between them. Many have closely adjoining polyps which effectively share a common wall. These form key diagnostic characters which help to distinguish between species.

In all cases, deposition of limestone causes the colony to expand radially as its polyps divide, but the different proportions in which different species do so results in the vast range of colony shapes and sizes that are seen. In species where division is even and regular growth is unimpeded, smooth, dome-shaped colonies result, and some of these, such as in the genus *Porites*, result in

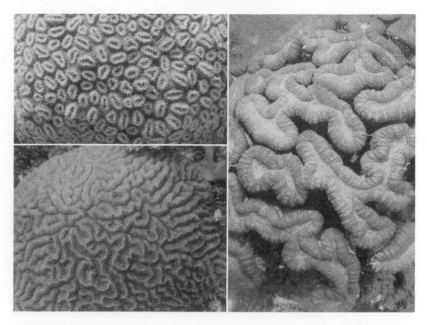

Figure 2.3   Top left: The Caribbean *Dichocoenia* contains polyps in cups, but its polyps are joined with interconnecting tissue called coenosarc, which also secretes limestone. Thus, limestone is deposited between the cups at a rate as great as in the cups. These polyps can be seen to elongate into oval shapes, which then pinch off to produce two daughter polyps. The fine tentacles can be seen emerged in the daylight, protruding just above the calices, or cups. Bottom left: The Caribbean coral *Meandrina*, which has meandering chains of polyps. There is no separation into individuals. The polyps occupy the valleys (but are tightly retracted and not visible in the daytime). Right: Indo-Pacific *Lobophyllia*, which shows meandroid groups of polyps, but in this case each series is separated from its neighbouring series. This coral is very fleshy, concealing the rows of large septa (the radial plates within all calices). In some, the small mouths of each centre can be seen.

huge colonies (Figure 2.4). Although the individual polyps are tiny—only 1–2 mm in diameter—such colonies can consist of millions of connected polyps after several centuries of growth. Others grow upwards very slowly or even not at all, yet the corallites around the perimeter bud profusely. This results in leafy colonies (Figure 2.5). These have probably the least amount of limestone per unit mass of polyp, and are commonly found in water where light is low, such as on deeper parts of reef slopes, where extensive monocultures can be found.

The most diverse growth form of all is the branching form, exemplified in the genus *Acropora*, and in many species of the family Pocilloporidae. Many other genera contain some branching species, but the former are almost exclusively so. *Acropora* species differ from most corals in that they have two kinds of polyps. An apical polyp is found at the ends of each branchlet, which is usually considerably larger than the other polyps and which leads to the elongation of the branchlet. All the rest are radial polyps: smaller, usually fairly crowded and covering the sides of the branches (Figure 2.6). These branching corals may dominate large parts of the reef and, to a considerable

Figure 2.4   A giant Indo-Pacific *Porites* colony. Each polyp is about 1 mm in diameter. The colony grows outwards at about 1 cm y⁻¹, so colonies like this are several centuries old, with millions of polyps. The fissures in the surface are occupied by sponges or bivalve molluscs which settle onto dead portions and then become progressively buried as the colony grows.

degree, they are responsible for the three-dimensional structure which lies behind much of the high biodiversity characterizing coral reefs.

Reproduction in corals may be sexual and asexual. About three quarters are hermaphrodites, while the remainder have separate sexes. In the case of some species of the free-living, or 'mushroom corals', *Fungia*, it has recently been shown that adults can change sex as they age; the two species known to do this begin as males and turn to females later in their lives. In one of the species, this may again later reverse back to being male again (Loya and Sakai 2008). Most species are 'broadcast spawners', that is, they release sperm and eggs into the water column, where the latter become fertilized. Well-described cases occur where most species broadcast their gametes on the same night, saturating the predators and ensuring sufficient numbers survive. Broadcasting can be so profuse that thick slicks of gametes and fertilized eggs form on the surface of the water, with much of these being occasionally being washed up onto the beach if there is an unfavourable change in wind direction.

The larvae are planulae, which remain in the water for periods usually of just a few days to as long as a few weeks. Planulae are elliptical and can swim only limited distances, so are dependent on currents for dispersal to any significant extent. They can control to some degree their vertical movement in the water column. They show a degree of chemotaxis and phototaxis too, so that, when they are ready to settle, they can show a considerable degree of site

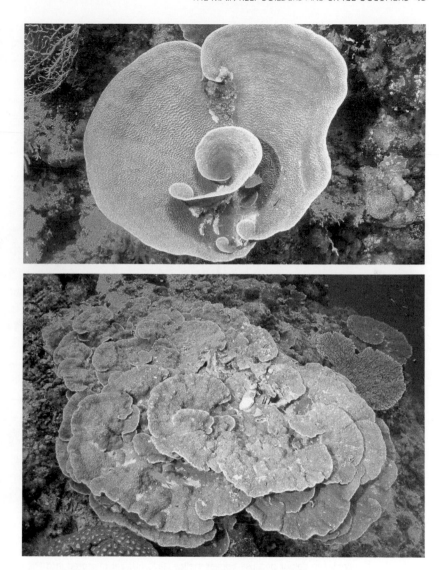

Figure 2.5   Two Indo-Pacific leafy corals. Top: A vase shaped *Turbinaria*, Bottom: A foliaceous *Montipora*. In both cases, living polyps are on the upper surface only. The undersurfaces of many are covered in coral tissue, perhaps entirely, perhaps only around the rim, but few undersurfaces have plankton-catching polyps.

selection (Baird and Moorse 2003). After settlement, the planula develops into a polyp, which then initiates colony formation.

In their mass spawning events, broadcasting corals show considerable synchronicity, which, of course, is necessary if sperm and eggs are going to even meet. Daylight length and water temperature are likely to be triggers involved in the process, as is the stage of the lunar cycle, and, in the short term, sunset.

Figure 2.6   Two species of the very speciose genus, *Acropora*. Top: The Caribbean *A. cervicornis*, forming branches that are up to 2 or 3 cm thick and which can reach over a metre long. Bottom: A shrubby Indo-Pacific form. In this, the clear axial polyps can be seen. These are lighter in colour because zooxanthellae have not yet migrated into the rapidly growing apical tissue. The apical polyps and corallites are much larger than the axial polyps.

Most corals that broadcast are the reef-building corals. The other method is used by the 'brooders'. These release only sperm, retaining their eggs in the parent body. Fertilization occurs internally, and developed planulae are released when suitably developed; when they are released, they may settle almost immediately. Most species that have adopted this system are ahermatypic, those which have no symbiosis with algae, although some common and important reef builders have brooding species on the reef also.

Asexual reproduction is commonplace in corals too, most commonly amongst branching species. Some use fragmentation and subsequent dispersal of fragments by wave action as the main method of propagating. *Acropora cervicornis*, shown in Figure 2.6 (top), is one such species. The result can be many colonies of the same species, which are in fact the same genetic individual, covering large expanses in favourable locations.

Variations of fragmentation occur, and an important one in lagoonal areas is that shown by *Goniopora stokesi*, an abundant species, members of whose genus are readily identified because of their long, extended polyps capped with rings of relatively small tentacles. This species makes 'polyp balls' (Figure 2.7), which are small spheres of skeleton which develop within the mass of polyps but detached from the main skeleton. At a certain size, these detach and roll away from the parent colony. While being an obviously successful means of propagating and dispersing, this species thrives in sedimented and sandy conditions and so also has the important effect of extending the hard limestone substrate over areas of sediments, literally extending the reef.

Most reef corals are hermatypic, that is, they contain large numbers of dinoflagellates, single-celled algae, called zooxanthellae. Each polyp contains large numbers of these, and the photosynthetic products of these algal cells are an important part of the coral's nutrition. The role of this symbiosis is a key aspect of corals and of reef growth, and Chapter 4 describes this in detail. All corals, whether they have this symbiosis or not, have tentacles which contain many batteries of stinging darts called nematocysts with

Figure 2.7   The Indo-Pacific *Goniopora stokesi*. This genus is abundant in calm areas, and has very long polyps, crowned by rings of small tentacles. It can reproduce asexually by the production of 'polyp balls' (arrowed). These will later detach, roll away and develop a new colony on soft sediments.

which they capture zooplanktonic prey (see Chapter 5, Figure 5.10). There is differentiation in the nematocysts: some seem to be primarily concerned with stinging and injecting toxin, while others appear to be more like 'hooks' which catch on to the prey and retain it. The quantity of zooplankton caught by these batteries of stinging cells provides some additional energy, but is thought to be important mainly in the provision of additional compounds which are essential to the needs of the coral. Coral species with larger polyps can sense zooplankton through amino acid residues and capture them using their tentacles. Larger food particles that cannot fit through the mouth can be digested externally using mesenterial filaments that move through the mouth or temporary body wall openings by ciliary action. This mechanism is also used to destroy corals competing with it for space (see 'Aggressive interactions amongst corals').

Aggressive interactions amongst corals

Competition amongst corals for space on the reef is vigorous. Many species are very aggressive, and will kill the tissue of neighbouring corals. The results of this are commonly seen where two colonies of different species grow close to each other, where one—the subordinate species—will exhibit a killed band of skeleton adjacent to the dominant coral. Mechanisms which corals use to attack another species fall into four categories. First, dominant species can extend digestive mesenterial filaments onto a neighbour (Figure 1). This is a short-range but rapid mechanism, which usually takes place at night, and the digestion of the target colony tissue that lies within range takes just a few hours. Second, sweeper tentacles may be developed which are loaded with stinging cells, and which, again during night-time, are swept over and onto the nearby subordinate coral (Figure 2). These sweepers may take several days or a couple of weeks to develop, following the coral's detection of a neighbour, but they have a longer range. Within any one pair of species, one is usually consistently dominant over the other for a given mechanism (Sheppard 1979). On the reef, the longer-range sweeper tentacles may act first but, if the same pair of species is artificially juxtaposed (or, presumably, if one is tumbled onto the other), the dominance pattern may be reversed because of the faster-acting but shorter-range mesenterial filaments deployed by the other.

A variation on sweeper tentacles is shown by *Goniopora*, which develops very elongated 'sweeper polyps', to similar effect. The third mechanism is a histological response but, for this to have effect, two corals must almost or actually touch. The fourth form of attack is by use of toxic chemicals which kill a neighbour, a method most commonly seen in soft corals.

Families of corals vary in their position in the aggressive hierarchy. The free-living mushroom corals (which are quite readily moved by waves or fish) are very aggressive. The coral *Galaxea* is likewise aggressive and uses this to dominate large expanses of lagoons. The Mussidae are aggressive in the Caribbean but much less so in the Indo-Pacific. Many families contain species showing a mix of dominant and subordinate species. But aggression is not the only characteristic to affect the overall abundance of a coral. Species that

Continued

Aggressive interactions amongst corals (*Continued*)

Figure 1. A polyp of *Lobophyllia* (right) has extended mesenterial filaments through its body wall and over a faviid colony. The right-hand half of the faviid is being digested away. This photo was taken 4.5 hours after placing the corals beside each other. **Figure 2.** A species of *Euphyllia* (right) has cast about eight sweeper tentacles over the coral on the left. The tentacles are packed with nematocysts, and part of the target colony nearest to the aggressor was killed. The sweeper tentacles took 2 weeks to develop, in this case.

are subordinate are likely to show other traits that enable them to stay on the reef, such as fast growth or high fecundity.

The continual aggressive interactions ensure that there is greater spatial movement and change than might be expected: while a quarter or a third of a typical reef slope might contain 'bare space', the precise location of each patch of bare space is continually changing as one coral slowly destroys another (Sheppard 1985).

Prof. Charles Sheppard, University of Warwick, UK.

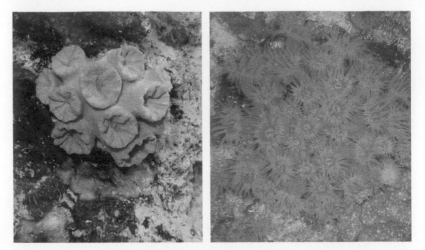

Figure 2.8   Possibly the commonest genus of azooxanthellate coral found on reefs is the genus *Tubastraea*. Left: An Indo-Pacific species, showing extratentacular budding found in many corals, where polyps bud from outside the tentacle rings. Polyps are tightly retracted. Right: The Caribbean *Tubastraea*, showing its bright yellow extended tentacles, photographed at night.

Corals that lack zooxanthellae are as numerous in terms of numbers of species as are the species that contain them, but these grow very much more slowly than those that do, although they usually do have a denser skeleton. Because they have no photosynthetic requirements, they can occur down to near-abyssal depths, where there is no light, and can grow close to the polar regions. They are a minor component of tropical reefs, however, as they are generally out competed in sunlit areas, but they commonly are profuse on vertical slopes, on the undersides of overhangs, and in other similarly dimly lit areas (Figure 2.8). These azooxanthellate corals occur throughout the world's oceans. Many are solitary or small, but some species can form large, dense aggregations which are also called reefs. These are briefly described in 'Cold-water corals' (Chapter 1). Even though these corals depend entirely on zooplankton capture by their tentacles for their nutrition, those that live on tropical coral reefs commonly retract their tentacles during the day, as do the hermatypic corals, presumably because of threat from grazing, although in dim locations they may be extended in daytime also.

2.2 Soft corals and sea fans

Soft corals and sea fans are grouped under the order Alcyonacea. There are as many different genera of soft corals as there are stony corals (Fabricius and Alderslade 2001) and several are abundant, conspicuous and very important, while others are much less so. Their skeleton consists of an organic matrix, in which are embedded numerous spicules, made from aggregates of calcite which, in a few areas, contribute substantially to reef building (Cornish and DiDonato 2003) but, generally, they are not major reef builders.

There is a fundamental difference in appearance between most of those of the Indo-Pacific region and those of the Caribbean. In the former, the typical soft coral is encrusting, lobed or, if branching, commonly fleshy while, in the Caribbean, branching forms predominate. For many years, a more definite taxonomic division was thought to exist between the 'soft corals' and the gorgonians or sea fans but, as more species were described and more anatomical work was done, it became clear that these exist in a continuum from simple soft corals to the elaborate gorgonians (Fabricius and Alderslade 2001). Nevertheless, there remains a marked difference between the Caribbean and Indo-Pacific in the appearance of the dominant forms found on the coral reefs (Figure 2.9).

Polyps of most are of a single kind, the autozooid, distinguished by having eight tentacles, with pinnules along both edges (see Figure 2.1, right). Some soft corals, mostly the larger forms in the Indo-Pacific, have a second tiny kind of polyp (siphonozooid) which has no, or rudimentary, tentacles. Siphonozooids are very numerous, and are thought to be involved in irrigating the colony.

Most octocorals have colonies of separate sexes, although some genera are dominated by hermaphrodites. Most are broadcasters, releasing eggs and sperm into the water in a synchronized fashion, triggered by temperature and lunar phase, although some are brooders in which eggs are retained and only sperm are released, such that development into well-developed planulae occurs within the mother colony. Larvae remain viable for days to weeks and may eventually settle out of the plankton and develop into adult colonies many kilometres from the parents. Reproduction, therefore, is broadly similar to that seen in the stony corals but, amongst the brooders, there is one variation not seen in stony corals, termed 'external brooding'. In this, the fertilized eggs are retained in mucus pouches on the outside of the mother colony, where they carry out their initial growth.

A settlement location is actively sought to some degree, using chemotaxis and phototaxis sensory systems. Most choose a hard substrate, and several select for crustose coralline algae. After developing the initial polyp, zooxanthellae may be taken in through the mouth; in others cases, such as the brooders, they may already contain zooxanthellae taken from the mother colony. Polyp division then occurs.

Asexual reproduction is common, and varied in some genera. The important genera *Lobophytum* and *Sarcophyton* (Figure 2.9, top) can divide into two by constricting their adult colonies, splitting the adult colony into separated individuals. The genus *Efflatounaria* forms runners which extend many centimetres away from the parent, attach to new substrate and then translocate

Figure 2.9   Two soft corals typical of their regions. Top: The Indo-Pacific *Sarcophyton*, lobed or cushion shaped. Bottom: The branching Caribbean *Pseudopterogorgia acerosa*.

tissue through the runner to the newly establishing colony, following which the runner is reabsorbed or disintegrates (Figure 2.10). Still another type involves a process of pinching off small parts of the edges of the adult colony, which then roll away, attach and settle, while the attractive, branching genus *Dendronephthya* can drop small balls of polyps which also roll away, attach and settle. All these methods have analogues amongst the stony corals.

Figure 2.10   The genus *Efflatounaria* can reproduce asexually by sending out runners, which attach to adjacent rock. After some weeks, the connecting tissue disappears, leaving a new, isolated clone.

Most colonies are suspension feeders, which capture plankton in a manner similar to the stony corals. The nematocysts are generally fairly weak, but are able to capture larvae of other animal groups such as molluscs and crustaceans.

Most importantly, many of them host zooxanthellae, again in a similar manner to the stony corals, but in most cases to a much lesser extent, and rates of photosynthesis may be insufficient to cover the soft corals' total respiratory needs. Fabricius and Alderslade (2001) make the point that, whereas most zooxanthellate stony corals have a thin layer of living tissue over a white, light-reflecting skeleton and so can easily cover their respiratory needs by photosynthesis, in typical Indo-Pacific soft corals, the amount of tissue biomass is greater, with a lower surface area. This may account for the relative inefficiency of photosynthesis in such soft corals, and it has been noted that some expand their colony size (by pumping their colony with water) in daylight to increase the surface area exposed to sunlight.

Caribbean reefs are heavily dominated by genera of branching soft corals that are so characteristic of the region but, when this occurs, less space is available for stony corals and other reef builders (Figure 2.11).

Figure 2.11   An assemblage of many species of typical Caribbean soft corals. These do not help to build a reef, although they may support a rich diversity. In such areas, stony corals may struggle to survive (See Plate 3).

2.3 Sponges

Sponges belong to the phylum Porifera and are the most primitive metazoan (i.e. multicellular) animals of all, with ancestors being traced as far back as the Cambrian era (505–570 Ma). They are enormously varied in shape; some, indeed, appear shapeless. Some, in the Caribbean especially, form huge vases (Figure 2.12), while others are encrusting or burrowing, consist of solitary pipes or a series of cones or pipes (Figure 2.13, left) or are tangles of flexible rope (Figure 2.13, right). They are important on the reef for several reasons. They are filter-feeders, several of them are vigorous eroders of the limestone reef and they occupy significant amounts of substrate, especially in the Caribbean region. Coral reef sponges largely belong to two classes: the Calcarea (calcareous sponges) and the Demospongiae (siliceous sponges), which includes the vast majority of species. Sponges range in size from a few millimetres to more than a metre in diameter.

Sponges have no true organs or tissues, and instead consist of a unicellular outer layer called the pinacoderm which encloses a collagen matrix that contains an organic skeleton composed of spongin fibres, and/or an inorganic skeleton composed of mineral spicules. This matrix, called the mesohyl, also

Figure 2.12   A large species of Caribbean sponge, *Verongula gigantea*.

Figure 2.13   Two morphologically different sponge forms. Left: The tubes of Caribbean *Callyspongia longissima*. In these, the outside contains thousands of very small inhalant pores, and the single exit aperture is at the top of each tube. Right: The rope-like *Aplysina fulva*. In this, there are relatively large exhalant pores at intervals along each 'rope'. The inhalant pores are minute.

contains various different types of cell, with functions such as phagocytosis, skeletal secretion and even 'muscular' contraction and, in many coral reef sponges, it houses symbiotic bacteria and cyanobacteria. Being filter-feeders, sponges have developed an elaborate aquiferous system. This consists of numerous entry pores (ostia) which lead to a series of canals and to one or more chambers in the sponge's body and, finally, to far less numerous but larger exhalent pores (oscula). The chambers contain flagellated 'collar cells' (choanocytes), whose beating generates water currents which continuously percolate through the sponge. Moderately sized particles (5–50 μm) are trapped in the canals, which get ever more narrow as they approach the chambers, where fine particles are trapped (<5 μm). Particles that are too large to enter the canals may be phagocytosed on the sponge's surface (Ruppert et al. 2004).

The calcareous and siliceous skeletons of sponges are only a minor component of their bodies and so sponges contribute relatively little to coral reef accretion, although in the geological past (e.g. 550 Ma in the early Cambrian and 210–290 Ma in the Permo-Triassic) sponges were the dominant reef builders and created 'sponge reefs' several hundreds of metres thick (Wilkinson 1998). Today, the functional importance of sponges to modern reefs is considerable, although the abundance and biomass of sponges are highly variable between geographic regions and between different parts of individual reefs. In the Caribbean, sponge abundances are on average fivefold greater than in the Indo-Pacific (Wilkinson 1998). There, they are commonly visually dominant, and can occupy as much space as corals, with many conspicuously large forms at all depths, although they may favour deeper waters, where there is a rich supply of plankton (Lesser 2006). On some Caribbean reefs, sponges can equal or exceed the biomass and abundance of corals (Rützler 1978).

Reflecting the availability of planktonic food, sponge populations decrease about fourfold from inshore to offshore regions of the Great Barrier Reef (GBR), while, on oceanic reefs of the Indo-Pacific, sponges are relatively uncommon (Wilkinson 1998). Despite this difference in abundance, species diversity in the Indo-Pacific is much greater than in the Caribbean, as is the case with most marine groups.

Over 5,000 species of sponge have so far been identified worldwide, with perhaps three times as many species awaiting identification. A large proportion of these live on coral reefs. In the Atlantic, >80 species are known from the Florida Keys, and about 300 species are known from the Bahamas (Lesser 2006) while, in the Indo-Pacific, about 600 species, 71% of which are endemic, are thought to inhabit the reefs and surrounding waters of New Caledonia, and nearly 830 species are known from the biodiversity hotspot of Indonesia (Hooper and van Soest 2002).

Sponges have a number of functional roles (reviewed by Bell 2008) that, where sponges are common, have significant ecological implications for coral reefs. The most important of these roles are (1) water filtration, (2) bioerosion and (3) sediment consolidation (i.e. binding). Other roles include the provision of habitat and food for other reef organisms; nitrogen cycling; and, where photosynthetic symbionts such as microalgae, cyanobacteria or macroalgae are present, primary production too.

Much of a sponge's diet consists of picoplankton (0.2–2.0 μm) such as heterotrophic bacteria and cyanobacteria (Pile 1997; Turon et al. 1997), and there is recent evidence that sponges even have the capacity to capture viruses, such that the coral reef sponge *Negombata magnifica* can remove 23%–63% of viruses from the water passing through its filtration system (Hadas et al. 2006). The speed and efficiency with which sponges filter particulate matter from the water column, enabled by their unique aquiferous system, is remarkable. For instance, in the Caribbean a range of sponge species were found to remove 65%–93% of the available particulate matter from the water column (Lesser 2006), while sponge communities at 25–40 m on the reef slope in Jamaica have been estimated to filter the entire water column above them every day (Wilkinson 1998)! This capture of plankton by sponges represents a key link between the water column and the benthos (known as 'bentho-pelagic coupling') and provides an important route through which particulate carbon and nitrogen are channelled to predators further up the food chain (Lesser 2006). Sponge predators include various nudibranch molluscs, sea urchins, starfish, crustaceans and fish, as well as the hawksbill turtle (Wulff 2006). Furthermore, the efficient uptake of suspended matter by sponges provides a valuable mechanism for the retention of essential nutrients by coral reefs in tropical seas that are typically nutrient poor.

Sponges of the families Clionidae and Spirastrellidae are major bioeroders of coral reefs, where they bore into calcium carbonate structures with the aid of chemicals produced by specialized 'etching cells' (Figure 2.14). Susceptible structures include coral skeletons, with the skeletons of dead corals being colonized within a few weeks of initial contact and the skeletons of live corals being colonized within a few months (Schönberg and Wilkinson 2001); the sponge then spends most of its life within the skeleton and spreads through it. Ultimately, this activity produces carbonate rubble and sand. Competitive species such as the Caribbean *Cliona caribbaea* and *Cliona lampa* can erode 8–24 kg m^{-2} y^{-1} of calcareous substrate in colonies that they have occupied (Acker and Risk 1985). Thus, these bioeroding sponges destabilize corals, by making their skeletons more fragile.

The creation by bioeroding sponges of fine sediment also acts as 'packing material' in the reef framework, and sponges also play a more direct role in stabilizing the coral reef, by binding unconsolidated framework material together. The stabilization of large corals by encrusting sponges has long been recognized and was experimentally tested by Wulff and Buss (1979), who removed sponges from a number of Caribbean patch reefs. These authors measured a 40% loss of coral colonies after 6 months, whereas only 4% of colonies were lost from untouched (i.e. control) reefs after this same period.

Their symbioses with bacteria and algae are both important and varied, and are described in Chapter 4.

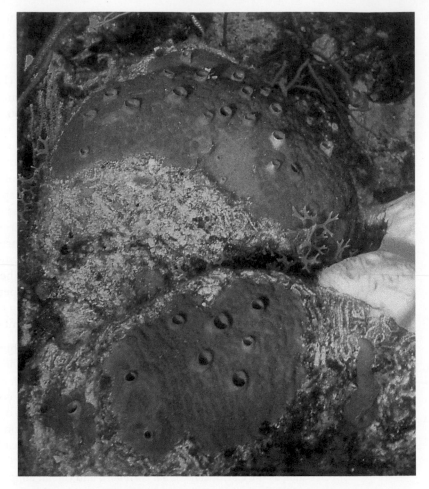

Figure 2.14 The boring sponge *Cliona* is very abundant in the Caribbean region. This species, *Cliona delitrix*, forms a thin surface film over a massive coral, visible only by the red film and the oscules with which it pumps water. Most of its mass is inside the greatly eroded coral skeleton.

2.4 Other animal species

2.4.1 Molluscs, echinoderms, crustaceans, polychaetes

These four groups, it can be justly argued, are each far too big, diverse and important to include into one short summary. However, one book on reef biology for so many important groups means that several such descriptions must be truncated or ignored. Also, the emphasis is on the whole, integrated reef system, not on each aspect of it, no matter how interesting, so only the very dominant 'space occupiers' are featured in much of this chapter.

The biomass of these four very important groups is a significant part of the whole. Many of them in addition carry a large, inorganic mass of shell of one kind or another (Figure 2.15). Mainly, these shells have arisen for the obvious reasons of protection or trying to overcome the protection of a prey species—the Red Queen at work, mentioned in Chapter 1—but, even so, some molluscs bore holes into others, starfish prise open shells of molluscs, and crustaceans use their armoured claws to crack open prey as well as to defend themselves. Even many polychaete worms secrete calcareous tubes, although most are simply cryptic (see Section 2.4.2). Several thousand species of each group occur on reefs or around them on the associated soft substrates. They, like other groups, play key roles in the nutrition and food chains of any reef. Many filter-feed and many more are detritus feeders, while many others also erode the reef fabric.

2.4.2 Cryptic and bioeroding species

The total mass and numbers of cryptic, eroding species is substantial, and has even been estimated to exceed that of visible biota (Glynn 1997). Bioerosion of corals is a natural and continuous event equally or more important

Figure 2.15   Representatives of three major invertebrate groups with high diversity on coral reefs: echinoderms, represented by the brilliantly coloured starfish *Protoreaster lincki*; crustaceans, represented by a mantis shrimp (Stomatopoda); and molluscs, represented here by the predatory cone shell *Conus geographus*. This is a piscivorous cone and so has a very strong and fast-acting toxin. (See Plate 4).

Source: Mantis shrimp photo by Charles Delbeek. Cone photo by Anne Sheppard.

to reefs as physical erosion is. The extent is very variable and is influenced by location, depth and pollution gradients but, overall, bioerosion is continuous and substantial. Both live and dead corals are subject to it, although different groups of organisms are responsible in different places. Important bioeroding groups are the parrotfish (scarids), sea urchins, polychaetes, sponges and molluscs. Relevant to this are major coral predators, such as crown-of-thorns starfish and several molluscs, which do not directly erode limestone but kill living tissue and so expose large expanses of denuded limestone which then become more easily invaded by bioeroders.

Attempts have been made to relate amounts of bioerosion to 'health' of a coral reef, perhaps as a proxy for numerous other methods, although correlations are mostly too weak, and variability is too high, for any clear relationship to be seen. Compounding the variability is that, when bioerosion takes place, the surface area of exposed limestone is further increased, leading to even greater potential for both direct physical and chemical solution of limestone (Hutchings et al. 2005). In addition, it is difficult to separate natural or background rates of bioerosion from those which are increased by anthropogenic influences such as increased run-off of water from the land, and several of these anthropogenic influences commonly act together (Chapter 9).

Estimates of net loss of limestone from bioerosion (i.e. loss minus accretion) may occasionally exceed 7 kg m^{-2} of reef each year, although values of 1–4 kg m^{-2} are more common in sites of the GBR, some parts of which are exposed to terrestrial effects (Hutchings et al. 2005). This should be contrasted with many rates of total reef accretion of about 1 kg m^{-2} because it shows that net growth of coral reefs can be easily and substantially reversed. In Eastern Australia, large differences in rates of erosion were attributed to proximity to river outflows, because freshwater deterred grazers, so that greater grazing and hence erosion was seen at greater distances from river plumes. At other sites, such as Tahiti, grazing causing erosion was very high close inshore (about 7 kg m^{-2}), which was attributed to very high densities of echinoids, whose numbers were encouraged by the overfishing of their predators (Pari et al. 2002). Thus, multiple factors and indirect causes strongly influence the amount of bioerosion, but its measurement may not easily find correlation with reef condition, since even healthy reefs undergo considerable bioerosion.

2.5 Macroalgae

Macroalgae are also species rich, productive and functionally important members of the coral reef benthic community. It is thought that, globally, about 2,000–3,000 species of macroalgae live on coral reefs with all the major groups being well represented: (1) Chlorophyta (green algae), (2) Phaeophyta (brown algae) and (3) Rhodophyta (red algae).

These algae can be further placed into several functional groups such as fleshy macroalgae, turf algae, crustose coralline red algae, branched coralline red algae and brown algal crusts. Some are very important in constructing the reef, while others are important components of food chains. Still others, especially the larger brown algae, do not appear to be eaten at all, even where they occupy substantial areas of reef.

Reef constructors are found mainly in the red and green algal groups. In the reds, the common stony reds that thrive on reef crests and which build the algal ridges and spurs are most important (described in Chapter 1 and Figure 1.12). Of the greens, a particularly important coralline species is *Halimeda* (Figure 2.16), on which much reef construction and sand production depends. These chains of calcareous discs can grow at the rate of one new disc per day on each chain, with each forming a coarse particle of sand upon detachment and death. On the Bahamas Banks, for example, much of the sand is composed of *Halimeda* discs. In the Indo-Pacific too, *Halimeda* occurs in huge 'bioherms', many of which have been studied on the shelf off Eastern Australia.

Figure 2.16   In both the Indo-Pacific and the Caribbean, reef walls may be heavily festooned with the calcareous green alga *Halimeda*. Each frond is a chain of small discs, made of limestone secreted by the disc's enveloping green tissue. Each chain may produce a new disc every day in good growing conditions, producing large quantities of coarse sand. Inset: Close-up of *Halimeda* discs, with several dead discs now forming part of the sand.

The brown algae are generally less important than the reds and greens in this respect, with a few exceptions such as the fleshy frondose brown alga *Padina*, which also creates fine sand and is important in many areas, like those parts of the Persian Gulf where salinity is too high to support much coral growth. The brown algae tend to be the largest. *Sargassum* forms dense thickets over 1 m tall in many places, especially where conditions are marginal for corals. The southern Red Sea is one notable location where this is the case. Another genus of large fleshy browns is *Turbinaria*, recognized by their triangular blades or fronds, which likewise can densely cover reef surfaces and thus exclude significant coral growth.

The greens and reds are numerous in terms of numbers of species, although most are smaller in size. The fleshy green alga *Caulerpa* from Australia and the Caribbean has achieved notoriety as an invasive species in the Mediterranean, where it is blanketing the native biota but, in its home range, it is a minor component, recognized by its chains of vivid green blades. The reds are extremely diverse, and there are many species of finely branching forms especially in sheltered areas, such as the red turf alga *Ceramiales*.

Often, coral reef macroalgae are not immediately obvious to casual observers, generally because much of the macroalgal biomass is removed by grazers as fast as it is produced. There are reports of 85%–100% of fleshy macroalgal biomass being removed by grazing fish in periods of just 8–12 hours (Lewis 1986; Mantyka and Bellwood 2007). This is not always the case, though, as macroalgae are opportunists and often thrive on reefs that are subject to disturbance, such as where terrestrial run-off and elevated concentrations of nutrients occur. For instance, at two inshore regions of the northern GBR, Fabricius et al. (2005) found that percentage cover of turf algae, coralline algae and fleshy macroalgae was positively related to nutrient levels, with turf alga cover showing an especially marked difference between sites of 50.7% versus 20.6%; most of the change in algal cover, abundance and taxonomic richness was related to red and green algae rather than brown algae. Reduced grazing pressure on some reefs may also lead to a greater abundance of macroalgae. For example, less fish grazing on nearshore than on mid-shelf reefs of the GBR is the dominant reason why the brown fucoid seaweed *Sargassum siliquosum* is most abundant at nearshore sites (McCook 1997).

Clearly, macroalgae have an important functional role as a food source for grazers such as fish and sea urchins, although they also have other hugely important roles on coral reefs. This especially applies to crustose coralline red algae that bind sediment and so help consolidate the reef, and all coralline algae that contribute calcium carbonate to reef sediments and assist the growth of reefs. On many reefs in the western Indo-Pacific, such coralline algae are the dominant benthic organisms and may contribute more towards reef building than do corals. A further function of macroalgae is their great capacity to capture nutrients and so assist the nutrient budget of the reef (Raikar and Wafar 2006).

A small number of macroalgae form associations with invertebrates. Perhaps the best known of these is the symbiosis between the Indo-Pacific red seaweed *Ceratodictyon spongiosum* and the sponge *Haliclona cymiformis*, which covers the alga's branches. The sponge benefits from the provision of photosynthetic compounds and by having a robust substrate to attach to in its otherwise unstable habitat of sandy lagoons and reef flats; in return, the alga receives the sponge's nitrogenous waste, which promotes its growth (Davy et al. 2002).

2.6 Seagrasses and mangroves

Two groups of angiosperms are also associated with coral reefs. Seagrasses are found on soft sediments from the tropics to subpolar regions, and may be abundant on reef flats and in lagoons. Seagrasses are certainly not restricted to coral reefs. Plants of one genus, *Thalassodendron*, can grow on hard substrates and hence may live in close association with corals, although, most often, reef-associated seagrasses live in soft substrates (Figure 2.17, top). They are true flowering plants (with roots, not holdfasts) and are hugely productive.

Very few reef species graze on seagrasses, but notable exceptions are found amongst two vertebrates in particular. One kind of seagrass is called turtle-grass, because it is food to green turtles, while many kinds are grazed by herds of dugong, marine mammals that depend entirely upon them. Mostly, however, the blades of seagrasses are not eaten directly but are decomposed by microbial activity when they die, so the organic material enters the food chain via the microbial food web (Chapter 5).

Their preferred habitats range from fine mud to coarse sand, depending on species. In reef areas, one role of greatest importance for seagrasses is their role in binding soft substrates, mainly those in 'backreef' areas, in lagoons and in sheltered locations where large quantities of sediment are continuously pumped by wave action from the nearby reef. Approximately half of the biomass of a seagrass bed, or even more, is contained in the extensive root system beneath the surface, and this stabilizes the otherwise highly mobile substrate. The extent of this can be seen where physical damage has killed a patch of seagrass, showing the relative quantity of root system compared with their leaves (Figure 2.17, bottom). Where seagrass has been removed or killed, erosion rapidly removes the substrate.

Seagrasses are commonly found in lagoons of atolls, behind barrier reefs or on sheltered reef flats of fringing reefs, within a few kilometres of reefs. Where this is the case, they interact with the reef in several important ways. They receive and stabilize sediments from the reef, as noted, but they may also be sites of deposition of fine sediments which enter as run-off from the land,

Figure 2.17   Top: Two seagrasses in a mixed stand, on a shallow reef in the Seychelles. The species with cylindrical, straw-shaped leaves is *Syringodium isoetifolium*; the leaf shape is the diagnostic feature of this species. The flat-bladed form is *Thalassia hemprichii*. Bottom: A bank of seagrass being eroded away. The extent of the root network is clearly visible.

and so trap fine sediments which do not therefore end up on the corals further offshore. But they also exchange species with reefs, and several species of fish move between seagrass beds and reefs, according to the time of day.

The second group of angiosperms, the mangroves (Figure 2.18), is also associated with coral reefs, particularly in coastal regions, which are easily colonized by mangrove seedlings, although mangroves can be found on offshore reefs also. Mangroves comprise a diverse group of halophytic trees and shrubs with little taxonomic relation to each other, sharing only the ability to live in salt water. Like seagrasses, they favour soft sediments in calm areas such as coastlines behind fringing and barrier reefs. Like seagrasses, mangrove forests are often not associated with coral reefs and, most strikingly, form complex forest ecosystems which progress into estuaries. There, clear zonation exists, with species arranged according to elevation within the intertidal zone and along the salinity gradient from 100% sea water to progressively lower salinity. They may then grade into true tropical forest at the highest end of the estuary.

Mangroves are largely distributed from the tropics through to subtropical regions, although a few species (e.g. the widespread Indo-Pacific species *Avicennia marina*) reach warm–temperate regions such as southern Australia and northern New Zealand. Where seagrass beds and mangrove forests do occur in close proximity to coral reefs, these three ecosystems can interact in several ways, the most important of which are (1) the stabilization of soft substrate by mangrove and seagrass roots, thus preventing sediment from dispersing across the coral reef; (2) the dissipation of wave energy by the coral reef, providing the calm, hydrodynamic regime favoured by mangroves

Figure 2.18   Left: A typical mangrove stand at high tide, in this case, *Rhyzophora* in the lagoon of Aldabra Atoll, Indian Ocean. Right top and bottom: Caribbean *Rhizophora* (San Salvador, Bahamas) showing the thickets of prop roots, forming 'nursery cages' which provide protection for countless juvenile and small fish from larger carnivorous fish.

and seagrasses; and (3) the provision, by mangroves and seagrasses, of sheltered, food-rich nursery habitat for coral reef organisms.

Regarding the first two, those of substrate stabilization and energy, their importance is most obviously seen when mangroves are removed, for timber or for land development. In such cases, rapid and massive scouring of the sediments becomes evident. The same effect occurs if an offshore breakwater, such as a coral reef, is excavated or otherwise killed, permitting greater wave energy to break in the mangroves. Many mangrove restoration attempts have been made, and many of those have been unsuccessful because of the difficulty of stabilizing the sediments sufficiently for mangrove seedlings to take root.

The provision of shelter for reef and other species is also crucial. The root systems of all mangroves are complex, designed for living in sediments which are easily shifted and which, in many cases, produce anaerobic conditions beneath the top layers, requiring elaborate systems of vertical shoots, pneumatophores, to permit oxygen to reach the root tissues. Root systems consisting of tangles of prop roots act in the manner of a cage, providing protection for many species of fish (Figure 2.18, right, top and bottom). Juvenile stages of several reef species exploit this protection before growing large enough to increase their chances of survival in more exposed locations and on coral reefs.

2.7 Rates of coral growth, rates of reef growth

Rates of growth of calcifying reef organisms are several orders of magnitude greater than rates of reef growth. That the first leads to the second is the reason behind the existence of many nations found in all three tropical oceans. Although reefs may grow more or less continuously for millennia, it is largely the accumulated growth of rock, rubble and sand over the last few thousand years which has led to the existence of coral islands.

Corals grow at rates that vary considerably according to genus, or even species. The largest boulder corals from the genus *Porites* form large domes, reaching a radius of several metres (see Figure 2.4). Growth of these is approximately 1 cm y^{-1} in terms of radius expansion, that is, a colony diameter expansion of 2 cm y^{-1} and a height increase of perhaps 1 cm y^{-1}. These corals are likely to be the largest and oldest living corals, and many are remarkably uniform in their hemispherical shape even after 400 or 500 years.

Large colonies of *Porites* are favoured for a range of geochemical work. When such colonies are drilled to extract a core, growth rings similar in principle to tree rings can be seen in the extracted cylinder of skeleton (see 'Corals as archives of past climate' in Chapter 8). Each growth ring of about 1 cm long represents 1 year's growth; the rings are due to differences in the density of limestone laid down in different seasons. A core of 300 cm long, therefore, might show about 300 pairs of lighter density–heavier density bands and will be 300 years old. The top band, where living tissue exists, is obviously that of the current year, and each band below that contains material deposited in successively earlier years. The value of this is that samples of limestone can be taken of known ages. From each age, measurements can be made of various parameters such as deposition of pollutants and, based on ratios of different isotopes of oxygen (^{16}O and ^{18}O), the temperature that bathed the colony at the time that particular layer was deposited.

Some coral skeletons are much denser than those found in *Porites*. Other massive-shaped corals expand more slowly, and several faviid corals, *Leptastrea* for example, grow to relatively small size but deposit skeletons that are extremely dense. Others have skeletons that are much more porous, none more so than *Alveopora*, which has skeletons so porous that dried ones will float.

Most colonies of *Acropora* are fairly porous; because of this, together with their branched colony shape and method of growing rapidly in a direction dictated by the axial corallites, they can expand each of their branches by 10 cm y^{-1}, and even a few faster rates have been estimated. Many bushy species may increase in height (assuming no breakage, which is usually considerable) by 10 cm y^{-1}, therefore. In the case of a tabular-shaped colony such as those shown in Figure 1.1, the diameter of the table may expand by 20 cm y^{-1}. The tips of the branches or tables contain limestone that is relatively very weak and even crumbly to the touch, and this material hardens up with more deposition of limestone by the radial polyps and coenosarc tissue as time progresses.

Clearly, the rate of reef growth is nothing like as fast, and vertical reef accretion figures of 1–10 mm y^{-1} are commonly cited, sometimes even 20 mm y^{-1} (Glynn 1997). The numbers vary because of several factors, including the location on the reef where measurements were taken and whether only modern growth or an average accretion taken throughout the entire vertical extent of a reef over millions of years is considered. In such cases, the rate per year is, of course, the average rate at which that reef *did* grow, not the rate at which it *could* have grown, and the rate is unlikely to have been the same throughout its life. All these rates, which will differ in any case between reefs, are one to two orders of magnitude slower than the rate of extension of coral colonies. Rapid coral growth may even be seen to occur on reefs that are actually suffering net erosion, such as when impacted by pollution of various kinds (Edinger et al. 2000).

Reef cementation is a major element of reef growth. Fine sediments are forced into the cavities in porous limestone, and these change gradually into solid rock. The corals (and other calcareous organisms) are the main source of this sediment. There is complex chemistry involved, as the limestone particles are solidified again by rain or pH changes surrounding the particles in the presence of microbial organisms, but the initial production of this limestone by corals is crucial. Estimates of total limestone deposition have been derived in several ways, for example, by chemistry changes in water surrounding corals on reefs closed off with inverted glass domes, by increases in weight of coral colonies, by pH and alkalinity sensors floating across a coral reef and by using geological records in which whole-reef growth is measured against dated sections. Quantities of limestone produced by a healthy reef range from ~1–5 kg m^{-2} y^{-1} (Barnes and Devereux 1984; Barnes et al. 1986; Beanish and Jones 2002; Hallock 2005).

Much of this deposited limestone, however, degrades. The progression of limestone rock degradation is important. In simplified form, coral rock becomes reduced to rubble, then to sand and then to fine silt (Wright and Burgess 2005; Hopley et al. 2008). Agencies driving each stage differ, from mechanical breakage down to processes of constant rasping, mixing and stirring by marine life, some of it very small. Areas in a lagoon where each stage dominates also differ. The first products of breakdown (boulders and rubble) do not move very far, of course. Sand size particles do move, and sand is piled up to form shores, but it moves decreasingly with greater depth, where wave energy is less. Much sand moves across lagoons to sedimented areas, where much reworking takes place. These final products of fine sand and then even finer sediment may become suspended, and can remain in suspension for several hours or days. Values of suspended sediment levels are known for many coral reef areas, where values range widely according to conditions. Clear water usually has about 2.5 g m^{-3} of solids in suspension; inshore reef water commonly contains 5–20 g m^{-3}, while values of 200 g m^{-3} or more are common after storms (Larcome et al. 1995; Hoitink 2004). Importantly, too, about 0.3 kg sand from each square metre each year dissolves back into the water in lagoons because of constant movement and bioturbation from organisms (Tudhope and Risk 1985).

Thus, there is a relatively rapid dynamic between growth and erosion, much more rapid than is the case with almost all other geological materials. In part, this is because limestone is a relatively soft rock, a factor which permits a large number of burrowing and rasping organisms to have developed niches within it, and in part this is because the rock it is alkaline, easily etched by changes in pH.

2.8 Soft substrates

The sand produced from physical erosion and bioerosion is one of the key building components of the reef. Behind most reefs there are large deposits of sand, pumped into drifts or mounds, or into extensive lagoons, where they contain a biota which may interlink with that of the reef but which has a characteristic of its own. The area of sandy expanse associated with a reef and derived from it may exceed that of the reef itself by orders of magnitude.

Whatever the origins of the sand, an important grading and sorting process takes place. Small particles are moved more easily than larger ones so, wherever there is a current, there will be gradients of sand and sediment particle size. Where a mixture of suspended particles exists in a current, such as that travelling over the reef from seaward, water velocity will drop as the flow travels into deeper water. Larger particles fall out first, followed by progressively lighter particles. Different species favour soft sediments of different grain size, so the result is a zonation also of different forms of life. Usually, the species living on sand are adapted exclusively to a soft substrate existence. Several forms of algae form films over more stable expanses, which themselves provide food for numerous grazers, notably echinoderms and molluscs. Burrowing organisms create tunnels and ensure a steady mixing of layers, which results in continued oxygenation of the sediment down to nearly half a metre depth in many locations. Molluscs, especially micro-molluscs, may be both diverse and abundant, as may holothurians and echinoids, with many more living and hunting beneath the sand than may be seen on the surface. An important group of animals that are commonly overlooked is the Foraminifera, which comprises protozoan animals that range in size from smaller than a millimetre to a couple of centimetres across. These make a skeletal case which, again, is made of limestone. Some are planktonic forms which sink to the sand after death, while others live in high density on soft substrates and make most of the sand itself. The whole community of micro-organisms is itself food for the sea cucumbers, which ingest huge quantities of it daily, extract the nourishment and excrete the sediments again, greatly reducing the amount of standing crop of organic material.

The biota of reef-associated sand patches is high, and the productivity likewise is high; given that the area of sand usually greatly exceeds that of the coral reef from which much of it derives, this region is a key, integrated part of the reef system as a whole. Its seagrass and algal beds provide habitat for many reef species, and the sediment itself both is derived from, and by later consolidation, may become part of the reef matrix itself.

3   The abiotic environment

3.1 Controls on coral distribution

Coral reefs are restricted to shallow tropical seas. The abiotic factors that determine the survival of reefs within these regions are quite specific and, in particular, coral reef growth is controlled by (1) salinity, (2) temperature, (3) light (both quality and quantity), (4) nutrients, (5) exposure and other hydrodynamic factors, (6) sediment and (7) seawater carbonate chemistry. These abiotic factors control the global distribution of coral reefs. Cold temperatures, high sediment loads and both reduced and elevated salinities lead to reduced reef growth. On the west coast of the Americas, the upwelling of cold, nutrient-rich water associated with the Californian and Peruvian currents limits coral reef formation and promotes the growth of extensive kelp forests; similarly, the Canary Current and Benguela Current upwelling systems limit coral reef formation on the coast of West Africa, as does the seasonal upwelling system on the Arabian coast (Figure 3.1). On the northeast coast of South America, major rivers such as the Amazon and Orinoco massively lower the salinity and transport light-blocking particulate matter into coastal regions, prohibiting reef growth. Clearly, a sound knowledge of the abiotic environment in which coral reefs live is essential for understanding their evolution, function and biogeography, as well as their susceptibility to environmental disturbance (e.g. global warming, land-based sediment discharge). Both the average and the broad limits of abiotic conditions in which coral reefs grow are summarized in Table 3.1.

3.2 Salinity

Because corals are intolerant to wide ranges in salinity, most coral reefs exist where salinity is stable and where average salinity is that of normal seawater (34–36 ppt). Marked dilution of shallow reef waters may occur after heavy periods of rainfall, particularly when these coincide with low tide or the

The Biology of Coral Reefs. Second Edition. Charles Sheppard, Simon Davy, Graham Pilling, and Nicholas Graham, Oxford University Press (2018). © Charles Sheppard, Simon Davy, Graham Pilling, and Nicholas Graham (2018). DOI 10.1093/oso/9780198787341.001.0001

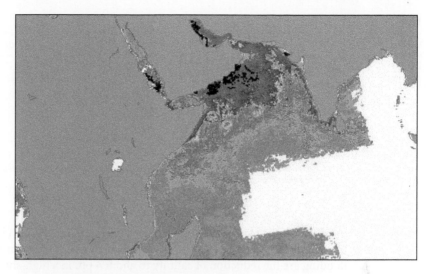

Figure 3.1   A satellite image of the Indian Ocean showing oceanic productivity. This image shows the Arabian upwelling, which takes place every year between about May and September, bringing cool and nutrient-rich water to the shores of Arabia; this explains why reef development in this area is substantially inhibited, even though the reef is located in warm water. The darker the colour, the greater is the chlorophyll concentration in the water. (Note that the white part in the extreme bottom right corner of the image is data deficient, not necessarily poor in plankton.)

Source: Image from the Coastal Zone Color Scanner.

Table 3.1 Average and extreme (minimum and maximum) values of abiotic variables on coral reefs around the world

Variable	Minimum	Maximum	Average	Standard Deviation
Sea surface temperature (°C) (weekly values)				
Avg.	21.0	29.5	27.6	1.1
Min.	16.0	28.2	24.8	1.8
Max.	24.7	34.4	30.2	0.6
Salinity (ppt) (monthly values)				
Min.	23.3	40.0	34.3	1.2
Max.	31.2	41.8	35.3	0.9
Nutrients (μmol L^{-1})				
Nitrate	<0.001	3.34	0.25	0.28
Phosphate	<0.001	0.54	0.13	0.08
Aragonite saturation				
Avg.	3.28	4.06	3.83	0.09

Continued

Table 3.1 (*Continued*)

Variable	Minimum	Maximum	Average	˙ Standard Deviation
Maximum depth of light penetration (m)				
Avg.	9	81	53	13.5
Min.	7	72	40	13.5
Max.	10	91	65	13.4

Note: Values do not include coral assemblages which do not form reefs.

Source: Data from Kleypas, J.A., McManus, J.W. and Meñez, L.A.B. (1999). Environmental limits to coral reef development: Where do we draw the line? *American Zoologist* 39:146–59.

flooding of coastal rivers, and such events may dilute reef waters for periods of minutes to hours, or days to weeks. Examples of short-term freshwater flooding are the reduction of salinity to about 4 ppt in a high-shore reef-flat tidepool on Onotoa Atoll (Kiribati) after a day of heavy rain (Cloud 1952) and to 10 ppt on shallow reefs of the Inner Gulf of Thailand, into which five major rivers discharge during the rainy season (Moberg et al. 1997). In such cases, these extreme levels are removed by subsequent tidal flushing and currents.

Many reefs exist at latitudes 7°–25° north and south of the equator where hurricanes (also called cyclones) occur. Over larger tracts of reef, periods of reduced salinities can follow hurricane events. When Hurricane 'Flora' hit Jamaica in October 1963, it reduced surface salinity (≤2.5 m depth) to as little as 3 ppt immediately after the hurricane, and salinities of less than 30 ppt persisted for more than 5 weeks (Goreau 1964). Cyclone 'Joy' on the east coast of Australia in late 1990/early 1991 caused considerable coastal flooding and, over the Great Barrier Reef (GBR), plumes of low-salinity water that persisted for up to 3 weeks (Van Woesik et al. 1995). In an extreme case, around Keppel Island, salinities during the flood peak were 7–10 ppt at the surface, 15–28 ppt at 3 m, 31–34 ppt at 6 m and 33–34 ppt at 12 m (Brodie and Mitchell 1992). In addition, some coral reefs are exposed to low salinities on a regular basis, with the record being a monthly minimum of 23.3 ppt for coral reefs around the Moscos Islands of Burma; other notable examples are at localities in the Bay of Bengal and the Eastern Pacific, where the monthly minimum may be as little as 27 ppt (Kleypas, McManus, et al. 1999).

At the other end of the salinity spectrum, some coral reefs exist in regions of elevated salinity, such as particular areas off Western Australia, some restricted lagoons of Pacific atolls and some semi-enclosed waters of the Middle East (Jokiel and Maragos 1978; Sheppard 1988; Coles and Jokiel 1992). In the Middle East, coral reefs exist where the average salinity exceeds 40 ppt; these areas include the Persian Gulf, the central and northern Red Sea, the Gulf of Aqaba and the Gulf of Suez (Sheppard et al. 1992; Kleypas,

McManus, et al. 1999; Coles 2003). The Persian Gulf is perhaps the best known of these high-salinity reef environments. There, high rates of surface evaporation and low rates of freshwater input lead to an average salinity of about 42 ppt in open water, with salinities off the coast of Saudi Arabia and in embayments of the Gulf of Salwah (between Saudi Arabia and Qatar) reaching extremes of 50–70 ppt (reviewed by Coles 2003). Such extreme salinities are, however, too high for the development of coral reefs, and the upper salinity for extensive coral coverage is about 45 ppt (Table 3.2; Sheppard 1988). Similarly, extensive coral communities develop in the central Red Sea at 40–45 ppt (Sheppard and Sheppard 1985; Piller and Kleemann 1992). Such coral communities are usually less diverse than those living at oceanic salinities, such that coral diversity decreases sharply, by about one species per every unit rise in salinity, in salinities of 41–50 ppt (Sheppard 1988). This reflects the differing salinity tolerance of coral species, and very few species can tolerate salinities above 45 ppt. Sheppard (1988) listed ten coral species that can survive salinities of at least 46 ppt for periods of 1–3 months or more, with three of these (*Siderastrea savignyana*, *Porites nodifera* and *Cyphastrea microphthalma*) being able to survive up to 50 ppt, perhaps continuously. One of these very hardy corals, *P. nodifera*, is largely responsible for forming reefs at 43–45 ppt (and, indeed, at temperatures of more than 35 °C too) in the Persian Gulf, where it builds high-diversity 'oases' amongst the seagrass (Sheppard 1988).

Table 3.2 The relationship of salinity and turbidity with five different coral-community types around Bahrain and in the neighbouring Gulf of Salwah (in the Persian Gulf)

	Type 1 Community	Type 2 Community	Type 3 Community	Type 4 Community	Type 5 Community
Salinity (ppt)	<43	<42	43	43–45	>45
Turbidity	Normal	Normal	High	Normal	High
Coral diversity	Medium	Rich	Medium	Poor	Poor
Per cent cover					
Total	30–90	35–70	1–10	12–70	2–5
Acropora spp.	20–75	5–15	–	–	–
Porites compressa	–	8–20	–	–	–
Porites nodifera	–	–	–	10–65	–
Brown algae	–	–	+	–	++

'Rich' and 'Poor' diversity refers to >15 species per site and <5 species per site, respectively. Values are per cent cover of total corals or named common coral taxa. For brown algae, '+' indicates tall but scattered plants, <1 every 5 square meters; '++' indicates tall plants, >1 every square meter and forming a good canopy. For turbidity, levels are expressed as 'normal' or 'high'. Environmental measurements were taken concurrently with the per cent cover estimates. Note that 45 ppt is the upper limit for extensive coral coverage (at least 70% cover), except where high turbidity levels limit corals, and the success of brown algae in areas where corals are unable to thrive.

Source: Data, and reef community types, from Sheppard, C.R.C. (1988). Similar trends, different causes: Responses of corals to stressed environments in Arabian seas. *Proceedings of the 6th International Coral Reef Symposium* 3:297–302. Note these data were obtained before massive mortality from sedimentation and warming in the 1990s.

The susceptibility of corals, and, indeed, other invertebrates and algae on reefs, to changes in salinity classes them as 'osmoconformers', where they become iso-osmotic with the surrounding seawater. Corals can exert a degree of control over their internal water content and the intracellular concentration of osmotically active molecules ('osmolytes'), and must do so in order to prevent cellular damage and optimize biochemical reactions when their external salinity fluctuates. Very little is known about the cellular mechanisms involved in the osmotic balance of corals, but they are likely to synthesize so-called compatible organic osmolytes (COOs) to prevent cellular dysfunction (Mayfield and Gates 2007; Yancey et al. 2010). Typical COOs include free amino acids, glycerol, glycine, betaine, and dimethylsulfoniopropionate. These compounds are synthesized by corals and/or their microalgal symbionts (dinoflagellates of the genus *Symbiodinium*, known as zooxanthellae), are abundant within their cells and may serve multiple metabolic functions; hence they are relatively 'cheap' in energetic terms. Furthermore, there is evidence that corals may use heat-shock proteins, such as Hsp60, to limit cellular damage under salinity stress (Seveso et al. 2013).

However, corals still have osmoregulatory limits and are unable to tolerate extreme or quickly changing salinities. At very low salinities (hypo-osmotic stress), the tissues of corals swell and become damaged (Van Woesik et al. 1995) because uptake of water by osmosis occurs too quickly to be countered by the coral's regulatory mechanisms. Likewise, algal cells may swell and rupture under hypo-osmotic stress (Lobban and Harrison 1994) and even small changes in salinity can induce a loss of photosynthetic capacity in some species. For instance, when the coral *Stylophora pistillata* was collected from the Gulf of Aqaba (the Red Sea), acclimated to 38 ppt for several years and then exposed to salinities of 34–40 ppt for 3 weeks, photosynthetic capacity at 34, 36 and 40 ppt was 50% less than that seen at 38 ppt; indeed, at 40 ppt, photosynthesis was insufficient to meet the metabolic demands of the coral colonies and they died (Table 3.3; Ferrier-Pagès et al. 1999). Ultimately, osmotic stresses on corals and their algal symbionts cause expulsion of the symbiotic algae (i.e. bleaching) (Coles 1992; Kerswell and Jones 2003), although how this is triggered is uncertain (Mayfield and Gates 2007).

As highlighted by their ecological distributions, not all corals are as physiologically susceptible to environmental changes as are others, and some show remarkable tolerance to fluctuating and extreme salinities. For instance, the coral *Porites furcata* collected from Biscayne Bay in Florida, where it encounters low and variable salinities due to terrigenous freshwater inputs, suffers reduced photosynthetic capacity when exposed to suboptimal salinities of 20, 25 and 45 ppt for 2–24 hours, but it still photosynthesizes enough to meet its metabolic demands and does not suffer any tissue degradation or death (Manzello and Lirman 2003). Similarly, *Siderastrea siderea*, another coral that frequents the coast of Florida, as well as the Caribbean, can tolerate

Table 3.3 Physiological measurements for the widespread Indo-Pacific coral *Stylophora pistillata* when maintained for 3 weeks at four different salinities (34–40 ppt)

Physiological parameter	Salinity (ppt)				Significant differences
	34	36	38	40	
$P_{grossmax}$ (µmol O_2 mg^{-1} chl h^{-1})	496 ± 19	665 ± 42	1,103 ± 62	318 ± 6	38 > 34, 36, 40 ppt 36 > 34, 40 ppt
$P_{grossmax}$ (µmol O_2 mg^{-1} protein h^{-1})	5.5 ± 0.8	5.2 ± 0.3	8.5 ± 0.1	6.3 ± 0.3	38 > 34, 36, 40 ppt
R (µmol O_2 mg^{-1} chl h^{-1})	171 ± 16	269 ± 12	363 ± 28	184 ± 20	38 > 34, 36, 40 ppt
R (µmol O_2 mg^{-1} protein h^{-1})	1.9 ± 0.1	1.6 ± 0.2	2.8 ± 0.2	3.9 ± 0.3	38 > 34, 36 ppt 40 > 34, 36, 38 ppt
P:R	1.3 ± 0.1	1.2 ± 0.1	1.5 ± 0.1	0.9 ± 0.1	38 > 40 ppt

Note: $P_{grossmax}$ = maximum rate of gross photosynthesis, and R = respiration rate, with rates normalized to either coral protein or chlorophyll content. P:R is the ratio of gross photosynthesis at an irradiance of 300 µmol photons m^{-2} s^{-1} over a period of 12 hours, divided by respiration over a period of 24 hours; P:R >1 indicates complete autotrophy, where photosynthetic production exceeds metabolic consumption. Values are means ± standard error, *n* = 5 corals per salinity.

Data from Ferrier-Pagès, C., Gattuso, J.-P. and Jaubert, J. (1999). Effect of small variations in salinity on the rates of photosynthesis and respiration of the zooxanthellate coral *Stylophora pistillata*. *Marine Ecology Progress Series* 181:309–14.

sudden increases or decreases in salinity of <9 ppt and <10 ppt without any loss of photosynthetic or respiratory function, respectively, but rates of both photosynthesis and respiration decrease when salinity changes suddenly by more than this (Muthiga and Szmant 1987). Furthermore, some coral species can acclimate to extreme salinities so long as the change in salinity happens over a prolonged period. For example, when Muthiga and Szmant (1987) gradually increased ambient salinity from 32 to 42 ppt over a period of 30 days, *S. siderea* showed no loss of photosynthetic or respiratory function, whereas a sudden increase to 42 ppt was potentially fatal even for this hardy species. The basis of this ability to tolerate variable salinities and to acclimate is unknown, but it is reasonable to think that interspecific differences in the production of COOs are at least partly responsible.

3.3 Temperature

Since the early 1800s, temperature has been regarded as one of the primary environmental determinants of coral reef global distribution. The distribution is very roughly bound by the 18 °C minimum monthly isotherm (see Figure 1.3), although this is greatly modified by both warm and cold currents, especially along continental shelf slopes, which cause reef development to extend or contract poleward, respectively (see 'High-latitude reefs: South Africa'). The average long-term minimum temperature conducive to reef growth is approximately 21 °C (Kleypas, McManus, et al. 1999), which largely limits coral reef development to between 30° N and 30° S. The world's southernmost coral reef at Australia's Lord Howe Island (31°33' S) has an average minimum temperature of about 17.5 °C, although rates of coral growth and reef formation, and coral species diversity (about 80 species compared to more than 350 on the GBR), are relatively low (Harriott et al. 1995; Harriott 1999). Thermal requirements are also responsible for major reef formations occurring mainly on the eastern sides of continents, which are influenced by warm western boundary currents; coral reefs on the western side of continents are generally less extensive and more fragmentary, and often exposed to cold, upwelled water. One exception is seen in Western Australia, where warm water flows southward from Indonesia. At a local scale, strong stratification of the water column in otherwise suitable tropical regions may produce a shallow thermocline below which corals do not survive even in sunlit water. Such a phenomenon restricts coral reefs in the north-western Hawaiian Archipelago to depths of 20 m or less (Grigg 1981).

High-latitude reefs: South Africa

High-latitude coral reefs are found in localities where conditions for the survival of this fauna would normally be limiting, for example, the Houtman Abrolhos Islands, Lord Howe Island and Bermuda. Lower temperatures, light attenuation and reduced aragonite saturation would normally preclude coral growth at these latitudes but local conditions conspire for their success. In this way, southern African coral communities form a continuum from the more typical, accretive reefs in the tropics of Mozambique to their marginal, southernmost African distribution in the province of KwaZulu-Natal, South Africa. Their occurrence here is mediated by the Agulhas Current, which originates primarily from the East Madagascar Current (but also Mozambique Channel eddies), forming one of the strongest of the warm, western boundary currents (mean velocity ~2 m s^{-1}, seasonal transport ~70 × 10^6 m^3 s^{-1}).

Corals are thus found in the IsiMangaliso Wetland Park, a World Heritage Site, within what is known as the Delagoa Bioregion. Despite being marginal, the coral communities are rich in biodiversity (Schleyer 2000). In terms of prominent biota, they comprise some 40 species of soft and 90 species of hard corals, with at least 30 species each of sea squirts and sponges. Overall, the reef communities consist of an admixture of tropical and temperate Indo-Pacific fauna and include numerous endemic species, of both invertebrates and fish. Because of the marginal conditions, the coral communities are non-accretive and grow as a veneer on sandstone reefs that developed as Late Pleistocene dune rock and beach rock, becoming reefs with the Holocene submersion of earlier coastlines. Soft corals are preponderant on most of the reefs, this again being attributable to their marginal nature and the turbulence found on reefs at these southern latitudes. The Alcyonacea appear to tolerate these conditions and form particularly extensive carpets on the flat, turbulent reef tops.

Continued

High-latitude reefs: South Africa (*Continued*)

The reefs are only 40 km² in extent and are clustered in three complexes, with some scattered reefs between the northern and central reef complexes (Figure, left). Corals are found in their shallowest reaches, which rise to a depth of 8 m, and extend down to 27 m. Surveys involving point intercept analysis of digitally imaged reef quadrats (e.g. Schleyer and Celliers 2005; Celliers and Schleyer 2008) revealed that there are 18 distinct benthic communities on the reefs, at a similarity level of 55%. When mapped in a GIS, it emerged that these occur in a gradient from north to south, with only one community type found in all the reef complexes, and none common to all of the reefs. This was taken into account in drafting management guidelines for the reefs, incorporating the addition of new sanctuary areas to existing sanctuaries to meet biodiversity protection targets in the IsiMangaliso Wetland Park. Accessible reefs were also zoned for recreational use according to their carrying capacity and the sensitivity of their coral communities to damage (Schleyer and Tomalin 2000).

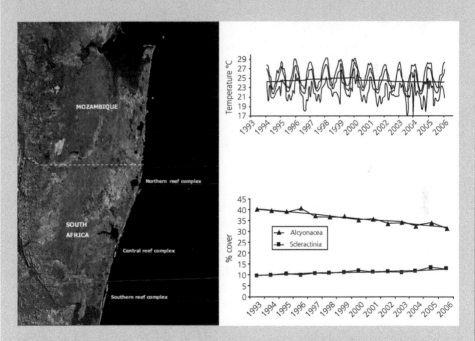

Figure. Left: Landsat image of the north-east coast of South Africa and southern Mozambique, with reefs in the region delineated in an overlay. Right, top: Mean, minimum and maximum monthly sea temperatures recorded at a depth of 18 m at the South African long-term coral reef monitoring site from 1994 to 2006. Mean temperatures from 1° × 1° satellite data were used to fill a gap in data recording from November 2002 to January 2003; comparisons with underwater temperature recorder data on either side indicated a good fit. Right bottom: Overall changes in alcyonacean and scleractinian cover measured at the South African long-term coral monitoring site from 1993 to 2006.

Source: From Schleyer, M.H., Kruger, A. and Celliers, L. (2008). Long-term community changes on high-latitude coral reefs in the Greater St Lucia Wetland Park, South Africa. *Marine Pollution Bulletin* 56:493–502;reproduced by permission of the *Marine Pollution Bulletin*.

Continued

High-latitude reefs: South Africa (*Continued*)

Long-term monitoring on the reefs in view of climate change (Schleyer et al. 2008) commenced in 1993, using temperature logging and image analysis of high-resolution photographs of a representative reef within fixed quadrats. Sea temperatures rose by 0.15 °C p.a. at the site up to 2000 but have subsequently been decreasing by 0.07 °C p.a. (Figure, right top). Insignificant bleaching was encountered in the region during the 1998 El Nino Southern Oscillation event, unlike elsewhere in East Africa. However, quantifiable bleaching occurred during warming in 2000 (Celliers and Schleyer 2002). Peak temperatures on the South African reefs thus appear to have attained the regional coral bleaching threshold. This has resulted in relatively little bleaching so far, but increased temperatures appear to have had a deleterious effect on coral recruitment success, as other anthropogenic influences on the reefs are minimal. Recruitment success diminished remarkably up to 2004 but subsequently improved. This change appears to constitute a 'silent' effect of the earlier temperature increases, the lag in its appearance being the grow-out period needed before detection of new recruits in the monitoring photographs.

Throughout, the corals have manifested changes in community structure, involving an increase in hard coral cover and a reduction in that of soft corals, resulting in a 5.5% drop in overall coral cover (Figure, right bottom). Warming may thus be encouraging hard coral growth at this stage, at the expense of the soft corals. The South African high-latitude reefs may thus provide a model for the study of many of the stresses and their effects on globally threatened coral reef systems.

Prof. Michael H. Schleyer, Oceanographic Research Institute, Durban, South Africa.

Limits to coral reef distribution imposed by low temperature are directly related to coral physiology and survival. For example, under experimental conditions, some Hawaiian and Australian corals can only survive 1–2 weeks at 18 °C or less (Jokiel and Coles 1977; Crossland 1984). Low temperature also reduces coral feeding, which in some cases may stop below 16 °C (Mayer 1915). The inability of many corals to tolerate even short-term exposure to low temperatures is best demonstrated by the devastating impacts of cold snaps on coral mortality. Heavy mortality has occurred in the Florida Keys, Persian Gulf and Mexican Pacific, and on relatively high-latitude reefs that are occasionally subjected to temperatures of 15 °C or less for a few days (Porter et al. 1982; Walker et al. 1982; Kemp et a. 2011; Lirman et al. 2011; Colella et al. 2012; Rodriguez-Troncoso et al. 2014). Likewise, on the southern GBR, intertidal corals have been seen to bleach (an indicator of physiological stress) when exposed to air temperatures as low as 12 °C during winter low tides (Hoegh-Guldberg and Fine 2004; Figure 3.2).

Resilience to low temperature is well illustrated by reefs in the Persian Gulf, where the major reef-forming coral *Porites harrisoni*, as well as various faviid corals, survived temperatures of just below 11.5 °C for four consecutive days

(a)   (b)

Figure 3.2   Cold-weather bleaching of corals. (A) *Acropora aspera* and (B) *Porites* sp. Both corals are on the reef flat of Heron Island, southern GBR, with the bleaching occurring during July (winter) 2007. Note that the corals only show signs of bleaching (i.e. whitening) when they are exposed to cold air at low tide; parts of the corals that remain submerged in the seawater at all times, and hence are protected from the extreme air temperatures, are not bleached.

Source: Photos by S. Davy.

and mean daily temperatures of 13 °C or less for more than 30 days (Coles and Fadlallah 1991); all of these Gulf corals showed signs of sublethal stress (indicated by bleaching) but ultimately recovered. However, not all coral species fared so well, with most suffering considerable mortality and algal overgrowth.

Elsewhere in the tropics, at reef sites such as the Galapagos and Pearl Islands (Gulf of Panama) in the Eastern Pacific, where temperatures fluctuate considerably from warm to cold because of upwelling deep water, coral growth occurs during warm periods but may slow or cease entirely when it becomes too cold. As a result, corals are less able to compete with other benthic organisms, such as macroalgae, and may suffer from increased bioerosion, limiting the rate of reef growth (Glynn and Stewart 1973; Glynn and Ault 2000). When low temperatures are encountered on a more persistent basis, corals may survive and form richly productive communities but do not form reefs due to reduced rates of calcium carbonate deposition and other, still unknown factors (Buddemeier and Smith 1999). Locations showing this include the far north of New Zealand, southern Australia and the mainland of Japan (Schiel et al. 1986; Veron 1993; Davy et al. 2006). Similar decoupling of reef growth from coral growth occurs in some tropical areas also, as described and illustrated in Chapter 1.

In contrast, development of many coral communities is not limited by a single maximum monthly isotherm. In the Persian Gulf and Red Sea, over 50 species of corals survive maximum mean weekly temperatures of about 34–36 °C (Coles and Fadlallah 1991). Although this is a very poor subset of the

Indo-Pacific fauna, it is still richer than most parts of the Caribbean, so coral diversity bears no relationship to reef construction. Such high temperatures occur because of the semi-enclosed nature of these water bodies, although some more open regions, such as the Andaman Islands in the Bay of Bengal, may approach these temperatures also (Kleypas, McManus, et al. 1999). The average temperature for reefs is 27.6 °C and the average *maximum* temperature is 29.5 °C, while the highest latitude reefs such as those around Lord Howe Island have maxima as low as 24 °C. Indeed, the majority of the world's coral reefs would be destroyed by temperatures as high as those seen in the Persian Gulf and other embayed areas.

The geographic range of many hardy coral species extends across both warm and cool waters (Hughes et al. 2003), which illustrates both the physiological plasticity of some and their capacity to adapt to local conditions. This adaptation is most likely linked to phenotypic and genetic diversity of both the coral and its zooxanthellae (Brown 1997a; Howells et al. 2013). Considerable genetic diversity is now recognized within coral species (Ayre and Hughes 2000) and zooxanthellae, each with different physiological characteristics. Consequently, corals are able to survive in more extreme thermal regimes by, in part, housing a more heat-tolerant algal partner (Rowan 2004; Berkelmans and van Oppen 2006; Silverstein et al. 2015)—see Chapter 4. The precise physiological mechanisms that form the basis of thermal tolerance in the coral and its zooxanthellae are not fully understood, but include (1) heat-shock proteins that re-fold denatured cellular and structural proteins and (2) antioxidant proteins that deactivate toxic oxygen radicals produced when thermal stress interferes with photosynthesis and other physiological processes (Fitt et al. 2009; Rosic et al. 2011; Krueger et al. 2015).

3.4 Light

Sunlight penetration of water on coral reefs is sufficient to support high densities of photosynthetic organisms, including seagrasses, macroalgae and microalgae, as well as the numerous invertebrates that harbour symbiotic algae, including reef-building corals. Benthic photosynthesis on coral reefs contributes 90% or more to total carbon fixation (Delesalle et al. 1993), and light also enhances rates of calcium carbonate deposition (and hence reef formation) by corals and coralline algae. Therefore, light availability and water quality profoundly influence coral reefs. Light is largely responsible for the depth distribution and geographic distribution of coral reefs too.

The depth of reef development is, in part, dictated by the amount of photosynthetically active radiation (PAR; light of wavelength 400–700 nm) reaching the sea surface, which is a function of the angle of the sun and atmospheric attenuation. The maximum depth of coral reef formation has been reported as 30–50 m (Grigg and Epp 1989; Grigg 2006), although, in very clear water, it has been seen at depths of 60–75 m (e.g. Jarrett et al. 2005). Corals may grow even deeper, although reef formation is unlikely. Maximum growth rates in branching corals, which do not have dense skeletons, are typically about 10 mm y^{-1} (Grigg and Epp 1989), although higher rates can occur in optimal environments (Buddemeier et al. 1974; Grigg 2006), while more solid, massive forms generally expand their radii by about 1 cm y^{-1}. These fast rates typically occur at about 5–10 m depth. Lower light limits growth in deeper water, while growth in the shallowest few metres is typically slowed by a number of factors, including harmful solar UV radiation and increased exposure to turbulent water. Growth rates drop dramatically with depth-related light attenuation so that, by 30 m in clear water, where only 30%–40% of surface irradiance remains, coral growth rates are only 15%–40% of those in the shallows (Buddemeier et al. 1974; Dustan 1975).

In deeper water, growth becomes insufficient to withstand rates of mechanical and biological erosion, so that net positive accretion (i.e. reef building) ceases. The major reef-building coral *Porites lobata* in the Au'au Channel off Maui in the Hawaiian Islands (Grigg 2006) grows at its maximum rate of 13.5 mm y^{-1} at 6 m depth but, below 50 m depth, growth slows to <3 mm y^{-1} (Figure 3.3). Water at that location is exceptionally clear, allowing *P. lobata* to live to 80–100 m deep, where irradiance is 5%–10% of that on the surface, but the coral's growth rate below 50 m is insufficient to counter the activity of endolithic boring organisms, such as that from various clionid sponges and bivalves (Grigg 2006), so that reef building does not occur.

Attenuation of light in the water is a function of water clarity; irradiance decreases exponentially with depth. The vertical attenuation coefficient of PAR, referred to as $K_d(PAR)$ is typically 0.03–0.04 m^{-1} in clear tropical seas (Gattuso et al. 2006; Grigg 2006). This parameter increases when water clarity is reduced by suspended particulate matter. It may be as high as 0.15–0.40 and 0.35–13.0 m^{-1} in temperate coastal seas and estuaries, respectively (Kirk 1994). The depth to which light penetrates on reefs is therefore highly variable, with the deepest reefs forming in equatorial, oceanic sites, such as atolls, and those with shallowest extent being in inshore regions and at high latitudes, where light penetration falls seasonally. Assuming a maximum depth of 30–40 m in clear waters and an associated $K_d(PAR)$ of 0.04 m^{-1}, Gattuso et al. (2006) estimated that reefs require a minimum irradiance of 400–600 μmol photons m^{-2} s^{-1} over the course of a day; this compares to a typical maximum surface irradiance in the tropics of about 1,500–2,000 μmol photons m^{-2} s^{-1}. Similarly, by modelling global reef distribution as a function of PAR, Kleypas (1997) estimated that reef formation is limited to regions that receive a minimum irradiance of 250–300 μmol photons m^{-2} s^{-1} (or 7–8 mol photons m^{-2} d^{-1}). Kleypas et al. (1999) subsequently used this minimum requirement to identify light-limited reefs, where enough light for reef formation occurs only at depths shallower than

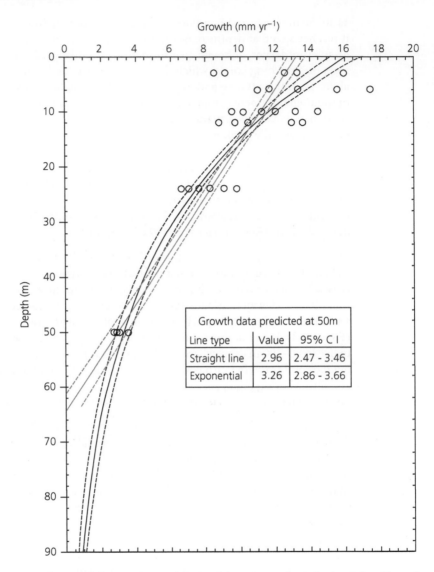

Figure 3.3   Growth rate of the massive coral *Porites lobata* versus depth in the Au'au Channel, Hawaii. Open circles are the mean growth rates (in millimetres per year) for coral colonies along the depth transect, with plots (solid lines) being the best-fit regressions with 95% confidence intervals (dotted lines). Regressions have been fitted with either a straight line (grey) or an exponential line (black). Growth rates predicted at 50 m depth from these two regression lines are given in the box. The Au'au Channel is characterized by oceanic circulation, producing maximum water clarity and optimal conditions for coral growth. Below 50 m in the Au'au Channel, coral colonies do not remain attached to the substratum, as bioerosion on the undersurface of colonies exceeds the rate of linear extension; 50 m is therefore the maximum depth for coral reef accretion at this site.

Source: Reprinted from Grigg, R.W. (2006). Depth limit for reef building corals in the Au'au Channel, SE Hawaii. *Coral Reefs* 25:77–84, with kind permission from Springer Science + Business Media.

15 m. Such reefs occur at a number of tropical inshore sites that have high turbidity; these include Bowden Reef on the central GBR, India's Gulf of Kutch, and Tunku Abdul Rah in Borneo. In many places, the depth of light penetration may exceed 15 m but still be too low for reef development. This is particularly obvious at high-latitude sites, such as Lord Howe Island and the Ryukyu Islands of Japan (Harriott et al. 1995; Kan et al. 1995), where seasonal impacts on turbidity and day length are more marked than those seen in the tropics. Indeed, while temperature may be primarily responsible for imposing latitudinal limits on coral reef development, light availability most likely explains why coral reefs living at these latitudinal extremes are restricted to relatively shallow depths.

3.5 Nutrients

Nutrients—nitrogen, phosphorus and various trace elements—are essential for the production of organic matter by photosynthesizing organisms such as corals, macroalgae and seagrasses. These nutrients can be in dissolved inorganic, dissolved organic and particulate organic (detritus and plankton) forms. Dissolved inorganic nitrogen (DIN) and dissolved inorganic phosphorus (DIP) in the forms of nitrate (NO_3) and phosphate (PO_4) are particularly important, with dissolved organic nitrogen (DON) also providing an important source of nitrogen. For instance, on the outer shelf of the GBR, nitrate and DON comprised 50% and 43% of total nitrogen, respectively, while phosphate comprised 97% of all phosphorus. Ammonium (NH_4), nitrite (NO_2), particulate nitrogen, dissolved organic phosphorus and particulate phosphorus made up the remaining 7% and 3% of total nitrogen and phosphorus, respectively (Furnas and Mitchell 1996).

Despite the crucial role of nutrients in primary production, coral reefs somehow support massive production and biodiversity in nutrient-poor (oligotrophic) water, providing a paradoxical situation that has been debated by reef scientists for decades. Typical concentrations in reef waters are 0.6 µM for nitrate and 0.2 µM for phosphate (compared to concentrations of 10–40 µM and 1–4 µM in deep waters, respectively) with nutrient concentrations being particularly low in the North and South Pacific gyres, the central Indian Ocean and the Coral Sea (Kleypas 1994). Algae on reefs are nutrient deficient as a result of these low ambient supplies of nitrogen and phosphorus (Cook and D'Elia 1987; Lapointe et al. 1997) while iron, present in trace concentrations on reefs, may limit algal growth further (Entsch et al. 1983). Indeed, an increase to just 1.0 µM DIN, for instance, may be enough to destabilize some reef communities and cause macroalgal overgrowth of corals and seagrasses (Lapointe et al. 1993, 1997). Low nutrient concentrations limit phytoplankton productivity; typical values for oligotrophic reefs are 0.3–0.7 µg L^{-1} chlorophyll *a* (Van Woesik et al. 1999; Fabricius et al. 2005).

The survival of coral reefs under such conditions is greatly facilitated by the efficient recycling and conservation of nutrients by symbioses between various animals and phototrophs (i.e. algae, cyanobacteria) (Wang and Douglas 1998) as well as within the reef system as a whole (D'Elia and Wiebe 1990; Furnas et al. 2011). Bacterial degradation of organic matter plays an essential role in this recycling as it regenerates nutrients within and between interstitial pore-water of coral heads, which have both a high porosity and permeability, and reef sediment (Rougerie et al. 1992). Indeed, the accumulation and bacterial mineralization of organic matter in sediment results in much higher concentrations of nutrients than exist in seawater. For instance, sediment on Davies Reef in the central region of the GBR was found to contain concentrations of up to 8.4 μM nitrate, 7.6 μM soluble reactive phosphorus (SRP) and 25.2 μM ammonium, in marked contrast to overlying reef water, which contained only ≤0.6 μM nitrate, ≤0.14 μM SRP and ≤0.2 μM ammonium (Entsch, Boto, et al. 1983). The availability of these recycled nutrients helps to explain the success of benthic algae and seagrasses in the coral reef ecosystem. Furthermore, resuspension of sediment by storms can increase nutrient concentrations in the seawater, at least for short periods (Ullman and Sandstrom 1987).

Nutrient recycling is crucial to the success of coral reefs in oligotrophic waters; yet, in most cases, the recycling is not 100% efficient, and coral reefs are net exporters of inorganic and organic nutrients (Hatcher 1985; Hallock and Schlager 1986). Exported material is lost to the surrounding ocean via turbulent transport. There must, therefore, be a regular net input of nutrients to sustain the coral reef ecosystem, which, given the oligotrophic nature of the surrounding ocean, may at first seem a difficult task. However, there are a number of natural mechanisms by which 'new' nutrients can be supplied to reefs, in addition to replenishment by regular ocean currents.

(1) *Nitrogen fixation*: Fixation of atmospheric nitrogen occurs across coral reefs as a result of the activities of abundant free-living (benthic and pelagic) and symbiotic microbes, most conspicuously cyanobacteria (Wilkinson et al. 1984; Larkum et al. 1988; Furnas et al. 1997b; den Haan et al. 2014; Cardini et al. 2015). At One Tree Island on the GBR, benthic cyanobacteria fixed 8–16 kg N ha^{-1} y^{-1} (Larkum et al. 1988). Similarly in Tikehau Lagoon, French Polynesia, nitrogen fixation rates of 0.4–3.9 mg N m^{-2} d^{-1} were measured, with nitrogen fixation contributing 24.4% of the total nitrogen required for benthic primary production (Charpy-Roubaud et al. 2001). Rates of benthic nitrogen fixation tend to be highest on highly disturbed surfaces that provide space for colonization by free-living cyanobacteria. These include substrates that have been heavily grazed by fish (Wilkinson and Sammarco 1983), and coral skeletons exposed by loss of overlying tissue due to crown-of-thorns starfish predation (Larkum 1988) or coral bleaching (Davey et al. 2008). Pelagic cyanobacteria also contribute nitrogen: Furnas et al. (1997b)

estimated that the pelagic colonial cyanobacterium *Trichodesmium* may fix anywhere between 2 and 72 kmol N m^{-1} per year along the central GBR (compared to 0.5 kmol N m^{-1} for benthic cyanobacteria), making this perhaps the biggest source of new nitrogen in the system. Input from all other sources, both natural and anthropogenic ones such as sewage, was 24–27 kmol N m^{-1} (Table 3.4).

(2) *Oceanic upwelling*: Upwelling of nutrient-rich deep water can carry nutrients to the reef. On the outer shelf of the northern and central GBR, nutrient-rich water upwells several times each summer as a result of phenomena such as internal tides and internal waves interacting with the bathymetry of the shelfbreak (Wolanski and Pickard 1983; Wolanski et al. 1988). Large upwelling events may displace one-third of the outer-shelf water, import substantial quantities of inorganic nutrients (e.g. nitrate and phosphate), and displace large quantities of organic nitrogen and phosphorus. This nutrient-rich water moves perhaps 60–80 km shoreward over the reef from the shelfbreak (Furnas and Mitchell 1996). Thus, Furnas et al. (1997b) estimated that, on the central GBR upwelling contributed about 3%–19% and 7%–18% of the total annual input of nitrogen and phosphorus, respectively (Table 3.4). However, it should be remembered that, as described at the start of this chapter, in regions of particularly strong and persistent upwelling, low seawater temperatures and elevated nutrient levels can slow or inhibit reef growth altogether.

(3) *Geothermal endo-upwelling*: Upwelling of nutrients through the coral reef framework has been proposed as an important source of nutrients for reefs in geothermally active regions, such as Pacific atolls and high-island barrier reefs (e.g. Tahiti) that have a volcanic foundation. Interstitial water (pore-water) of the skeletal framework and sediments is nutrient rich, as well as CO_2 rich, compared to the surrounding oligotrophic seawater as a result of bacterial mineralization and metabolism. Furthermore, at oceanic sites with volcanic foundations, there is input of nutrients and CO_2 by deep water currents. Typically, pore-water in the upper 20 m or so of the skeletal framework tends to be well oxygenated as a result of wave turbulence; anoxic conditions prevail below this, as they do below the top few centimetres in reef sediments. The well-oxygenated water supports rapid rates of aerobic microbial activity and nutrient mineralization (especially nitrate regeneration by nitrifying bacteria). In comparison, the anoxic pore-water is home to denitrifying bacteria, which contribute atmospheric nitrogen for fixation (Wiebe 1985), and contains a high concentration of reduced compounds such as ammonia (Sansone et al. 1990; Rougerie et al. 1992). Nutrients present in larger cavities of the reef are readily made available to the overlying reef by tidal currents, while geothermal heat aids the release of nutrients held in smaller cavities and deeper within the reef matrix. The cumulative build-up of this heat causes a lowering in the density

of the pore-water and the establishment of a slow convective circulation system that transports nutrients upwards through the porous limestone framework of the reef (Rougerie et al. 1992). The release of this nutrient-rich water onto the reef is inhibited in calm regions where sediment clogs the reef, forcing the water up in shallower areas such as the algal crest and spur zone where currents and wave action limit sedimentation. Rougerie et al. (1992) assumed a vertical flow of pore-water through the reef matrix of 1 cm h^{-1} and hence estimated that this geothermal endo-upwelling may cause pore-water to leak out of the reef at a rate of 10 L m^{-2} h^{-1}.

(4) *Rainfall and terrestrial run-off*: Rain provides a modest source of nitrogen and phosphorus for coral reefs, for instance contributing 0.8%–3% of nitrogen and 0.7%–0.9% of phosphorus to the central GBR (Furnas et al. 1997b; Table 3.4). Terrestrial run-off is far more important. Except where pelagic nitrogen fixation is locally elevated, run-off is the biggest contributor of nitrogen and phosphorus to coastal reefs. By far the most studied coral region with respect to run-off is the GBR, where about 42 km^3 of freshwater flow onto the reef each year from terrestrial catchment areas (Furnas et al. 1997a). Huge quantities of particulate material and nutrients are transported. Quantities fluctuate seasonally and annually, and are affected by flood events, meaning that nutrient input to the reef is not constant. Concentrations of particulate nitrogen and phosphorus vary directly with water flow, reaching a peak during major seasonal flood events (Mitchell et al. 1996). However, concentrations of dissolved inorganic nutrients (most commonly nitrate, the most abundant form of DIN in rivers; less commonly phosphate) tend to peak with the first major run-off event of the summer wet season after the build-up of oxidized nutrient stocks (e.g. nitrate) in catchment soils during the dry season. For instance, DIN and phosphorus concentrations in run-off may exceed 40 µM and 3 µM, respectively, at this time. After this peak, concentrations of dissolved inorganic nutrients tail-off as stocks are exhausted. In contrast, concentrations of dissolved organic nutrients, particularly nitrogen in the form of amino acids and urea, tend to decline with flood events, indicating dilution of a relatively constant input from watershed stocks (Mitchell and Furnas 1997; Furnas 2003). Nutrient concentrations also vary with the nature of biological and chemical processes in catchment soils, the extent and rate of soil erosion and, where present, input from sewage and agricultural fertilizers (Brodie et al. 2007). Thus, nutrient inputs from different rivers or catchment regions can be quite distinct (Furnas 2003). Whatever the differences, though, rivers are a globally important source of new nutrients for inshore reefs as illustrated by the fact that, on the central GBR, it has been estimated that rivers contribute about 21.5%–82.0% and 71.0–91.0% of the overall annual input of nitrogen and phosphorus, respectively (Furnas et al. 1997b; Table 3.4).

Table 3.4 Input, export and recycling fluxes of nitrogen and phosphorus on the central Great Barrier Reef

	Nitrogen	Phosphorus
Inputs		
Upwelling	2.7–5.0	0.2–0.4
Rivers	21.3	2.0
Sewage	0.14	0.02
Rainfall	0.84	0.02
Nitrogen fixation by *Trichodesmium* (a planktonic cyanobacterium)	2.0–72.0	—
Reefal nitrogen fixation	0.5	—
Demand	165	10
Recycling		
Zooplankton excretion	7.5	0.4
Microbial mineralization	61.0	?
Benthic mineralization	14.0	1.8
Resuspension	230	10.0
Sinks		
Upwelling displacement	1.9–3.7	0.1–0.2
Shelf denitrification	13.4–19.6	—
Reefal denitrification	0.003	—
Burial	0.61	0.13
Shelfbreak exchange	?	?

Note: All values are annual fluxes in kilomoles per metre of shelf.

Data from Furnas, M., Mitchell, A. and Skuza, M. (1997). Shelf-scale nitrogen and phosphorus budgets for the Central Great Barrier Reef (16–19°S). *Proceedings of the 8th International Coral Reef Symposium* 1:809–14.

While coral reefs are generally considered to exist in nutrient-poor water, they in fact exist in seas containing a wide range of nutrient levels. Some coral reefs live at relatively high nutrient concentrations (>2 μM nitrate; >0.4 μM phosphate; Kleypas, McManus, et al. 1999). These include a number of atolls (e.g. Christmas Island) along the equatorial belt in the Central Pacific, where wind-driven divergence of surface waters leads to relatively consistent upwelling of nutrient-rich water. High-nutrient coral reefs exist but with very limited development and extent in the Arabian region, where peak nutrient concentrations may reach 3.34 μM nitrate in the Strait of Hormuz and 0.54 μM phosphate in the Arabian Sea, while the non-reefal coral communities of the Galapagos Islands are exposed to concentrations of 5.61 μM nitrate and 0.54 μM phosphate as a result of the deep oceanic water that upwells around the islands. In such places, corals may live alongside macroalgae (Figure 3.4).

The fact that corals can tolerate a wide concentration range of these nutrients may indicate that the distribution of reefs is relatively independent of these parameters. However, other nutrients are also important, and some data are available for iron, which may limit production on reefs as it does in some open ocean regions (Entsch, Sim, et al. 1983; Kolber et al. 1994).

Figure 3.4   In marginal areas, corals live amongst macroalgae. In this case, *Acropora* tables are seen in southern Oman amongst the macroalgae *Sargassum* at the top and (foreground) the more kelp-like *Nizamuddinia zanardinii*, which is a regional endemic (See Plate 5).
Source: Photo by Rob Baldwin.

High nutrient concentrations enhance growth of macroalgae and phytoplankton, which compete for space and light, respectively, and hence nutrient enrichment limits coral cover indirectly (Lapointe et al. 1997; Costa et al. 2000).

3.6 Exposure and other hydrodynamic factors

Strength and direction of waves and local currents, tidal range and frequency and magnitude of storm impacts also affect formation of coral reefs. Many in the Caribbean (e.g. Jamaica, the Cayman Island) are sheltered from ocean swells and exposed to low-energy, trade-wind-generated waves for much of the year; moreover, they only encounter small tidal ranges. In contrast, reefs in the Pacific are more likely to be exposed to forceful ocean swells that, in combination with relatively large tidal ranges, increase the threat of wave damage to reefs. This capacity to inhabit a wide range of hydrodynamic settings highlights that, unlike factors such as temperature or aragonite saturation state (see Sections 3.3 and 3.8), hydrodynamic regime plays little role in dictating the global limits of coral reef distribution, although on a more localized scale, waves, currents, tides and storms are important for determining (1) reef morphology; (2) reef community structure; (3) sediment distribution patterns; (4) zones of early reef diagenesis; and (5) larval transport pathways.

Coral reefs tend to favour high-energy sites (e.g. the windward side of islands and some atolls), where waves and currents deliver planktonic food and nutrients to the reef from offshore sources, keep reef water well flushed and oxygenated (Jokiel 1978) and prevent smothering by sediment (Murray et al. 1977). The ability to withstand considerable wave action is enabled, in part, by the spur-and-groove formation that is typical of the windward side of coral reefs in both swell- and trade-wind-dominated regions (Chapter 1). This spur-and-groove formation consists of parallel linear ridges (spurs) of reef separated by depressions (grooves). In high-energy regions of the Pacific, the spurs are generally formed by coralline red algae of the genera *Porolithon* and *Lithothamnion* and low-relief encrusting corals whereas in the lower-energy Caribbean, the spurs are dominated by a diverse community of higher-relief corals such as *Acropora palmata* (Roberts et al. 1992). Similarly, a conspicuous red coralline algal ridge (the 'reef crest') forms on high-energy Indo-Pacific reefs, where it is generally elevated a little above low tide and runs parallel to the edge of the reef; by comparison, algal ridges are relatively uncommon in the Caribbean. These trends highlight the effect of wave action on not only reef morphology but also community composition, with increasing wave action and hydrodynamic stress in general favouring increased structural strength and hence a shift from fine-branching corals to massive forms.

The amplitude and spacing of the spurs and grooves affects the frictional attenuation and scattering of waves. Indeed, interaction with the forereef causes a 20%–47% reduction in wave height, depending on reef configuration and water depth, while wave breaking on the reef crest accounts for 50%–90% of the remaining wave energy (Lugo-Fernández, Roberts and Wiseman 1998). Overall, coral reefs attenuate 72%–97% of incident wave energy (Roberts et al. 1992, Sheppard et al. 2005). Thus, these regions protect the backreef from potentially damaging hydrodynamic forces. Similarly, flow interaction with the rough seabed and reef wall, and, in particular, the spur-and-groove topography, weakens currents greatly. For instance, at Grand Cayman Island in the Caribbean, current speed is reduced by 60%–70% between the edge of the forereef shelf (22 m depth) and a shallower (8 m depth) forereef site, a distance of just 400 m (Roberts et al. 1977). The relationship between reef structure/roughness and the capacity to absorb wave energy means that healthy reefs with high coral cover and complex coral reef matrices, consisting of multiple offshore reefs with spaces in between (e.g. the GBR), are especially effective at dissipating wave action (Gallop et al. 2014; Monismith et al. 2015).

The extent to which wave energy is dissipated by the forereef and reef crest is influenced further by the tidal range, with tidal reduction of water depth causing a shift in the wave-breaking point seaward from the reef crest, and broken waves to travel longer distances as a bore-like wave. This means an intensification of wave breaking and hence greater attenuation of wave energy at low tide than at high tide, even in regions with small tidal ranges such as the Caribbean. For instance, at St. Croix in the US Virgin Islands, it was estimated that 15% and 6% more wave energy was attenuated at low tide than at high tide between the forereef and reef crest, and between the forereef and lagoon, respectively (Lugo-Fernández, Roberts and Wiseman 1998). The intensification of wave breaking at low tide also intensifies the associated current surges that are directed over the reef crest and into the lagoon. These surges can travel at speeds of 180 cm s^{-1} (Roberts and Suhayda 1983) and at Davies Reef on the central GBR were estimated to account for 40% of the transport across the reef crest (Pickard 1986). Hence, they are an essential mechanism for flushing lagoon waters, helping maintain temperature, salinity, oxygen and nutrients at favourable levels. Current surges also transport sediment from the reef crest into the lagoon, with this transport intensifying at low tide. Current surges are, therefore, important for preventing smothering of corals and other reef organisms.

Wind, waves and tides are major driving forces of across-reef flow, although oceanic currents may also play a part around atoll and platform reefs, as may pressure-gradient forcing over the adjacent shelf (Andrews and Pickard 1990; Lugo-Fernández et al. 2004; Kench et al. 2009). The direction of across-reef flow at a particular point in time is the result of these factors combined with aspects of reef morphology. For instance, where large breaks in the reef crest or across-reef channels occur, they provide an opportunity for tidal reversal of across-reef flow, which otherwise tends to be driven towards the lagoon by waves (Suhayda and Roberts 1977). This is particularly the case where large tidal ranges and light winds co-occur; in contrast, when tidal ranges are small and winds strong, continuous lagoonward flow is more likely. Very strong seasonal and monsoon winds can dominate even strong tidal fluctuations and dictate the direction of across-reef flow at certain times of year (Yamano et al. 1998; Hoitink and Hoekstra 2003). The physics which underlie the oceanographic processes operating across coral reefs and thus determine factors such as the direction and velocity of across-reef transport are complex (Andrews and Pickard 1990; Roberts et al. 1992; Lugo-Fernández, Roberts and Suhayda 1998; Fernández, Roberts and Wiseman 1998; Wolanski 2001).

Wave and storm surges generated by hurricanes (cyclones) are especially destructive (Done 1992), with seaward sides of reefs being most impacted. Forereefs may be subjected to waves of 20 m or more in height, while storm surge can be as much as 5 m above the normal tidal level (Scoffin 1993). These conditions at the forereef can smash corals to depths of 20 m, with branching forms being particularly vulnerable. During storms, sand can be washed away or swept across the reef flat, and debris from the forereef and reef crest can accumulate (Scoffin 1993). While not the everyday situation on coral reefs, storms are thought to play an important role in determining the density, structure and local distribution of coral assemblages (Massel and Done 1993; Madin and Connolly 2006; Perry et al. 2014). The extent to which coral reefs are impacted by hurricanes is not uniform. For instance, on the GBR, reefs at 21° latitude are most likely to encounter and be damaged by cyclones, with the likelihood of damage decreasing both north and south of 21° latitude (Massel and Done 1993).

3.7 Sediment

Many of the world's coral reefs inhabit coastal regions that are subjected to terrestrial run-off. Coastal waters are relatively dynamic and shallow, with wind-driven waves and tidal currents causing resuspension of sediment (Kleypas 1996; Hoitink and Hoekstra 2003; Orpin and Ridd 2012). Reef waters are often loaded with varying proportions of clay, calcareous or silicate sand, organic detritus and plankton. When river water enters the sea, nutrients adsorbed to particulate matter become available to biota, leading to the colonization of the particles by microbes whose mucus secretions in turn increase particle stickiness and aggregation (Wolanski and Gibbs 1995; Fabricius and Wolanski 2000). The reality in many nearshore regions is far removed from the classic magazine image of offshore coral reefs surrounded by crystal-clear blue waters (Figure 3.5).

Variations in the particulate load of reef waters, from nearshore to offshore systems, is perhaps best illustrated by the GBR. Furnas (2003) estimated that an average of 14.4×10^6 tonnes of terrestrial sediment are discharged onto the GBR shelf from the Australian continent each year, with 42% of this originating from two major drainage basins (Burdekin and Fitzroy River basins) in the central and southern regions. This considerable annual input of sediment, combined with greater wave action and shallower water depths close to shore, leads to a marked gradient in median suspended sediment concentrations from a peak of 800–3,300 µg L^{-1} in inshore waters, to 700–1,500 µg L^{-1} in mid-shelf waters and 500–600 µg L^{-1} in outer-shelf waters. Peak levels reach much higher concentrations than median values, however, with peak levels in nearshore water of just 2 m depth being up to

Figure 3.5   A sediment-impacted coral reef. This photograph shows a community of corals and sponges smothered by sediment on Sampela reef, off Kaledupa in the Wakatobi Marine National Park, SE Sulawesi, Indonesia. The sediment is thought to originate from a neighbouring mangrove forest (including deforested areas) and/or human inputs such as sewage.

Source: Photo by J. J. Bell.

1,000 mg L^{-1} (Wolanski and Spagnol 2000). Annual sediment export can also vary 20-fold depending upon the degree of flooding and terrestrial run-off with, for example, just 3×10^6 tonnes being exported in 1987 versus 59×10^6 tonnes being exported in 1974 (Furnas 2003). More locally, sediment plumes vary with the nature of the catchment area (Neil et al. 2002; Table 3.5). Small but high run-off catchments in the consistently humid 'wet tropics' may suffer frequent but short-duration flooding, whereas large catchments of the more seasonal and variable 'wet–dry tropics' (e.g. the Burdekin and Fitzroy River catchments) undergo infrequent but more lengthy periods of flooding. Sediment concentrations from these large wet–dry tropical catchment areas are high due to relatively low vegetative cover and high rates of soil erosion.

Typically, sediment from rivers contributes much less to reef waters than does sediment resuspended by wave action. On a fringing reef of the GBR it was estimated that, annually, <10% of suspended sediment came from river inputs (Neil 1996), and it is likely that this figure is far lower in most high-energy wave environments, even when extensive sediment-laden river plumes are located nearby (Larcombe and Woolfe 1999). River plumes are still important, as they have the capacity to transport significant quantities of sediment, and alter turbidity levels across wide tracts of coral reef. This

Table 3.5 Sediment yield and flow-weighted sediment concentration for five climatic regions along the Great Barrier Reef, at both undisturbed and anthropogenically disturbed (largely agricultural) locations

Region	Catchment	Run-Off (ML km^{-2} y^{-1})	Area (km^2)	Sediment Yield (kt km^{-2} y^{-1})		Flow-Weighted Sediment Concentration (mg L^{-1})	
				Undisturbed Locations	Disturbed Locations	Undisturbed Locations	Disturbed Locations
Northern, monsoon	Eastern Cape York	29	43,300	1.00	3.00	32.0	99.0
Northern, humid	Daintree–Murray	61	11,965	0.84	1.90	11.28	21.00–76.0
Northern, dry	Herbert–Don	71	150,515	9.50	37.60	29.0–225.0	100.0–906
Central, humid	Proserpine–Pioneer	21	6,410	0.54	1.95	23.0–26.0	87.0
Southern, dry	Fitzroy–Burnett	25	202,610	5.80	23.0	43.0–461.0	147.0–1,800

Source: Data from Neil, D.T., Orpin, A.R., Ridd, P.V. and Yu, B. (2002). Sediment yield and impacts from river catchments to the Great Barrier Reef lagoon. *Marine Freshwater Research* 53:733–52.

is because their relatively low salinity (and hence density) means that they are buoyant and so can travel considerable distances, increasing turbidity at the sea surface rather than close to the seabed; the opposite is true of resuspended sediment. This means that plumes reduce light reaching the coral reef below, even though the reef itself may be in relatively clear water. On the other hand, plume sediment does not blanket the reef community directly in the way that resuspended sediment can do (Neil et al. 2002).

Associated ecosystems such as seagrass beds and lagoonal soft-bottom communities rely heavily on sediment for habitat and nutrition (e.g. Lee Long et al. 1993; Alongi et al. 2007). Riegl and Branch (1995) modelled energy budgets for a range of South African hard and soft corals under different levels of light and sedimentation (Figure 3.6), concluding that sediment not only reduces coral photosynthesis and increases the relative rate of respiration, but also increases the energy required by corals for mucus production from 35% to 65% of daily respiration. Mucus is used to slough off sediment, so corals have to limit or shut down most of their normal metabolic functions in order to produce sediment-shedding mucus. Not surprisingly, sediment has substantial negative impacts on coral tissue growth, calcification, health and, ultimately, survival (Stafford-Smith 1992; Stafford-Smith and Ormond 1992).

Sediment also discourages settlement of coral larvae (Hodgson 1990; Babcock and Davies 1991). Not all coral species, or indeed individuals of the same species, are equally susceptible, however, while seasonal variability in temperature can also influence sediment clearance rates (Ganase et al. 2016). In their study of 42 Indo-Pacific coral species, Stafford-Smith and Ormond (1992) identified four widespread mechanisms for actively

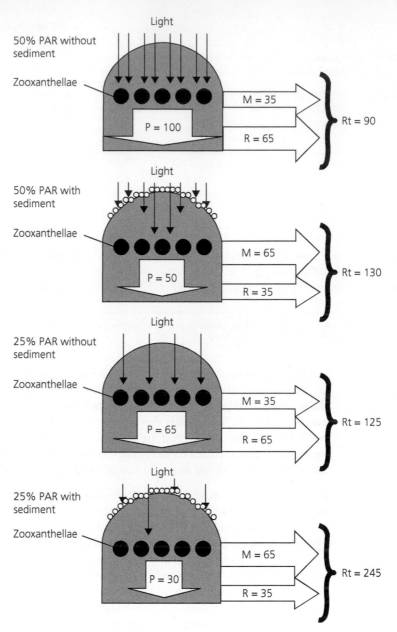

Figure 3.6   The effect of sediment on the energy budgets of corals at two different irradiances (50% vs 25% of photosynthetically active radiation (PAR)). Grey domes represent corals containing symbiotic algae (zooxanthellae) (●), either with (o) or without surface sediment. Thin vertical arrows indicate downwelling light while the broad vertical arrows represent photosynthetic productivity (values are the percentage of productivity when compared to 50% PAR/no sediment treatment). Outwards-facing arrows represent respiratory use of photosynthetic production for mucus synthesis (M) and other energetic requirements (R), with values for these parameters being the percentage of total production allocated to these energetic sinks. Rt is total respiration, and the value given is the percentage of photosynthetic production required for respiration (e.g. for 50% PAR/no sediment treatment, the equivalent of 90% of photosynthetic production is needed to support the coral's respiration).

Source: Redrawn from Riegl, B. and Branch, G.M. (1995). Effects of sediment on the energy budgets of four scleractinian (Bourne 1900) and five alcyonacean (Lamouroux 1816) corals. *Journal of Experimental Marine Biology Ecology* 186:259–75. Copyright 1995, with permission from Elsevier.

removing sediment: (1) ciliary activity, (2) mucus entanglement, (3) expansion of the polyp or coenosarc tissue and (4) tentacular manipulation. The effectiveness of each mechanism differed between coral species. Corals with large calices of >10 mm in diameter (and hence large polyps; e.g. *Symphyllia recta, Lobophyllia hemprichii*) or smaller calices (3.5–10.0 mm) but with strong ciliary activity (e.g. *Turbinaria peltata, Gardineroseris planulata*) are especially good at removing sediment, while corals with calices of <2.5 mm in diameter and hence very small polyps (e.g. the common massive corals *Porites lobata* and *P. lutea*) are poor rejecters of sediment. Some corals have upright, finely branched morphologies that negate the need for active sediment removal and are able to shed sediment passively. Such corals include the widespread *Pocillopora damicornis* and species of *Acropora*. However, removal efficiency alone is not an indication of tolerance to smothering by sediment. For instance, *Gardineroseris planulata* suffers tissue death within 6 days of smothering, while the relatively inefficient *Porites lobata* and *P. lutea* exhibit tissue bleaching but not death after this time, and recover once the sediment is removed (Stafford-Smith 1993).

Given these species-specific differences, sediment clearly has the potential to determine species distributions, and to drive community composition. On Kenyan reefs exposed to a gradient of riverine sediment input, the lowest sedimentation reef (Watamu National Park) was dominated by sediment-intolerant and intermediate-tolerance coral genera, while the highest sedimentation reef (Malindi National Park) had an abundance of sediment-tolerant and intermediate-tolerance genera, as well as soft corals; there was no evidence for an overall loss of coral diversity or ecological health at the high sediment reef (McClanahan and Obura 1997).

Ultimately, all corals have a limit to the amount of sediment they can tolerate. In regions with high sediment concentrations, a loss of coral cover, density and diversity may occur, with dominance by those species and morphologies best able to deal with the prevailing conditions (Kleypas 1996; Dikou and van Woesik 2006). In the most extreme cases, sediment prevents coral reef development completely. Indeed, tropical coasts subject to high levels of terrestrial run-off are far less likely to possess coral reefs than are coasts situated away from such influxes (McLaughlin et al. 2003).

3.8 Seawater carbonate chemistry

Carbonate, particularly in the form of aragonite, is essential for coral calcification and hence reef formation. Other marine organisms deposit calcium carbonate in slightly different crystalline forms (Table 3.6). The capacity to deposit a calcium carbonate skeleton is related to the carbonate saturation state of seawater. That is, if the saturation state is too low, then carbonate

Table 3.6. Forms of calcium carbonate deposited by different taxonomic groups

Group	Photosynthetic	Crystal Form of Carbonate	Ecological and Main Calcification Role
Corals	Reef forms have symbiotic algae	Aragonite	Reef builders
Soft corals	Some	Calcite spicules	Some contribute to reef building
Macroalgae (greens and browns)	Yes	Aragonite, Mg calcite and calcite	Sand production
Calcareous red algae	Yes	High-Mg calcite	Reef crest construction
Echinoderms	No	High-Mg calcite	Minor role in limestone production
Molluscs	Not usually (some clams have symbiotic algae)	Calcite and aragonite	Minor role in limestone production
Crustaceans	No	Skeleton mainly organic, but some with mineral salts especially High-Mg calcite	Negligible role in limestone production
Foraminifera	Some	Calcite	Benthic and planktonic, sand production
Pteropods	No	Aragonite	Planktonic food
Coccolithophores	Yes	Calcite	Planktonic food

Note: High-Mg calcite is commonly 3%–15% MgO.

remains in solution and cannot be deposited. The aragonite saturation state (Ω_{arag}) is calculated from the product of the concentrations of calcium and carbonate ions at the in situ temperature, salinity and pressure, divided by the equilibrium constant for aragonite, that is, $\Omega_{arag} = [Ca_2^+][CO_3^{2-}]/K_{arag}$. The close relationship between Ω_{arag} and coral skeletogenesis was demonstrated by Marubini et al. (2003), who incubated a range of coral species under both 'normal' ($\Omega_{arag} = 4.4$) and 'low' ($\Omega_{arag} = 2.3$) aragonite ion concentrations, and observed a 13%–18% reduction in calcification rate and a weaker skeletal structure at the lower concentration. It has also been found (Leclercq et al. 2000) that reef community calcification (corals, calcareous algae, crustaceans, gastropod molluscs and echinoderms) decreased by about 60% as Ω_{arag} was lowered from 5.4 to 1.3. In addition, Ω_{arag} is at least as important as temperature in dictating the global distribution of coral reefs (Kleypas, McManus, et al. 1999; Kleypas et al. 2006).

The average Ω_{arag} in the tropics is about 4.0; the 'optimal' Ω_{arag} for reef growth is >4, while an Ω_{arag} of 3.5–4.0 is considered 'adequate' for reef growth (Guinotte et al. 2003). Some coral reefs exist at lower saturation states than this, with those of the Southwest Pacific (e.g. Lord Howe Island) and the Houtman Abrolhos Islands off Western Australia (the southernmost coral reefs in the Indian Ocean) growing under 'marginal' conditions where Ω_{arag} values of 3.28–3.35 and 3.36, respectively, have been recorded (Kleypas, McManus, et al. 1999). Similarly, there is recent evidence that inshore regions of the GBR persist in waters where Ω_{arag} values may regularly be less than 3.3, as a result of both thermodynamic effects and anthropogenic run-off (Uthicke et al. 2014). Non-reefal coral communities exist at a number of other locations where similarly marginal Ω_{arag} values have been recorded, such as the Galapagos Islands, Easter Island, mainland Japan and northern New Zealand, where Ω_{arag} is about 3 (Kleypas, McManus, et al. 1999). In contrast, high Ω_{arag} values of as much as 5–6 have been reported for the Red Sea (Kleypas, Buddemeier, et al. 1999).

Geographical patterns in aragonite saturation state are reflected in coral calcification rates. For instance, the relatively high Ω_{arag} of the Red Sea supports coral calcification rates that are higher than those seen on the coral reefs of Hawaii, despite the two areas being at similar latitudes and having comparable light and temperature regimes (Heiss 1995; Kleypas, McManus, et al. 1999). At the other end of the spectrum, the poorly cemented and thin coral accretions of the Eastern Pacific and Galapagos might be a result of the relatively low Ω_{arag} in that part of the world (Kleypas, McManus, et al. 1999). However, such conclusions are confounded by the low seawater temperatures and high rates of bioerosion at these locations, as these factors also limit rates of carbonate accretion.

The importance of Ω_{arag} to coral reef distribution is now widely recognized and this parameter is currently receiving considerable attention (see 'Reefs and changing seawater chemistry'). This is largely because, as atmospheric CO_2 emissions continue, the Ω_{arag} of coral reef waters will decrease and, ultimately, at higher latitudes at least, may drop below the threshold required for reef accretion by the end of the twenty-first century (Guinotte et al. 2003; Hoegh-Guldberg et al. 2007). This is discussed more fully in Chapter 8.

Reefs and changing seawater chemistry

Modern scleractinian corals secrete skeletons of aragonite, a metastable form of $CaCO_3$. Living coral reefs are restricted to shallow-water tropical and subtropical environments characterized by warm temperatures, high light intensities and high aragonite supersaturation (where concentrations of calcium and carbonate ions exceed the thermodynamic mineral solubility product) (Figure 1). Corals and other calcifying organisms precipitate carbonate ions, forcing a re-equilibration of the bicarbonate-dominated marine inorganic carbon system, which also creates a source of CO_2. In the short term, a coral reef therefore represents the net accumulation of $CaCO_3$ and a source of CO_2, which in turn probably leads to feedback controls on the global carbon cycle (Buddemeier 1997).

Continued

Reefs and changing seawater chemistry (*Continued*)

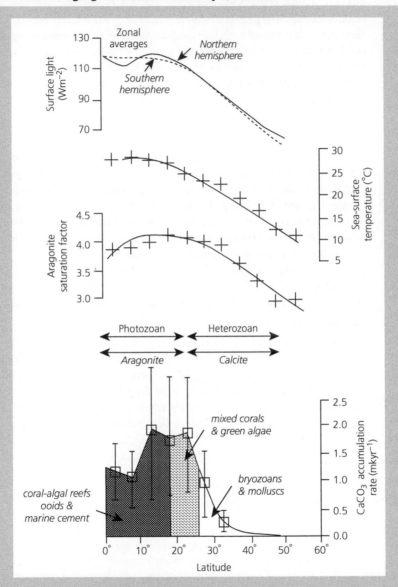

Figure 1. For all graphs, the x-axis is latitude. From top to bottom: photosynthetically active surface radiation (400–700 nm), surface temperature, surface water aragonite saturation ratio and (bottom) CaCO₃ accumulation rates and types, where shading indicates the relative importance of the major depositing communities.

Source: Modified from Buddemeier, R.W. (1997). Symbiosis: Making light work of adaptation. *Nature* 388:229–30.

Continued

Reefs and changing seawater chemistry (*Continued*)

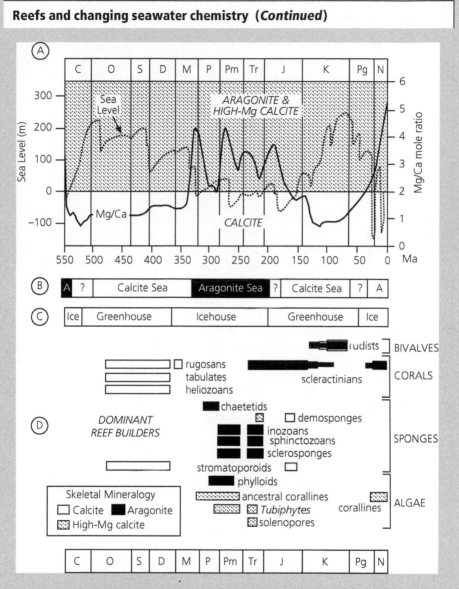

Figure 2. (A) Correspondence between changing ocean chemistry and carbonate mineralogy through time as a function of the Mg/Ca ratio in seawater and the general global sea level change. The boundary between fields of calcite (<4 mol% $MgCO_3$), and high-Mg calcite (>4 mol% $MgCO_3$) and aragonite is the horizontal line at Mg/Ca = 2. (B) The different dominant non-skeletal mineralogies precipitated in seawater through time (Sandberg 1983). (C) The different global climatic and oceanographic periods. (D) The mineralogy of different reef-building organisms.

Source: From James, N.P. and Wood, R.A. (2010). Reefs. In: R. Dalrymple and N.P. James (eds), *Facies Models: Response to Sea Level Change*. Geological Association of Canada, p. 421–47.

Continued

Reefs and changing seawater chemistry (*Continued*)

Coral reef calcification depends on the saturation state of the aragonite in surface waters and, if calcification declines, then reef-building capacity also decreases. Predictions suggest that, on present trends, this will be catastrophic for reefs and the modern carbonate system, as the timescale required for natural feedbacks to operate is far greater than the rate of greenhouse gas increase. Both experimental work and models suggest that the increased concentration of CO_2 is lowering pH, leading to ocean acidification and decreasing the aragonite saturation state in the tropics by 30%, and the biogenic aragonite precipitation by 14%–30% (Kleypas, Buddemeier, et al. 1999; Kleypas, McManus, et al. 1999). Indeed, biological calcification rates are already known to be 10%–20% lower than under pre-industrial conditions. Coral reefs are particularly threatened, because most modern reef-building organisms secrete a metastable form of $CaCO_3$.

The dominant form of precipitated crystalline $CaCO_3$ has oscillated during the geological past (Sandberg 1983), with both inorganic and organic production of aragonite and high-Mg calcite dominating carbonate formation during cool periods (icehouse, 'aragonite' seas), and low-Mg calcite predominating during warm periods (greenhouse, 'calcite' seas; see Figure 2). Such mineralogical shifts are interpreted as markers for major changes in seawater chemistry and climate.

It has been proposed that shifts in the Mg/Ca ratio have controlled the predominance of calcite versus aragonite producers, particularly reef builders, due to the inhibiting effect of high Mg^{2+} concentration on calcite secretion. The changing Mg/Ca ratio of seawater has probably been controlled by variations in the rate of production of oceanic crusts, as oceanic hydrothermal alteration is a major sink for Mg and an important source of Ca. Experimental work has subsequently confirmed the profound influence of Mg/Ca seawater ratios on modern reef builders, including scleractinian corals (Reis et al. 2006) and the calcareous green alga *Halimeda*.

Scleractinian corals were dominant reef builders in the Jurassic, but they did not build extensive reefs during the greenhouse period (calcite seas) of the Cretaceous. During this period, their species diversity remained high but with lower abundance on carbonate platforms than in the Jurassic and with a distribution shifted from platform edges to outer platform settings and higher latitudes (~35°–45° N). During much of the Cretaceous, carbonate platforms were dominated by rudist bivalves. Many hypotheses have been offered to explain these observations, including high temperatures and restricted circulation but, most persuasively, differences in seawater chemistry appear to have selected for the calcite-producing rudist bivalves over aragonite corals. Indeed, rudist bivalves with thick outer calcitic skeletons underwent a dramatic radiation in the late Cretaceous when the Mg/Ca ratio of seawater became exceptionally low (Steuber 2002).

Major shifts in the dominant composition of carbonate skeletal particles through geological time also mirror in part these proposed changes in seawater chemistry and climate, but mass extinctions also play a role by triggering changes in the predominant form of $CaCO_3$ produced by marine calcifiers (Kiessling et al. 2008). Changes in the abundance of aragonitic organisms following mass extinction events appear to have been predominantly driven by selective recovery rather than selective extinction.

Prof. Rachel Wood, School of GeoSciences, University of Edinburgh, UK.

It is clear that coral reefs have very particular abiotic needs which promote coral growth and net reef accretion and favour corals at the expense of their potential competitors such as macroalgae, so keeping the reef ecosystem in balance. Clearly, none of these abiotic factors works in isolation. For example, light penetration is influenced by suspended sediment which is itself influenced by the hydrodynamic regime. Similarly, high sea surface temperature is linked to high salinity through evaporation. Understanding how these various abiotic factors interact and hence how changes to them in the future (e.g. in response to climate change) might affect different species of corals, and the coral reef ecosystem as a whole, is of considerable importance for predicting the fate of coral reefs in our changing world.

4   Symbiotic interactions

4.1 What is symbiosis?

Symbiosis, as defined by Anton de Bary in 1879, is an association where two or more different species of organism live together for prolonged periods (Paracer and Ahmadjian 2000). A symbiotic organism may live inside the other (endosymbiosis) or outside it (ectosymbiosis), and may be wholly dependent upon the relationship for survival (obligate symbiosis) or able to also exist in a free-living state (facultative symbiosis). The three major forms of symbiosis are: (a) *parasitism*, where one partner derives benefit at the expense of the other; (b) *commensalism*, where one partner benefits and the other is unaffected by the relationship; and (c) *mutualism*, where both partners derive benefit, although not necessarily to the same degree. This chapter will focus predominantly on mutualistic symbioses, which are a dominant and extremely important feature of coral reefs, both in ecological and evolutionary terms.

The most conspicuous mutualistic symbioses on coral reefs are those between (a) eukaryotic microalgae and many invertebrates, including all reef-building corals and some sponges; and (b) prokaryotic cyanobacteria (= blue-green algae) and/or heterotrophic bacteria, and numerous sponges. In all these cases, the invertebrate acquires an important additional nutritional capability (Douglas 1994), such as photosynthesis or nitrogen fixation, that facilitates its survival in an environment that typically has limited supplies of nutrients and planktonic food. Indeed, corals and sponges, which together form the structural foundation of the coral reef ecosystem, could not exist in such abundance were it not for their mutualistic symbioses with various algae and microbes.

4.2 Algal–invertebrate symbioses in corals and soft corals

The vast majority of reef building (= hermatypic) scleractinian corals contain endosymbiotic dinoflagellate algae, called zooxanthellae (Figure

The Biology of Coral Reefs. Second Edition. Charles Sheppard, Simon Davy, Graham Pilling, and Nicholas Graham, Oxford University Press (2018). © Charles Sheppard, Simon Davy, Graham Pilling, and Nicholas Graham (2018). DOI 10.1093/oso/9780198787341.001.0001

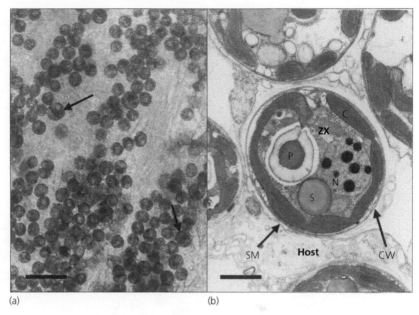

(a)   (b)

Figure 4.1   Symbiotic dinoflagellates (zooxanthellae) of the genus *Symbiodinium*. (a) Light micrograph of zooxanthellae (coccoid cells) in squashed coral tissue. Arrow points to a zooxanthella cell undergoing mitotic division to produce two daughter cells. (b) Transmission electron micrograph of zooxanthellae in a thin section of the sea anemone *Exaiptasia pallida* showing the cellular structures and host–symbiont arrangement also seen in corals; ZX, zooxanthella cell; Host, animal tissue; P, stalked pyrenoid surrounded by a starch sheath; N, nucleus containing dark, permanently condensed chromosomes; C, peripheral chloroplast; S, starch storage granule; CW, cell wall; SM, symbiosome membrane separating zooxanthella from host tissue. Scale bars represent 20 μm (A) and 2 μm (B).

Source: Photos: O. Hoegh-Guldberg (a) and K. Lee and G. Muller-Parker (b).

4.1). Evidence from the stable isotopic (δ^{15}N) signature of the skeletal organic matrix of a fossil coral (*Pachythecalis major*) suggests that the symbiosis occurred as early as the Triassic, 240 Ma (Muscatine et al. 2005); this signature differs in symbiotic versus non-symbiotic corals. Zooxanthellae also live in numerous other invertebrate taxa, including soft corals, sea anemones, jellyfish, giant clams, sponges, protists (e.g. foraminiferans) and nudibranch molluscs. In both hard and soft corals, the zooxanthellae are held in the animal's endodermis cells and bound by one or more membranes of animal origin, which form a vacuolar compartment known as the symbiosome. Zooxanthellae reach densities of 0.5 $\times 10^6$ to 5×10^6 cells cm^{-2} of coral (Porter et al. 1984; Hoegh-Guldberg and Smith 1989a, b) and the density may fluctuate over time in response to seasonal variables such as irradiance and temperature (Fagoonee et al. 1999; Fitt et al. 2000). Of note, symbiotic nitrogen-fixing cyanobacteria

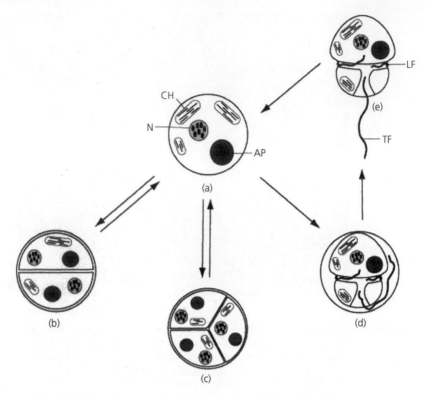

Figure 4.2   Life cycle of *Symbiodinium* (zooxanthellae) (a) Vegetative cyst (coccoid form); (b) dividing vegetative cyst forming two daughter cells; (c) dividing vegetative cyst forming three daughter cells; (d) developing zoospore (flagellated motile form); (e) zoospore; CH, chloroplast; N, nucleus; AP, accumulation product; LF, longitudinal flagellum; TF, transverse flagellum. Steps (a) to (b) and (a) to (c) occur both inside the host and in culture, while steps (d) to (e) occur only when zooxanthellae are isolated from the host.

Source: Reprinted form Stat, M. Carter, D. and Hoegh-Guldberg, O. (2006). The evolutionary history of *Symbiodinium* and scleractinian hosts: Symbiosis, diversity, and the effect of climate change. *Perspectives in Plant Ecology, Evolution and Systematics* 8:23–43; copyright 2006 with permission from Elsevier.

have been found to coexist with zooxanthellae in the Caribbean coral *Montastraea cavernosa* and two species of Pacific *Acropora* (Lesser et al. 2004, 2007; Kvennefors and Roff 2009); this symbiosis may function in a similar manner to the common sponge–cyanobacterial symbioses described in Section 4.10.

Zooxanthellae belong to the genus *Symbiodinium* (phylum Dinophyta, order Gymnodiniales). They are 5–10 μm in diameter, coccoid (spherical), usually lack flagella when inside the host (*in hospite*), and are surrounded by a cellulose cell wall. Internally, the cell possesses a large nucleus containing permanently condensed chromosomes, and a peripheral, lobed chloroplast(s) with a conspicuous stalked pyrenoid (a structure that houses the enzyme ribulose-1,5-bisphosphate carboxylase (RuBisCO), for CO_2 fixation) surrounded by a sheath of starch (Figure 4.1b). Like other photosynthetic dinoflagellates (note that some free-living dinoflagellate species are heterotrophic only), zooxanthellae contain chlorophylls *a* and *c* alongside the accessory pigments peridinin and diadinoxanthin, which give zooxanthellae, and hence their translucent coral hosts, a brownish hue. In contrast to the situation *in hospite*, when zooxanthellae are cultured outside their host, they periodically alternate between the stationary coccoid form and a motile dinomastigote form (Figure 4.2). This dinomastigote stage looks like a 'typical' dinoflagellate, with two distinct halves (the epicone and hypocone) and two flagella that propel the cell in a rapid, spiralling motion; the cell is protected by a series of thin thecal plates. The presence of free-living zooxanthellae in seawater and sediment was confirmed only quite recently (Carlos et al. 1999; Gou et al. 2003; Cunning et al. 2015; Nitschke et al. 2016), although it seems likely that these free-living cells also undergo regular alternation between stationary and motile forms. Zooxanthellae reproduce asexually, by mitotic division, when in the coccoid state only (Figure 4.2). Mitosis typically produces two to three daughter cells, each of which, *in hospite*, is ultimately housed within its own symbiosome. There is growing molecular evidence that sexual recombination also occurs in zooxanthellae, possibly when they are free-living, although gamete production and sexual reproduction have yet to be demonstrated conclusively (Stat et al. 2006; Wilkinson et al. 2015).

4.3 Diversity of zooxanthellae

Originally thought to be one species (*Symbiodinium microadriaticum*), morphological, biochemical and molecular studies have revealed zooxanthellae to be genetically diverse (see 'Molecular characterization of zooxanthellae'). A number of *Symbiodinium* species have been given full scientific names (e.g. Trench 1993; Jeong et al. 2014; Hume et al. 2015; Parkinson et al. 2015), although more commonly the numerous zooxanthella types have been placed into a series of major phylogenetic divisions, or clades, based upon their gene sequences (Figure 4.3; Coffroth and Santos 2005; Pochon and Gates 2010). Different types of *Symbiodinium* show various levels of specificity for their host corals (reviewed by Baker 2003), with some only being found in a single taxon of coral ('specialists') and others being found in a large number of coral taxa ('generalists'). For

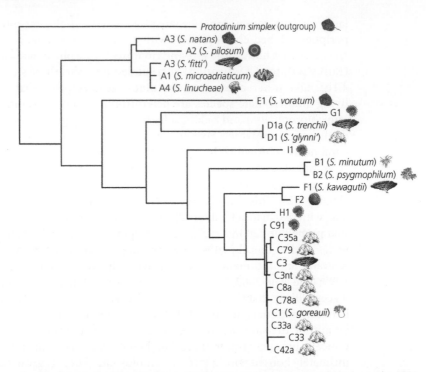

Figure 4.3 *Symbiodinium* diversity. Maximum-likelihood phylogenetic tree based on the ITS2 and D1/D2 regions of the large ribosomal subunit DNA. The *Symbiodinium* genus is currently divided into nine major clades (A–I), with several nested subclades either formally described as species, designated informal species-like names (*Nomena nuda*; shown here in quotation marks), or assigned alphanumeric nomenclature in the absence of sufficient morphological and genetic information for species description. Different *Symbiodinium* species/types have been found in the free-living state (*S. natans* and *S. voratum*) and/or in symbiosis with a variety of hosts, with examples including zoanthids (*S. pilosum*), giant clams (*S. microadriaticum*), jellyfish (*S. linucheae*), foraminiferans (several members of Clades G, I, H and C), sea anemones (*S. minutum* and *S. goreauii*) and various corals, including soft corals (*S. psygmophilum*), complex scleractinians such as table corals (*S. 'fitti'*, *S. trenchii* and *S. kawagutii*) and robust scleractinians such as the maze coral *Meandrina* sp. (F2) and the cauliflower, raspberry and bird's nest corals of the family Pocilloporidae (*S. 'glynni'*, and several types within Clades F and C). Images of these various hosts are shown on the right-hand side of each *Symbiodinium* type in the tree.

example, on the Great Barrier Reef (GBR), *Symbiodinium* type C8a has been found only in the coral *Stylophora pistillata*, while types C1, C3 and C21 have been found in a wide range of coral species (LaJeunesse et al. 2003). Similarly, some coral species harbour numerous types of *Symbiodinium* (often simultaneously), while others associate with a single type or a subset of closely related types (LaJeunesse 2002; Baker 2003; Putnam et al. 2012; Lee et al. 2016).

Molecular characterization of zooxanthellae

The symbiotic dinoflagellates (zooxanthellae) of corals and other marine invertebrates were initially considered to be a single pandemic species, *Symbiodinium microadriaticum* Freudenthal. However, early molecular studies revealed polymorphic enzyme isoforms amongst different *Symbiodinium* strains, with these isoforms corresponding to subtle differences in size, ultrastructure, division rates and host specificity (Schoenberg and Trench 1980a–c). This pioneering work provided some of the first clues that *Symbiodinium microadriaticum* might harbour a trove of hidden diversity. The increased availability of the PCR technique during the early 1990s brought about the detection of sequence variation in nuclear DNA encoding the ribosomal subunits (Wilcox 1998), and a region of the chloroplast *cp23S* gene (Santos et al. 2002). The phylogenetic congruence of multiple genes spanning different organelles enabled the organization of the *Symbiodinium* complex into well-supported major taxonomic clades, nine of which are currently recognized (Clades A–I; Pochon and Gates 2010) and see Figure 4.3. These clades are supported by genetic distances that approximate order- or even class-level divergence in free-living dinoflagellates and other algae (Blank and Huss 1989), and hence the current single-genus classification appears to be outdated. However, reaching a general consensus on *Symbiodinium* systematics remains a challenge, primarily due to the existence of multiple levels of within-clade genetic diversity.

The development of high-resolution genetic markers, including two spacers of the nuclear ribosomal DNA cistron (ITS1 and ITS2; LaJeunesse 2001), and a hypervariable non-coding region of the chloroplast *psbA* minicircle (Moore et al. 2003), revealed hundreds of new subcladal phylotypes, particularly within the highly divergent Clade C group. A number of microsatellite markers (repetitive short tandem repeats interspersed throughout the genome) were also developed for some clades, offering a phylogenetic signal at an even finer level of taxonomic resolution. These markers are sufficiently sensitive to reveal important processes occurring at the population level, including patterns of host–symbiont co-evolution (Baums et al. 2014), invasion dynamics (LaJeunesse et al. 2016) and even the occurrence of cryptic sexual recombination in *Symbiodinium*, whose reproductive mode was once thought to be confined to clonal cell division (LaJeunesse et al. 2014).

The field of *Symbiodinium* genetics is currently going through an exciting transformation with the advent of high-throughput sequencing (HTS) technology. Millions of DNA and RNA sequence reads can now be rapidly generated at a fraction of the cost of just a few years ago, enabling a wide range of new research questions to be addressed. Applications of HTS include extensive sequencing of *Symbiodinium* communities from host tissue and the surrounding environment (environmental DNA meta-barcoding; e.g. Cunning et al. 2015), quantification of intracellular RNA transcripts that can characterize molecular-level responses to environmental stimuli (transcriptomics; e.g. Palumbi et al. 2014), the identification of vast numbers of ancestry-informative markers (single-nucleotide polymorphisms, or SNPs), and landmark assemblies of entire *Symbiodinium* nuclear and organellar genomes (e.g. Shoguchi et al. 2013). These new capabilities offer valuable insights into *Symbiodinium* function, ecology and evolution, and their potential roles in coral acclimation and adaptation to warming, acidifying oceans caused by anthropogenic climate change.

Dr Shaun P. Wilkinson, School of Biological Sciences, Victoria University of Wellington, New Zealand.

The prevalence of a *Symbiodinium* type at a particular location is related to the environmental regime and its physiological suitability to this regime. For instance, types belonging to Clade C tend to dominate in 'normal', stable conditions, while those belonging to Clade B tend to be more prevalent in harsher, colder conditions. This physiological difference is thought to explain the biogeography of these two important clades, where Clade C dominates Pacific reefs, and Clade B is more common on Caribbean reefs, possibly because it was favoured over Clade C by the relatively tough conditions experienced in the Caribbean during the Pliocene–Pleistocene transition (Baker 2003; LaJeunesse et al. 2003). Likewise, in the widespread Indo-Pacific coral *Plesiastrea versipora*, a latitudinal shift from Clade C to Clade B is seen as one moves

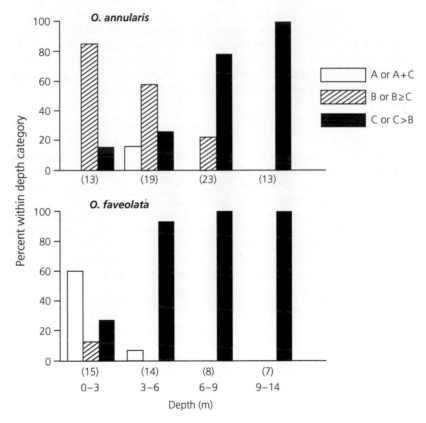

Figure 4.4   Occurrence of *Symbiodinium* Clades A, B and C in the Caribbean corals *Orbicella annularis* and *O. faveolata* at different depths (0–3 m; 3–6 m; 6–9 m; 9–14 m). Corals contained either Clade A or A plus C; B or B with an equal or lesser amount of C; or C or more C than B; no corals simultaneously harboured Clades A and B. Plots show the percentage of individual corals found to contain each clade (or combination of clades), and values in brackets show total number of coral colonies sampled within each depth range.

Source: Rowan, R. and Knowlton, N. (1995). Intraspecific diversity and ecological zonation in coral algal symbiosis. *Proceedings of the National Academy of Science of the United States of America* 92:2850–3. Copyright 1995 National Academy of Sciences, USA.

south from the tropical waters of the GBR to the temperate, more variable waters of southern Australia (Rodriguez-Lanetty et al. 2001). Clade D zooxanthellae tend to dominate in high-temperature and turbid environments (Berkelmans and Van Oppen 2006; Pettay et al. 2015; Silverstein et al. 2015), meaning that they are locally abundant in stressful sites such as the Persian Gulf where seawater temperatures regularly exceed 33 °C (Baker et al. 2004).

Physiological diversity can also be used to explain within-site distributions of different *Symbiodinium* types. In particular, there has been found to be a strong relationship between depth (i.e. light regime) and type. Rowan and Knowlton (1995) discovered that the relative proportions of Clades A, B and C in the Caribbean corals *Orbicella annularis* and *O. faveolata* changed with depth, with Clade C becoming more common in deeper water (Figure 4.4). This led to the suggestion that some zooxanthellae are 'shade lovers' while others are 'sun lovers'. It is important to note, though, that all members of the same *Symbiodinium* clade do not necessarily have the same environmental preferences. For example, the recently discovered Clade C zooxanthella *Symbiodinium thermophilum* is dominant in the extremely warm waters of the southern Persian Gulf (Hume et al. 2015). Furthermore, we are yet to fully understand the true relationship between genetic and physiological diversity, and the extent to which the environmental limits (e.g. depth or latitudinal ranges) of different coral species are determined by the physiology of the zooxanthellar type(s) with which they can form successful symbioses.

4.4 Physiology of coral calcification

The coral's skeleton is a two-phase composite of (a) an organic matrix that includes various mucopolysaccharides, proteins, glycoproteins and calcium-binding phospholipids and (b) fibre-like crystals of calcium carbonate ($CaCO_3$) in aragonite form (Goreau 1959; Constantz and Weiner 1988; Muscatine et al. 2005; Tambutté et al. 2011). The organic matrix is synthesized by the calicoblastic epithelium before being secreted into the underlying sub-epithelial space. The matrix then facilitates $CaCO_3$ nucleation and provides a framework for the aragonite crystals. Zooxanthellae are directly linked to the synthesis of the organic matrix through the provision of organic carbon from photosynthesis, although exogenous food supplies also provide some of the material needed. This incorporation of organic carbon has been traced using ^{14}C-labelled bicarbonate and food in the corals *Pocillopora damicornis* and *Fungia scutaria*, respectively (Muscatine and Cernichiari 1969; Pearse 1971). Zooxanthellae also assimilate inorganic nitrogen from seawater and synthesize and release organic nitrogen (i.e. amino acids) for matrix formation (Muscatine et al. 2005). However, exogenous food may be a more important source of some nitrogenous components such as aspartic acid, an acidic amino acid that is abundant in proteins of the skeletal matrix (Allemand et al. 1998; Houlbreque et al. 2004).

The mechanism of $CaCO_3$ deposition (i.e. calcification) in corals remains poorly understood, with the role of the zooxanthellae being an ongoing topic of controversy (see reviews by Allemand et al. 2011; Tambutté et al. 2011; Davy et al. 2012). The epidermis of all scleractinian corals is able to secrete $CaCO_3$; however, the rate of secretion is faster in symbiotic corals than in non-symbiotic corals and, in the vast majority (>90%) of cases, symbiotic corals secrete $CaCO_3$ more rapidly in the light than in darkness (Gattuso et al. 1999). For instance, [45]Ca incorporation in the coral *Manicina areolata* was nearly 6.5 times faster in the light than in the dark, and 16.5 times faster in the light when zooxanthellae were present than when they were absent, while [45]Ca incorporation in the coral *Stylophora pistillata* was 4 times greater in the light than in the dark (Furla et al. 2000). Calcification and photosynthesis are therefore clearly coupled, as shown in the following equations:

$$Ca^2 + CO_2 + H_2O \rightarrow CaCO_3 + 2H^+ \text{ (Equation 1, calcification)}$$

$$2H^+ + 2HCO_3 \rightarrow 2CO_2 + 2H_2O \text{ (Equation 2, bicarbonate conversion)}$$

$$CO_2 + H_2O \rightarrow CH_2O + O_2 \text{ (Equation 3, photosynthesis)}$$

$$Ca^{2+} + 2HCO_3^- \rightarrow CaCo_3 + CH_2O + O_2 \text{ (Equation 4, net of equations 1–3)}$$

However, the exact nature of this coupling is not fully understood and several different mechanisms have been proposed to explain why calcification is faster in the light than in the dark. The two best known are light-enhanced calcification and trans-calcification. In the light-enhanced calcification model (Goreau 1959, 1961), increased $CaCO_3$ saturation results from photosynthetic removal and a reduced partial pressure of CO_2 in the coral tissues (i.e. an equilibrium shift in favour of $CaCO_3$ secretion). One notable problem with this model is that calcification often happens most rapidly at the tips of coral branches, where the density of zooxanthellae is relatively low (Smith and Douglas 1987). In the trans-calcification model (McConnaughey 1991; McConnaughey and Whelan 1997), Ca^{2+}-ATPase delivers Ca^{2+} to the calcification site and carries H^+ released during $CaCO_3$ precipitation to the coelenteron, where the protons dehydrate HCO_3^- and hence generate CO_2. Trans-calcification would therefore enhance photosynthesis through increasing the availability of CO_2, and the suggestion is that the coral regulates calcification so that rapid $CaCO_3$ deposition in the light benefits zooxanthellae at a time when they are photosynthesizing; evidence in support of trans-calcification is equivocal, though, with inhibition of calcification in the coral *Stylophora pistillata* having no effect on photosynthetic rate (Gattuso et al. 2000). Other mechanisms proposed to explain faster calcification in the light than in the dark include (1) neutralization of H^+ (i.e. acidic) produced during calcification by OH^- (i.e. alkaline) liberated during photosynthesis, which produces an environment favourable to $CaCO_3$ deposition; (2) dark repression of calcification; (3) light-enhanced removal by zooxanthellae of compounds that interfere with calcification; (4) provision by zooxanthellae of photosynthetic products for organic matrix synthesis or to provide energy for active transport; and (5) maintenance of an oxic environment in the tissues through photosynthetic O_2-production (Gattuso et al. 1999; Holcomb et al. 2014). It may well be that several of these mechanisms play a concurrent role in coral calcification, but their relative contributions are unknown.

4.5 Photosynthesis and carbon fluxes

The coral and zooxanthellae derive a number of benefits from the symbiosis (Table 4.1). The major benefit derived by the coral is the capacity to obtain photosynthetically fixed organic carbon compounds ('photosynthate') from the zooxanthellae, which supplement the relatively small quantity of organic carbon obtained by feeding on zooplankton. In other words, the zooxanthellae enable the coral to be polytrophic (i.e. both autotrophic and heterotrophic), thus increasing the feeding options in an environment where planktonic food supplies are limited. There is little evidence for the digestion of healthy zooxanthellae; rather, the zooxanthellae use a portion of the photosynthate for their own metabolism, growth and storage, and then release surplus material to the coral (see 'Isotopic labelling').

Table 4.1 Benefits of the coral–zooxanthella symbiosis

Nutritional benefits
Autotrophy, arising from the translocation of energy-rich photosynthetic products (e.g. glycerol, glucose) from zooxanthellae to the coral; these products support coral respiration, tissue growth, gamete production and survival.
Enhanced availability and retention of nutrients, arising from:
– **capture of planktonic food** by the coral and the supply of excretory waste (containing nitrogen and other elements such as phosphorus) to the zooxanthellae.
– **nitrogen conservation**, due to the preferential use of energy-rich photosynthetic products rather than amino acids for coral respiration.
– **nitrogen recycling**, arising from uptake of the coral's nitrogenous waste by the zooxanthellae, its incorporation into amino acids and their translocation back to the coral.
Coral skeleton formation and reef accretion
Light-enhanced calcification (i.e. $CaCO_3$ deposition) and the provision of photosynthetically fixed carbon for the coral skeleton's organic matrix.
Other benefits
Fixed position in the water column and a favourable light regime for the zooxanthellae, due to the coral's physical properties (e.g. branch orientation, light scattering by the skeleton or fluorescent pigments in the tissues), which optimize harvesting of downwelling light.
Protection of zooxanthellae from planktonic grazers by the coral's tissues.
Detoxification of coral cells, via the uptake and utilization of potentially toxic nitrogenous (NH_4^+) waste by the zooxanthellae.
Supplementary O_2 for coral metabolism arising from zooxanthellar photosynthesis.
Supplementary CO_2 for photosynthesis arising from coral respiration.

Note: The importance of each of these benefits differs greatly, with those relating to nutrition, calcification and improved access to light for the zooxanthellae likely being the most significant.

Isotopic labelling

The translocation of photosynthetically fixed carbon from zooxanthellae to corals (or, indeed, from phototrophic symbionts to sponge hosts) is fundamental to the ecological success of coral reefs. Yet, not until the late 1950s was this movement of organic carbon demonstrated directly in a symbiotic invertebrate, by using ^{14}C (Muscatine and Hand 1958). This radioisotope is supplied as dissolved ^{14}C-carbonate or, more commonly, ^{14}C-bicarbonate and then incorporated via photosynthesis into organic compounds. These ^{14}C-labelled compounds can then be tracked as they move through the symbiosis, both by visualizing tissue sections with autoradiography and by separating the animal and algal partners and measuring the levels of radioactivity in each fraction (see Figure).

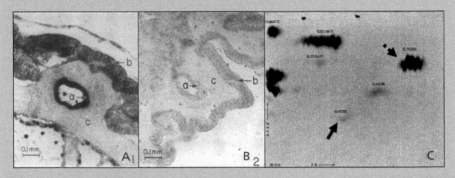

Figure. Transfer of radiolabelled (^{14}C) photosynthate from zooxanthellae in the symbiotic sea anemone *Anthopleura elegantissima*. (A) Section through the tissues of the sea anemone after 4 weeks of being incubated in ^{14}C. Dark regions (contrast with the unlabelled section in Panel B) indicate the presence of radioactive tracer, which has been incorporated by photosynthesis into the zooxanthellae, which are concentrated in the gastrodermis (a), and then translocated into the surrounding gastrodermal cells, epidermis (b) and mesogloea (c). (B) Anemone tissue incubated without a ^{14}C radiotracer; a, b and c as in Panel A. (C) A radiochromatogram, showing individual ^{14}C-labelled compounds released by freshly isolated zooxanthellae when incubated in a tissue homogenate of the anemone; two of the major release products, glucose and glycerol, are indicated by the solid and dashed arrows, respectively.

Source: Photos (A) and (B) from Muscatine, L. and Hand, C. (1958). Direct evidence for the transfer of materials from symbiotic algae to the tissues of a coelenterate. *Proceedings of the National Academy of Sciences of the United States of America* 44:1259–63. Photo (C) from Trench, R.K. (1971). The physiology and biochemistry of zooxanthellae symbiotic with marine coelenterates. II. Liberation of fixed ^{14}C by zooxanthellae in vitro. *Proceedings of the Royal Society of London. Series B, Biological Sciences* 177:237–50, with kind permission from R.K. Trench and The Royal Society of London.

The technique has proved useful in quantifying the translocation of photosynthate from algal or cyanobacterial symbionts to their host (Trench 1971; Muscatine et al. 1981; Wilkinson 1983) and for determining the fate of the released carbon in the symbiosis, for example, with respect to its use in gamete or mucus production (Crossland et al. 1980; Rinkevich 1989). Furthermore, it is possible to identify photosynthetic compounds released by symbiotic algae by incubating isolated algae in an homogenate of host tissue

Continued

Isotopic labelling (*Continued*)

and then performing a radiochromatographic analysis of ^{14}C-labelled products (Muscatine and Cernichiari 1969; Trench 1971); a similar protocol first demonstrated the existence of a 'host release factor' (HRF) in coral tissues (Muscatine 1967) and has since been employed in the characterization of this molecule(s) (Grant, Remond, et al. 2006; Grant, Trautman, et al. 2006).

More recently, stable isotopic labelling, for example with ^{13}C or ^{15}N, has been used to further our understanding of the nutritional fluxes between zooxanthellae and their host corals (Grover et al. 2003, 2008). Of particular note, stable isotopic labelling has been combined with the cutting-edge technology of nanoscale secondary ion mass spectrometry to visualize nutritional fluxes both between and within the symbiotic partners at the cellular level (Kopp et al. 2013).

Prof. Simon Davy, Victoria University of Wellington, New Zealand.

Zooxanthellae fix CO_2 by the C_3 (Calvin–Benson) pathway (Streamer et al. 1993) (also see Pulse amplitude modulation fluorometry and reefs). CO_2 can potentially be obtained from (a) coral and zooxanthellar respiration (Harland and Davies 1995); (b) coral skeletogenesis, where CO_2 is produced during the calcification process (Ware et al. 1991); and (c) seawater bicarbonate (HCO_3^-). In this latter case, HCO_3^- is converted to CO_2 in the coral and zooxanthellae by so-called carbon-concentrating mechanisms, such as via the enzyme carbonic anhydrase in the zooxanthellae and coral and via a vacuolar proton pump in the coral that acidifies the symbiosome that encloses the zooxanthellae, and hence creates an environment favourable for photosynthesis (Weis 1991; Bertucci et al. 2013; Barott et al. 2015). Photosynthesis shows a characteristic asymptotic relationship with irradiance: it increases linearly at first before steadily reaching a plateau (i.e. photosynthetic saturation), normally at an irradiance of about 200–300 μmol photons m^{-2} s^{-1}. This is a relatively low irradiance when compared to the surface irradiance that, on sunny days in shallow reef waters, can reach more than 2000 μmol photons m^{-2} s^{-1} (Muller-Parker and Davy 2001).

Irradiance decreases exponentially with depth. To counter low light levels, the coral–algal symbiosis employs a number of methods to optimize use of light. Foremost amongst these is photoacclimation by the zooxanthellae, where the size of the photosynthetic units (PSUs) and hence the amount of chlorophyll *a* per zooxanthella increase in response to reduced irradiance, resulting in greater light-harvesting efficiency; under bright conditions, PSUs are smaller but more numerous, meaning that light-harvesting efficiency is less but the maximum rate of carbon fixation is greater. Such photoacclimatory responses explain why shade-adapted corals appear darker than light-adapted corals (Figure 4.5). For instance, shade- and light-adapted colonies of *Stylophora pistillata* contain similar densities of

Figure 4.5 Photoacclimation in shade- versus high-light-adapted corals (*Stylophora pistillata*) from the Red Sea. The two corals contain similar densities of zooxanthellae, but visibly darker, shade-adapted colonies (right) contain more chlorophyll *a* per unit biomass than do pale, light-adapted colonies (left).
Source: Photo by Z. Dubinsky.

zooxanthellae, but visibly darker shade-adapted colonies contain over seven times more chlorophyll *a* per unit biomass than do pale light-adapted colonies (Falkowski and Dubinsky 1981); increasing chlorophyll concentration in the coral's tissues by increasing the density of zooxanthellae would cause self-shading. Another way that corals deal with declining light availability is by harbouring 'shade-loving' zooxanthellae as opposed to the 'sun-loving' zooxanthellae seen in shallower waters (Rowan and Knowlton 1995). In addition, corals may strategically position fluorescent pigments behind or alongside the zooxanthellae, to enhance light capture (Salih et al. 2000); the pigments transform light into useable wavelengths and scatter it back towards the zooxanthellae for photosynthesis.

In contrast, in high-light conditions, fluorescent pigments that emit light at non-useable wavelengths are situated above the zooxanthellae, where they reflect light away from the algal cells and so protect them from potentially damaging levels of photosynthetically active radiation (PAR) and ultraviolet (UV) light (Salih et al. 2000). UV-absorbing mycosporine-like amino acids in both coral tissue and zooxanthellae provide a further means of screening out the harmful wavelengths of sunlight (Dunlap and Shick 1998; Banaszak et al. 2000), while the coral also can withdraw its tentacles and contract its polyps for protection. Such photoregulatory mechanisms are of considerable importance.

Pulse amplitude modulation fluorometry and reefs

Since the early 1990s, chlorophyll fluorescence has been measured in studies of coral physiology, providing significant insights into mechanisms whereby symbiotic dinoflagellates of corals can contend with the potentially damaging light levels characteristic of shallow tropical seas. These techniques have provided valuable insights into the mechanism of coral bleaching, and how contaminants affect symbiotic algae and subsequently the host coral. These techniques are possible in part by the development of portable chlorophyll fluorometers (Maxwell and Johnson 2000; see Figure).

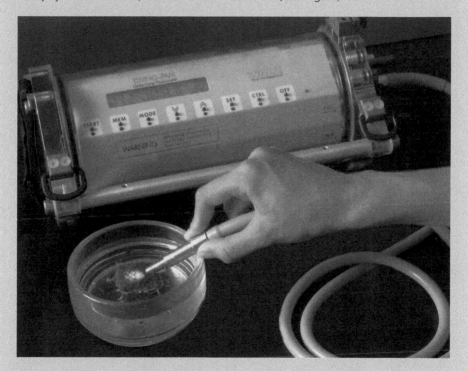

Figure. A DIVING-PAM underwater pulse amplitude modulation chlorophyll fluorometer (Walz GmbH, Effeltrich, Germany) with a fibre optic probe, measuring chlorophyll fluorescence of symbiotic dinoflagellates within the tissues (*in hospite*) of a small fragment of the hard coral *Orbicella franksi*. The unit is submersible to a depth of 50 m.

In photosynthesis, antenna pigments absorb light, and excitation energy is transferred to reaction centres of the two photosystems, where it drives the photochemical reactions of photosynthesis. About 3%–5% of excitation energy is dissipated by fluorescence, mainly from chlorophyll *a* of Photosystem II. Fluorescence emission competes with two other de-excitation processes that deactivate the excited chlorophyll states. These processes reduce (or quench) the amount of fluorescence, and are referred to as photochemical

Continued

Pulse amplitude modulation fluorometry and reefs (*Continued*)

quenching and non-photochemical quenching. Photochemical quenching reflects useful photochemistry (i.e. assimilatory electron flow). Non-photochemical quenching reflects photoprotective dissipation of excess absorbed energy as heat. Chlorophyll fluorescence displays characteristic changes upon illumination, reflecting changes in the properties of excitation and energy conversion at Photosystem II. Differentiating between the two main quenching components can provide useful insights into regulatory processes that occur within the photosynthetic apparatus, especially under stress conditions.

Measurement of chlorophyll fluorescence in the symbiotic algae uses highly selective modulation techniques (pulse amplitude modulation). This allows separation of the fluorescence signal from the much stronger excitation light. Light is provided from a modulated beam from a high-frequency LED. Fluorescence is measured by an amplifier which selectively picks up fluorescence signals against non-modulated and scattered light, allowing measurements to be made of the efficiency of Photosystem II electron transport, even in full sunlight (Schreiber 2004). Using a series of saturating pulses of light as well as actinic light, the maximum quantum yield (Fv/Fm) and effective quantum yield of photochemical efficiency can be measured.

Chlorophyll fluorescence techniques are rapid, allowing reliable estimates to be made within seconds. The method is non-destructive and non-intrusive, allowing direct assessment of the algae inside the coral host cells, without having to relocate organisms to the laboratory or to enclose them in a sealed container (as done with measurements of photosynthesis based on O_2/CO_2 exchange). Advances in the integration of various light sources, together with miniaturization, has resulted in compact, portable instrumentation which has extended the possibilities of real-time measurements of photosynthesis in corals in their natural environment. Other systems include machines which can examine individual symbiotic algae, suspensions of freshly isolated or cultured symbiotic algae at very low concentrations, and spatial variations in photosynthetic activity in coral branches or fragments (up to 20 or 30 mm in size) by capturing two-dimensional images.

Under natural conditions, dinoflagellates in corals can experience light intensities high enough to initiate downregulation or photoinhibition. This is a means of photoprotection which reduces long-term damage by partitioning less energy into photochemical reactions and more into non-photochemical quenching. This allows for rapid recovery once light levels return to lower levels. Work on cultured *Symbiodinium* spp. and from studies inside the host's tissues has shown that, under conditions of heat stress and light/temperature interactions, there is irreversible damage to the reaction centre of PSII, leading to bleaching. This may also be caused by cyanide, herbicides, effluent from the offshore oil and gas industry, low salinity, light shock and cold stress (Jones 2005).

Dr Ross Jones, Australian Institute of Marine Science, Perth, Australia.

Photosynthate released to the host largely consists of low-molecular-weight, energy-rich compounds, with the major components traditionally believed to be glucose and glycerol, although the significance of this latter compound has been questioned in recent years (Burriesci et al. 2012; Davy et al. 2012). Released photosynthate also contains almost all essential amino acids (often in the form of protein/sugar glycoconjugates), organic acids and lipids (Markell and Trench 1993; Trench 1993; Muscatine et al. 1994; Peng et al. 2011). Zooxanthellae in some corals have been calculated to release 78%–98% of their photosynthate to their coral host (Davies 1984; Muscatine et al. 1984; Edmunds and Davies 1986), in marked contrast to free-living microalgae, which typically leak out less than 5% of their photosynthate. The stimulus for this release has been the matter of much debate over the years, but one idea is that corals (as well as other zooxanthellate invertebrates such as giant clams and sea anemones) produce a chemical cue, a so-called host release factor (HRF) of uncertain identity (Hinde 1988; Gates et al. 1995; Cook and Davy 2001). HRF activity is demonstrated by placing isolated zooxanthellae in a homogenate of their host's tissues and labelling with a radiotracer (i.e. ^{14}C). Under these conditions, the release of photosynthate far exceeds that seen when zooxanthellae are incubated in seawater. Evidence from the coral *Plesiastrea versipora* suggests that HRF stimulates glycerol synthesis, hence producing a high concentration of glycerol inside the zooxanthellae, and that this, in turn, leads to the diffusion of glycerol from the alga to the animal (Grant et al. 1998). However, it is not yet known how the release of other photosynthetic compounds is induced, or whether the mechanism in *P. versipora* operates in other species of corals and invertebrates.

None of the released carbon compounds, with the exception of amino acids, which comprise only a minor fraction of the photosynthate, contain nitrogen; hence, they have been called 'junk food' (Falkowski et al. 1984). This means that they can directly support coral respiration, and be used to synthesize energy-rich storage products such as starch and lipid (especially triacylglycerols and wax esters), but can only be used for coral growth and reproduction when nitrogen is available, such as from a coral's predatory food capture. The importance of the photosynthate to coral respiration is considerable; under sunny conditions in shallow water, all of a coral's metabolic energy demands can be met by zooxanthellar photosynthesis but, under cloudy or turbid conditions, or in deeper water, respiratory demands may not be met. For instance, in the Indo-Pacific coral *Pocillopora damicornis*, the zooxanthellae supply 135% of the coral's respiratory carbon needs on a sunny day but only 79% on a completely overcast day, when the coral has to draw upon its lipid reserves (Davies 1991). Where more than enough photosynthate is available to support coral respiration, surplus photosynthate is free for use in coral growth and reproduction (both relatively small sinks for carbon utilization), storage or excretion as mucus lipid. Indeed, in well-lit, shallow waters, corals excrete copious quantities of mucus lipid, perhaps up to 45% of the carbon fixed in photosynthesis (Crossland et al. 1980; Davies 1991). A full carbon/energy budget for *P. damicornis* on sunny and overcast days is shown in Table 4.2.

Table 4.2 A daily (24 h) energy budget for the coral *Pocillopora damicornis* on both a sunny day and an overcast day in Hawaii at 3 m depth

	Zooxanthellae				Coral Animal		
	Photosynthesis	Respiration	Growth	Excess for Transport to Coral Animal	Respiration	Growth	Excess for Loss (e.g. mucus)
Sunny day	355.2 (100%)	29.5 (8.3%)	1.2 (0.3%)	324.5 (91.4%)	241.1 (67.9%)	12.4 (3.5%)	71.0 (20.0%)
Overcast day	224.7 (100%)	29.5 (13.1)	1.2 (0.5%)	194.0 (86.3%)	241.1 (107.3%)	12.4 (5.5%)	(−59.5) (−26.4%)

Note: Units are in joules of energy, and budgets are calculated for coral 'nubbins' of 10 g skeleton weight. Values in parentheses are the percentages of photosynthetic production utilized for the named process. Energy translocated from the zooxanthellae to the coral animal (in the form of photosynthetically fixed, energy-rich carbon compounds) is assumed to be that still available after utilization for respiration and growth by the zooxanthellae. Energy losses from the coral animal (e.g. as mucus) are calculated as the energy remaining (from the total translocated by the zooxanthellae) after utilization for coral respiration and growth. The budget shows that, on a sunny day, the coral animal has an excess of energy-rich photosynthate and so can meet all of its respiratory and growth demands by autotrophy while, on a cloudy day, it must draw upon its lipid reserves to meet its needs.
Source: Data from Davies, P.S. (1991). Effect of daylight variations on the energy budgets of shallow-water corals. *Marine Biology* 108:137–44.

4.6 Nitrogen acquisition and fluxes

Corals can acquire nitrogen from a number of different sources: (a) Dissolved inorganic nitrogen (DIN) in seawater; (b) dissolved organic nitrogen (DON) in seawater; (c) sediment settled on the coral's surface (particulate organic nitrogen (PON)); and (d) plankton and suspended particulate matter caught by the coral polyps. DIN is taken up from seawater as both ammonium (NH_4^+) (Grover et al. 2002) and nitrate (NO_3^-) (Wilkerson and Trench 1986; Grover et al. 2003), while DON is taken up from seawater as amino acids such as glycine and alanine (Ferrier 1991; Grover et al. 2008). Sediment contains a variety of PON sources, including detritus, bacteria, protozoans, microalgae, micro-invertebrates and sorbed organic matter (Lopez and Levinton 1987). Sediment settles on the coral surface and is trapped in mucus before being either ingested by the polyps or sloughed off; in both cases, the coral takes up nitrogen from the sediment, assimilating up to 30%–60% of that available (Mills and Sebens 2004). Corals consume numerous types of zooplankton, especially at night, when zooplanktonic organisms are most active in the water column. Ingested zooplankton includes holoplanktonic copepods and salps, and the meroplanktonic larvae of various turbellarians, polychaetes, barnacles, bryozoans and other corals; corals also capture animals that are stirred into the water column from the seabed (Porter 1974). All of these nitrogen sources support colony growth and reproduction, and assimilated nitrogen is lost from the colony as organic nitrogen in mucus, as is illustrated (alongside carbon inputs and outputs) for the Caribbean elkhorn coral *Acropora palmata* (Table 4.3).

Table 4.3 Mean annual carbon and nitrogen inputs/outputs for the Caribbean coral *Acropora palmata*

	Input			Output			
	Photosynthesis	Particulate Feeding	Dissolved Nutrient Uptake	Respiration	Tissue Growth	Gamete Production	Dissolved Organic Losses
Carbon	16.40 (91.1%)	1.60 (8.9%)	0.00 (0.0%)	11.50 (63.2%)	2.10 (11.5%)	1.60 (8.8%)	3.00 (16.5%)
Nitrotgen	0.00 (0.0%)	0.50 (71.4%)	0.20 (28.6%)	0.00 (0.0%)	0.30 (40.0%)	0.10 (13.3%)	0.35 (46.7%)

Note: Units are grams of carbon per year or grams of nitrogen per year, for a coral of 250 g initial skeletal weight; values in parentheses are percentages of the total input or output.

Source: Data from Bythell, J.C. (1988) A total nitrogen and carbon budget for the elkhorn coral *Acropora palmata* (Lamarck). *Proceedings of the 6th International Coral Reef Symposium* 2: 535–40.

However, like all nutrients, nitrogen (whether as DIN, DON or PON) is typically in short supply in many reef waters. Consequently, the capacity to scavenge nitrogen in various forms, and to conserve and recycle it, is a hugely important attribute of the coral–algal symbiosis and, along with the autotrophic supply of photosynthetic carbon, explains the paradoxical success of corals in nutrient-poor tropical seas. Most work on nitrogen assimilation by corals has focused on ammonium, whether it is sourced from exogenous seawater supplies or from coral excretion, where it is an end product of animal cell catabolism (or, more specifically, amino acid deamination). The assimilation and movement of ammonium nitrogen between the coral and its zooxanthellae is shown in Figure 4.6. Both the coral host and its zooxanthellae are able to assimilate ammonium, which diffuses readily into cells. In the animal, assimilation occurs largely via the NADP-glutamate dehydrogenase (NADP-GDH) pathway, where this enzyme catalyses the conversion of ammonium to the amino acid glutamate (Miller and Yellowlees 1989; Roberts et al. 2001); the glutamate can then be used to synthesize other amino acids. Zooxanthellae, on the other hand, assimilate ammonium by the glutamine synthetase/glutamine 2-oxoglutarate amido transferase (GS/GOGAT) pathway, where the end product is again glutamate (Roberts et al. 2001). Active uptake of nitrate occurs in both the coral and zooxanthellae, but only zooxanthellae can convert nitrate to ammonium for subsequent assimilation into amino acids; the conversion is through the action of nitrate and nitrite reductases (Miller and Yellowlees 1989; Grover et al. 2003; Leggat et al. 2007).

Given that the assimilation of ammonium involves amino acid synthesis, organic carbon skeletons must be available for assimilation to occur. This explains why the uptake of ammonium is dependent upon the presence of not only zooxanthellae, but light for photosynthesis; symbiotic corals placed in the dark show net release of ammonium, once all stored organic carbon supplies are used up, although these stores may be sufficient to support ammonium assimilation for several hours and even overnight (Muscatine and D'Elia 1978). Photosynthetic carbon fixation therefore not only supports

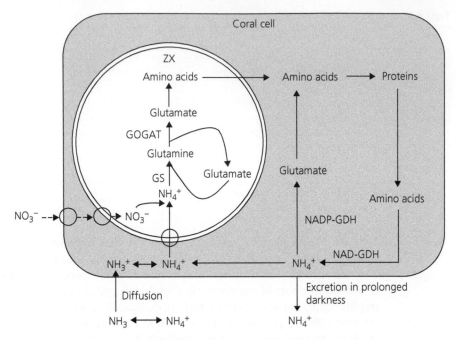

Figure 4.6   Nitrogen fluxes in the coral–zooxanthella symbiosis; ZX, zooxanthella cell; GS, glutamine synthetase; GOGAT, glutamine 2-oxoglutarate amido transferase; GDH, glutamate dehydrogenase. Small circles on cell surface indicate specific membrane transport mechanisms. Dashed arrows indicate inconclusive evidence for nitrate uptake route. Note that, for clarity, protein synthesis and catabolism pathways in the zooxanthella are not shown.

Source: Modified from Davies, P.S. (1992). Endosymbiosis in marine cnidarians. In: D.M. John, S.J. Hawkins and J.H. Price (eds), *Plant–Animal Interactions in the Marine Benthos*. Clarendon Press, pp. 511–40.

the metabolic and structural demands of the zooxanthellae and their coral host, but is also critical for the conservation of valuable nitrogen that would otherwise by excreted into seawater (Wang and Douglas 1998). The supply of exogenous carbon to support nitrogen assimilation is likely to be relatively small. Further conservation of nitrogen also arises because the coral uses zooxanthellar photosynthate in preference to amino acids as a respiratory substrate. This nitrogen conservation mechanism has been demonstrated in zooxanthellae by Wang and Douglas (1998) who found that cellular concentrations of amino acids increased and ammonium decreased, when the zooxanthellate animal (the tropical sea anemone *Exaiptasia pallida*) was placed in darkness but supplemented with exogenous organic carbon compounds.

Once nitrogen has been assimilated into amino acids, either by the coral or zooxanthellae, it can be incorporated directly into proteins and used for growth. Alternatively, organic nitrogen, with the various energy-rich non-nitrogenous compounds described earlier, is passed from the zooxanthellae to the coral (Davy et al. 2012; Kopp et al. 2013). For instance, the amino acid alanine has been identified in ^{14}C-labelled photosynthate released by zooxanthellae when incubated in tissue homogenates of a number of different host species (Lewis and Smith 1971; Sutton and Hoegh-Guldberg 1990), and immunolocalization techniques have shown glycoconjugates containing all essential amino acids to be translocated from zooxanthellae to host (Trench 1993; Markell and Wood-Charlson 2010); whether the zooxanthellae take up nitrogenous organic compounds such as amino acids from the coral is unknown, although there is some evidence for such 'reverse translocation' of organic nitrogen in other algal–invertebrate symbioses (e.g. Douglas 1983). Amino acid catabolism in the coral cells produces ammonium waste that again becomes available for assimilation by both the coral and its zooxanthellae. This 'nitrogen recycling' is one of the best-known characteristics of the coral–zooxanthella symbiosis, although whether it is as important as nitrogen conservation to the ecological success of corals is unclear (Wang and Douglas 1998; Piniak and Lipschultz 2004).

The potential for host assimilation and withholding of nitrogen may play a role in the control of zooxanthellar growth. However, zooxanthellar growth may be limited most by nitrogen deficiency arising from low ambient supplies of nitrogen. Indirect evidence for nitrogen deficiency comes from corals and other zooxanthellate invertebrates exposed to elevated concentrations of DIN or provided with a more ready supply of food, which show increased growth rates and densities of zooxanthellae (Hoegh-Guldberg and Smith 1989a; Muscatine et al. 1989), increased chlorophyll levels per zooxanthella (Cook et al. 1988; Dubinsky et al. 1990) and various ultrastructural changes to the zooxanthellae, such as denser packing of the thylakoids in the chloroplast and fewer stores of organic carbon (i.e. lipid and starch), which can be used for growth when nitrogen becomes available (Berner and Izhaki 1994; Muller-Parker et al. 1996). Direct evidence for nitrogen deficiency comes from elevated rates of dark carbon fixation by zooxanthellae from the Atlantic corals *Madracis mirabilis* and *Orbicella annularis*, when the zooxanthellae are isolated and exposed to an ammonium supplement (Cook et al. 1994); this assay is based on the ammonium-stimulated amination of Krebs-cycle acids and the condensation of CO_2 to replace the aminated acids (Cook et al. 1992). The limitation of zooxanthellar growth is important to prevent overgrowth by zooxanthellae of the coral tissues. It may also free up organic carbon for translocation to the coral host, although evidence for this is equivocal (Falkowski et al. 1993; Davy and Cook 2001a,b).

4.7 Phosphorus

As with nitrogen, phosphorus is essential for structural and functional needs. Corals and other zooxanthellate invertebrates can acquire phosphorus as dissolved inorganic phosphorus (DIP; mainly phosphate (PO_4^{3-})), dissolved organic phosphorus (DOP), and particulate organic phosphorus (POP) (reviewed by Ferrier-Pagès et al. 2016); PO_4^{3-} is the most accessible form and can be scavenged by corals at extremely low seawater concentrations of just 0.2–0.3 μM (D'Elia 1977). There is evidence that this uptake is primarily driven by the zooxanthellae, as aposymbiotic hosts (Muller-Parker et al. 1990) and non-symbiotic species of reef coral (Pomeroy and Kuenzler 1969) are not able to take up PO_4^{3-} from seawater, while cultured zooxanthellae are able to do so (Deane and O'Brien 1981). Moreover, PO_4^{3-} uptake is enhanced in the light, and zooxanthellate hosts excrete PO_4^{3-} when subjected to prolonged darkness (D'Elia 1977; Cates and McLaughlin 1979; Godinot et al. 2009). However, analysis of uptake kinetics in the coral *Stylophora pistillata* suggests that both the zooxanthellae and their coral host possess phosphate transporters; indeed, phosphate must be actively transported into the coral given that it travels against a concentration gradient, from the seawater to the coral cell cytoplasm (Jackson and Yellowlees 1990; Godinot et al. 2009). Nevertheless, while more research is needed to clarify the mechanisms of phosphorus acquisition by the symbiosis, it is clear that zooxanthellae play a crucial role in the uptake and retention of PO_4^{3-}, even though they still tend to be phosphorus limited when inside their host (Godinot et al. 2011; Ferrier-Pagès et al. 2016). The subsequent fate of phosphorus in the coral–algal symbiosis is not well understood, although phosphorus is a crucial constituent of ATP and phospholipids; these latter compounds help to stabilize membranes in animal tissues, and also contribute to the organic matrix of the coral skeleton (Allemand et al. 1998; Watanabe et al. 2003). Phosphorus is also incorporated into coral tissue phosphonates, a group of compounds that may enhance structural strength (Godinot et al. 2016; Ferrier-Pagès et al. 2016).

4.8 Symbiosis establishment and stability

When corals reproduce asexually, zooxanthellae are passed directly to the new coral colony in the parental tissues. The situation becomes far more complex when the coral reproduces asexually via embryonic development of unfertilized eggs (i.e. parthenogenesis) or sexually. In these situations, zooxanthellae must either be transferred from parent to offspring in the gametes ('vertical transmission') or be acquired anew by the coral from the surrounding seawater ('horizontal transmission'). In corals, vertical transmission is the less common of these two methods (Babcock et al. 1986). Vertical transmission involves movement of zooxanthellae from maternal tissue into the developing oocyte and their inclusion in the cytoplasm of the egg (Hirose et al. 2001); coral sperm are too small to contain zooxanthellae. In horizontal transmission, uptake of zooxanthellae by the planula or recently settled polyp is important for the continued survival of the coral, and relies on the availability of zooxanthellae in the water column. The zooxanthellae originate from a number of sources: healthy corals, which expel zooxanthellae into the seawater to control the density of their intracellular zooxanthellae (Hoegh-Guldberg et al. 1987); unhealthy corals, which expel zooxanthellae as a general stress response (i.e. coral bleaching; Jones 1997); the guts of zooplanktonic grazers that are subsequently eaten by corals (Fitt 1984); and the waste food and faeces of corallivorous fish and invertebrates (Muller-Parker 1984). Motile zooxanthellae may employ chemotaxis to locate potential hosts; this is suggested by experiments with cultured zooxanthellae, which swim towards ammonium, the excretory waste of corals and other marine invertebrates (Fitt 1984). When a zooxanthella comes into contact with a coral, it must be phagocytosed by the coral and avoid subsequent digestion or expulsion. The molecular and cellular events involved in these processes are only just beginning to be understood (Weis et al. 2008; Davy et al. 2012), but they may involve molecules on the algal cell that interact with receptor molecules on the animal cell and so trigger phagocytosis (Markell et al. 1992; Lin et al. 2000; Kvennefors et al. 2008; Logan et al. 2010), and signals that inhibit phago-lysosome fusion and so prevent digestion of the zooxanthella once it is inside the coral cell (Fitt and Trench 1983; Chen et al. 2004; Fransolet et al. 2012).

Once established, the stability of symbiosis under 'normal' conditions is maintained by the host controlling the proliferation of algal symbionts within its cells by actively slowing their growth (Smith and Muscatine 1999), and digesting and especially expelling excess algae (Titlyanov et al. 1996; Baghdasarian and Muscatine 2000). However, this stable relationship between host-cell and algal-symbiont growth can be perturbed dramatically by environmental stress (Weis 2008). Coral bleaching, one of the most important phenomena affecting corals today, involves the coral either losing its zooxanthellae or, at the very least, photosynthetic pigments from its zooxanthellae; other zooxanthellate organisms such as giant clams, sea anemones and sponges may also undergo bleaching. The end result is that the coral tissues become translucent, revealing the white $CaCO_3$ skeleton beneath (Figure 4.7). As a result of bleaching, the coral's tissue growth, fecundity, calcification and, ultimately, survival are severely compromised, and the health of the whole reef ecosystem is threatened. Stressors that induce bleaching include high or low temperatures; high or low levels of light (PAR); UV radiation; reduced salinity; microbial infection; and various marine pollutants, such as copper and cyanide (Hoegh-Guldberg 1999, 2004). Foremost amongst these stressors is elevated sea surface temperature related to global warming; this is thought to be primarily responsible for increasingly frequent episodes of *mass* bleaching, global events where large tracts of coral reef around the world bleach over a period of just a few months (Hoegh-Guldberg 1999; Hoegh-Guldberg et al. 2007); the effect of temperature is further aggravated by intense light.

Bleaching of corals at elevated temperatures has been widely linked to photoinhibition of the zooxanthellae (Jones et al. 1998; Smith et al. 2005; Weis 2008), although direct impacts of thermal stress on the coral host

Figure 4.7   Coral bleaching. The usual brown colouration of this *Acropora* coral (left) is lost under environmental stress when the zooxanthellae (inset) are expelled from its tissue, leaving the white coral skeleton visible through the translucent tissues (right) (See Plate 6).
Source: Courtesy of O. Hoegh-Guldberg.

may also play a significant role (Hawkins et al. 2013; Krueger et al. 2015). The photoinhibition model posits that temperature limits CO_2 acquisition and assimilation, meaning that the rate of excitation (i.e. light capture) exceeds the rate of light utilization in photosynthesis. Some of this excess energy may be dissipated as heat through the conversion of the xanthophyll diadinoxanthin to diatoxanthin (Brown et al. 1999; Venn et al. 2006), but remaining energy is passed to oxygen causing a build-up of reactive oxygen species (ROS). Under normal circumstances, the impact of ROS production is minimized by the activity of ROS-scavenging enzymes, but the overproduction of ROS associated with thermal bleaching causes proteins to denature and the photosynthetic apparatus to be destroyed. The photoinhibition model helps explain why high PAR levels exacerbate the influence of high temperature (Fitt and Warner 1995) and why the upper, more brightly lit surfaces of corals bleach before other parts of the coral colony (Figure 4.8).

While there is general agreement that photoinhibition of zooxanthellae occurs at high temperatures, the initial site of photosynthetic breakdown has yet to be identified unequivocally. Possibilities include dysfunction of Photosystem II and the associated D1 protein (Warner et al. 1999), loss of thylakoid membrane stability (Tchernov et al. 2004) and a loss of RuBisCO activity

Figure 4.8   Colony of *Goniastrea*, Indian Ocean. Many corals were killed only on their upper surface, where light was strongest, and more shaded areas remained alive.

(Jones et al. 1998) (Figure 4.9). RuBisCO is the enzyme responsible for fixing CO_2, and any loss of activity would interrupt the flow of energy through the dark reactions of photosynthesis. A temperature-dependent slowing of enzyme activity could also explain photoinhibition and bleaching at cold temperatures (Saxby et al. 2003; Hoegh-Guldberg and Fine 2004).

Photoinhibition may not only have direct impacts on zooxanthellar physiology, but also have consequent effects on host-cell health. Coral cells undergo apoptosis ('programmed cell death') at elevated temperatures, perhaps as a result of the increased production of toxic ROS by the zooxanthellae (Dunn et al. 2004). Furthermore, heat-stressed zooxanthellae produce nitric oxide (NO), a cytotoxic, membrane-permeable molecule with the potential to kill the host's cells (Hawkins and Davy 2012; Hawkins et al. 2014). Therefore, the overall bleaching response most probably results from the combined impacts on both the zooxanthellae and their host coral.

The potential for corals to withstand and recover from bleaching events is a hugely important subject. While the coral itself may undergo a degree of thermal acclimatization, so limiting the impact of heat stress (Berkelmans and Willis 1999; Williams et al. 2010), most attention has focused on the potential role of zooxanthellar diversity in dictating patterns of bleaching susceptibility and recovery. As noted, zooxanthellae are genetically and physiologically diverse. This diversity of zooxanthellae helps explain why some coral colonies are more prone to bleaching than others (i.e. their stress tolerance is a reflection of the zooxanthellar type they contain) and, in cases

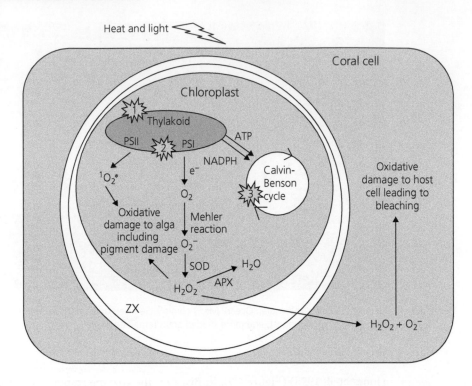

Figure 4.9   Schematic of the coral bleaching mechanism, showing three proposed sites for the primary impact of temperature on the photosystems of zooxanthellae. Proposed impact sites are (1) dysfunction of Photosystem II (PSII) and degradation of the associated D1 protein; (2) loss of thylakoid membrane integrity, and hence energetic uncoupling of the thylakoid membranes; and (3) damage to the Calvin–Benson cycle. Damage to the photosystems means that the rate of excitation (i.e. light capture) exceeds the rate of energy utilization in photosynthesis, leading to the reduction of O_2 by Photosystem I (PSI) to produce superoxide radicals (O_2^-) (the Mehler reaction). Some of these superoxide radicals are detoxified to hydrogen peroxide (H_2O_2) and then water by the combined action of the antioxidants superoxide dismutase (SOD) and ascorbate peroxidase (APX); however, when more superoxide radicals are produced than can be dealt with by these antioxidants, oxidative damage occurs to both the zooxanthellae and the coral host, leading to coral bleaching. Further damage can be caused by the production of singlet oxygen ($^1O_2^*$), a highly energized and toxic form of oxygen, which is formed at damaged PSII reaction centres. This singlet oxygen has the potential to denature proteins, such as the D1 protein in the PSII reaction centre, and cause the bleaching of chlorophyll and accessory pigments. Note that there is some evidence for direct impact of thermal stress on the host too; the potential sites of this impact are not shown here.

Source: Redrawn from Venn, A.A., Loram, J.E. and Douglas, A.E. (2008). Photosynthetic symbioses in animals. *Journal of Experimental Botany* 59:1069–80.

where corals harbour multiple types of zooxanthellae in different portions of the same colony, why patchy bleaching occurs. For instance, the Caribbean corals *Orbicella annularis* and *O. faveolata* simultaneously host multiple zooxanthellar types and under environmental stress lose the more shade-loving types from the most brightly illuminated parts of their intracolony distribution; the more resilient type(s) may then spread into these vacated regions (Rowan et al. 1997). Such adjustments in the symbiosis as a result of environmental change form the basis of the 'adaptive bleaching hypothesis' (ABH; Buddemeier and Fautin 1993). The ABH is founded on five assumptions (Fautin and Buddemeier 2004): (1) multiple host and symbiont types commonly coexist; (2) hosts can form symbioses with a diverse range of zooxanthellae; (3) different host–symbiont combinations differ physiologically; (4) bleaching provides an opportunity for repopulation by a different dominant symbiont type, either originating from the residual *in hospite* population or from the surrounding environment; and (5) host–symbiont combinations that are stress sensitive have a competitive advantage over more robust combinations under non-stressful conditions. There is evidence to support several of these assumptions, although the ABH remains highly controversial. In particular, the ABH assumes rapid adaptation to environmental change (LaJeunesse et al. 2003), yet this may be hindered by host–symbiont recognition and specificity (Hoegh-Guldberg 2004; Weis et al. 2008), at least in the case of acquiring novel symbionts from the surrounding environment, and the suboptimal function of some host–symbiont combinations (Starzak et al. 2014). Furthermore, given the increasing incidence of bleaching episodes around the world, adaptive bleaching may be too slow to keep pace with current rates of global warming (Hoegh-Guldberg 2004). This fascinating and contentious topic has been much debated (e.g. Baker 2001; Hoegh-Guldberg et al. 2002; Baker et al. 2004; Fautin and Buddemeier 2004).

4.9 Coral–microbial associations

Corals also associate with a large number of other microbes about which far less is known. Bacteria (including cyanobacteria) are known to inhabit the outer surface of corals, especially the outer mucus layer, as well as live inside coral tissue (Blackall et al. 2015; Thompson et al. 2015). The diversity and host specificity of bacterial communities associated with healthy corals was first demonstrated by the application of bacterial cultivation techniques to coral mucus samples (Ducklow and Mitchell 1979; Ritchie and Smith 1997), but it is the advent of DNA-based culture-independent techniques that has most strikingly illustrated the potential diversity of coral-associated microbial communities. For example, Rohwer et al. (2002) measured the bacterial diversity (both heterotrophic bacteria and cyanobacteria) associated with healthy colonies of the massive corals *Orbicella franksi*, *Diploria strigosa* and *Porites astreoides*, collected from Panama and Bermuda. From a combined total of just 14 tissue samples, these authors discovered 430 distinct bacterial ribotypes, with half of these probably representing novel bacterial genera and species. Moreover, it was speculated that more comprehensive DNA sequencing from these coral samples might produce a remarkable

6,000 bacterial ribotypes. Also of note, bacterial diversity from Panama was greater than that from Bermuda, paralleling diversity patterns seen in metazoans. These results highlight the huge diversity of novel and species-specific bacteria associated with healthy reef corals, with coral-associated bacteria belonging to at least 12 bacterial divisions (Knowlton and Rohwer 2003). These include the γ-Proteobacteria and the α-Proteobacteria, which were found to dominate the tissues and surface mucus, respectively, of the coral *Pocillopora damicornis* on the GBR (Bourne and Munn 2005).

In addition to bacteria, there is now strong evidence that other prokaryotes are associated with corals. Wegley et al. (2004) determined the presence or absence of archaea with a range of coral species from Panama, Puerto Rico and Bermuda. These coral-associated archaea are widespread, novel and diverse (93 archaeal ribotypes from just three coral species), but not species specific, in contrast to many coral-associated bacteria. Furthermore, archaea are frequently abundant, with densities of $>10^7$ cells/cm^2 being detected on *Porites astreoides*, where they comprised nearly half of the prokaryotic community.

Viruses are perhaps the least well known microbial associates of reef corals, although we are slowly beginning to appreciate their diversity and function on coral reefs (Vega Thurber et al. 2017). Davy and Patten (2007) were the first to study the viral community associated with the mucus of healthy corals in the field. They reported 17 subgroups belonging to a total of five morphological groupings in the surface mucus of the corals *Acropora muricata* and *Porites* spp. from the GBR. Viruses have since also been found in the healthy tissues and mucus of a number of other coral species (Patten, Harrison, et al. 2008; Nguyen-Kim et al. 2014; Lawrence, Davy, et al. 2015), while the advent of metagenomic, transcriptomic and proteomic approaches in particular has advanced our knowledge of the coral 'virome'. To date, about 60 virus families, equating to about 58% of all known virus families, have been found in corals and their symbiotic algae, with 9–12 of these families being especially dominant (e.g. Podoviridae, Myoviridae, Phycodnaviridae and Poxviridae). These dominant families belong to one of three viral lineages: Group I (double-stranded DNA (dsDNA) viruses, Group II (single-stranded DNA (ssDNA) viruses), or Group IV (retroviruses). Of particular interest with respect to coral disease and health has been the discovery of herpes-like viruses (Wood-Charlson et al. 2015; Correa et al. 2016; Vega Thurber et al. 2017).

Protozoans have been described from the surface tissues and mucus of healthy corals too. For example, various species of stramenopile protists form aggregates that sometimes cover extensive areas of the coral surface (Kramarsky-Winter et al. 2006).

Microbial communities are associated not only with the coral surface and tissues, but also with the coral skeleton. About 98% of reef-building corals contain filamentous green algae, as well as cyanobacteria, fungi and bacteria in their skeletons, where they form a series of distinct horizontal bands (Le

Campion-Alsumard et al. 1995; Bentis et al. 2000; Marcelino and Verbruggen 2016; Figure 4.10). The endolithic algae are adapted to very low light levels (<1% surface irradiance) and limited exchange (e.g. of gases and nutrients) with the water column, and hence have very low metabolic rates (Shashar et al. 1997; Ralph et al. 2007). The most common endolithic alga is the siphonalean chlorophyte *Ostreobium quekettii*, which is found in both the Pacific and Atlantic, while the closely related *O. constrictum* is only found in Caribbean corals; the major band of *Ostreobium* is typically about 5–30 mm from the skeleton surface. The conchocelis stages (very fine filamentous phases of the life cycle) of some red algae also survive within coral skeletons. These endolithic algae are not thought to be as important to reef production as are free-living forms, but they are still a valuable food source for grazers such as corallivorous fish (e.g. parrotfish) and sea urchins, and they also supply

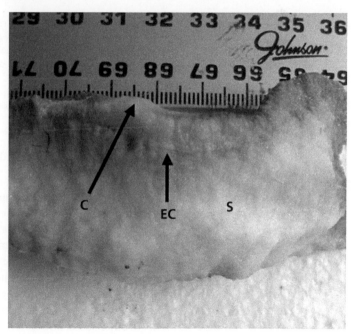

Figure 4.10   Cross section through the skeleton of an Indo-Pacific coral (*Porites* sp.), showing the endolithic community just below the surface. Such communities may contain a variety of microorganisms, including filamentous green algae, red algae, fungi, heterotrophic bacteria and cyanobacteria. The endolithic community forms distinct horizontal bands within the skeleton. In this specimen, the endolithic band is especially pronounced midway across the image, where the overlying coral tissue has been damaged, so allowing more light through to photosynthetic members of the endolithic community (e.g. the filamentous green alga *Ostreobium quekettii*); EC, endolithic community; S, skeleton; C, coral tissue on surface. Scale is in centimetres (See Plate 7).

Source: Photo by J. Davy.

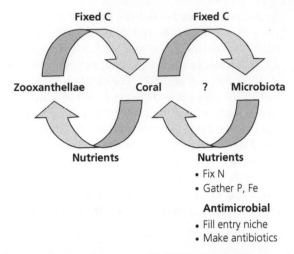

Figure 4.11   Proposed model of a coral colony as a 'holobiont'. The model shows the well-studied interactions between the coral and its zooxanthellae, as well as possible interactions between the coral and its prokaryotic microbial community. The holobiont also involves several other groups of organisms not shown in the model, including endolithic algae and fungi, protozoans and viruses.

Source: After Rohwer, F., Seguritan, V., Azam, F. and Knowlton, N. (2002). Diversity and distribution of coral-associated bacteria. *Marine Ecology Progress Series* 243:1–10; with kind permission of Inter-Research.

photosynthetic products to the overlying coral tissue and so support coral metabolism (Schlichter et al. 1997).

The interrelationships between most of these various microbes and their healthy coral hosts are unknown, but possible roles for prokaryotic microbes living on the coral surface, for instance, include (1) acquisition of novel metabolic functions by the coral, such as photosynthesis and nitrogen fixation, as may occur when cyanobacteria are present either in the coral's tissues or in the endolithic community inside the coral's skeleton (Tribollet et al. 2006; Lesser et al. 2007); (2) scavenging and recycling of nutrients (Ceh et al. 2013); and (3) protection from opportunistic pathogens through competition, the production of antibiotics or the occupation of 'entry niches' (Rohwer and Kelley 2004; Krediet et al. 2013; Raina et al. 2016). The suggestion that the prokaryotic community has a symbiotic relationship with the coral host led to the concept of the coral 'holobiont' (Rohwer et al. 2002), which encompasses the coral animal, the zooxanthellae and prokaryotic microbes (Figure 4.11), as well as endolithic algae and fungi, protozoans and viruses. The interactions between the various members of the coral holobiont are largely unknown, although there is increasing recognition that an improved knowledge of these interactions is critical for our understanding of coral reef function and health.

4.10 Sponge symbioses with non-photosynthetic bacteria, cyanobacteria and algae

Sponges also exist in symbioses with microorganisms, including bacteria, archaea and unicellular algae and fungi (Webster and Taylor 2012; Thomas et al. 2016). Of particular importance, it is likely that all marine sponges form symbioses with non-photosynthetic bacteria (Wilkinson 1984), while numerous coral reef sponges of the class Demospongiae form associations with algae or cyanobacteria (often of the genus *Synechococcus*) too (Figure 4.12a; Usher et al. 2004). For instance, about 50% of the sponge species that inhabit middle- and outer-shelf reefs of the GBR harbour intracellular cyanobacteria, dinoflagellates (i.e. zooxanthellae), diatoms or unicellular green algae (Wilkinson 1987; Trautman and Hinde 2001; Schonberg and Loh 2005; Webster et al. 2013). Moreover, some sponges form intercellular symbioses with macroalgae, especially red or green seaweeds (Figure 4.12b; Bergquist and Tizard 1967; Davy et al. 2002). These various algae and cyanobacteria can constitute up to 75% of the symbiosis (Wilkinson 1987; Davy et al. 2002) and on many reefs sponges may be second only to hermatypic corals in terms of space occupancy (Wulff and Buss 1979). The potential ecological importance of sponge symbioses on coral reefs is therefore high, as illustrated by the symbiosis between the sponge *Haliclona cymiformis* and the red seaweed *Ceratodictyon spongiosum* (Figure 4.12b) at One Tree Island on the GBR. On the sand/rubble of the reef flat at this location, primary production by the sponge–macroalgal symbiosis far exceeds that by corals, whose settlement and survival are not favoured by the loose substrate (Trautman and Hinde 2001). Sponges are also important for the consolidation of reef sediment and rubble, and bioerosion (Bell 2008), and algal partners can stimulate both. For instance, zooxanthellae in the bioeroding sponge *Anthosigmella varians* increase rates of both growth and bioerosion, and improve overall sponge fitness (Hill 1996); similar links have been made between sponge growth and the presence of symbiotic cyanobacteria (Wilkinson and Vacelet 1979). Moreover, having a photosynthetic partner may conserve energy by reducing the sponge's need to actively filter-feed (Pile et al. 2003).

Symbiosis is strongly linked to the huge success of sponges on coral reefs, but sponge–partner interactions are far less well understood than is the case with their coral–algal counterparts. The role of symbiotic heterotrophic bacteria is particularly poorly defined, although it may be that the bacteria assimilate dissolved organic matter from seawater and pass this to the sponge (Reiswig 1971). Furthermore, these bacteria are believed to play an important role in the nitrogen metabolism of sponges (Webster and Taylor 2012). For example, some heterotrophic bacterial symbionts have the capacity to nitrify amino nitrogen to nitrate, which can then be assimilated by cyanobacterial or zooxanthellar symbionts; nitrate can also be excreted into the surrounding seawater and so contribute a little to reef productivity (Corredor et al. 1988).

(a)   (b)

Figure 4.12   Sponge–phototroph symbioses. (a) Cyanobacteria (arrow) inside the Australian sponge *Chondrilla australiensis*. (b) Symbiosis between the red macroalga *Ceratodictyon spongiosum* and the sponge *Haliclona cymiformis* in the lagoon at One Tree Island, Great Barrier Reef. The morphology of the association reflects the profusely branched macroalga, with the intercellular encrusting sponge spreading around and between the alga's branchlets and enveloping most of the alga; only the branchlets at the apex of the thallus are free of sponge. Scale bars represent 200 nm (a) and 5 cm (b).

Source: (a) Photo: K. Usher; (b) photo: D. Trautman.

The photoautotrophic partners of sponges are a little better understood than the heterotrophic bacterial ones, and they may well interact with the sponge in a similar manner to that seen in the coral–zooxanthellar symbiosis. For instance, cyanobacteria release photosynthetic products to their host sponges, largely as glycerol, although the extent of this release may be small (5%–12% of photosynthate) when compared to that in corals (Wilkinson 1980). Similarly, the red seaweed *Ceratodictyon spongiosum* releases photosynthetic products to its partner sponge, *Haliclona cymiformis*, again at a low rate of <1.3% (Pile et al. 2003). These rather low release rates may limit the extent to which cyanobacteria and algae support the nutrition of their sponge partners, although photosynthetic rates generally exceed sponge respiratory rates in these associations, sometimes by as much as four times, indicating that the symbioses as a whole are net primary producers (Wilkinson 1983; Cheshire et al. 1997).

Parallels with coral–zooxanthellar symbiosis may also extend to nitrogen fluxes between the sponges and their photoautotrophic partners. In particular, excretory nitrogen from the sponge may be utilized by the photosynthetic partner, as illustrated by sponges releasing more ammonia to the seawater when their photosynthetic partners are either removed (Sara et al. 1998) or when photosynthesis is inhibited by darkness (Davy et al. 2002). Moreover, in the *H. cymiformis–C. spongiosum* symbiosis, the supply of excretory nitrogen by the sponge is sufficient to support the nitrogen demands of algal growth, although the potential for nitrogen recycling (i.e. release of assimilated waste nitrogen from the algal partner back to the sponge) in this symbiosis is limited (Davy et al. 2002). Nothing is known about nitrogen recycling in other sponge–photoautotroph symbioses, although cyanobacterial symbionts can fix nitrogen (Wilkinson and Fay 1979; Fiore et al. 2010) and may therefore translocate organic nitrogen to their sponge host.

In addition to nutritional benefits, sponges and their photoautotrophic partners may derive a number of other benefits. Best known of these is the production of toxic compounds, especially by cyanobacteria, that defend the sponge from predators (Unson and Faulkner 1993; Unson et al. 1994); in return the sponge's sharp spicules may deter organisms from grazing on cyanobacterial and algal partners (Scott et al. 1984). In some cases, particularly where a seaweed partner is involved, the sponge may benefit from the structural support afforded by the seaweed. For example, the sponge *Haliclona cymiformis* modifies the growth form of its intercellular macroalgal partner *Ceratodictyon spongiosum*, so providing the sponge with a rigid algal framework around which to grow; this is particularly of benefit in the unstable sandy lagoons frequented by species in this symbiosis (Bergquist and Tizard 1967; Trautman and Hinde 2001).

4.11 'Macro' or iconic symbioses (e.g. fish and sea anemones, shrimps and fish)

Numerous other endo- and ectosymbioses exist on coral reefs. Some of the most charismatic examples are between fish and various invertebrates. Fish that form endosymbioses with invertebrates include various species of small, eel-like members of the family Carapidae, known commonly as the pearlfish (Figure 4.13a). Pearlfish live inside echinoderms such as starfish and sea cucumbers, where they either feed on the echinoderm's tissues (i.e. parasitism) or find shelter when not hunting for prey outside (i.e. commensalism) (Parmentier and Das 2004; Parmentier and Vandewalle 2005). Various species of goby form exosymbioses with invertebrates. For instance, gobies may form a mutualistic association with shrimps, where the shrimp digs the burrow in which they both live and the fish flicks the shrimp's antennae with its tail to warn it of danger (Karplus 1979; Thompson 2004). Gobies also form obligate symbioses with scleractinian corals, with the species of goby being specific for just one or two species of hard coral, or even a particular colour morph of a single coral species (Munday et al. 1997, 2001).

(a)   (b)

Figure 4.13   Fish–invertebrate symbioses. (a) The graceful pearlfish (*Encheliophis gracilis*) with its echinoderm partner. This uncommon Indo-Pacific fish species lives inside and may parasitize starfish (*Culcita discoidea*) and sea cucumbers (*Holothuria scabra* and *H. argus*). (b) McCulloch's anemonefish (*Amphiprion mccullochi*) on the coral reef of Lord Howe Island, southeast Australia. This particular individual has formed a mutualistic symbiosis with the sea anemone *Entacmaea quadricolor. A. mccullochi* is found only at Lord Howe Island and a few other subtropical sites off the east coast of Australia.

Source: (a) Photo: E. Parmentier; (b) photo: J. Davy.

Probably the most iconic example of a fish–cnidarian symbiosis, however, is the mutualistic association between anemonefish and sea anemones (Figure 4.13b). Twenty-eight species of anemonefish are known, all from the Indo-Pacific region, while just ten species of sea anemone (i.e. only 1% of all known sea anemone species) form symbioses with these fish (Fautin and Allen 1997); of note, these sea anemones all contain zooxanthellae in their tissues. Some anemonefish are 'extreme specialists', forming a specific relationship with just one particular sea anemone species, while others are 'extreme generalists', forming symbioses with all ten sea anemone species (Mariscal 1970; Fautin and Allen 1997). Protection from the host sea anemone's sting is developed during the anemonefish's metamorphosis from a pelagic larva to a benthic juvenile (Elliot and Mariscal 1996), either through a behavioural strategy (e.g. acclimation during the initial, tentative stages of symbiosis establishment) and/or an innate chemical mechanism (Fautin 1991). The host sea anemone protects the anemonefish from predators (Fautin 1991) while, in return, the anemonefish improves sea anemone survival rates by deterring predators such as butterflyfish (Porat and Chadwick-Furman 2004) and by increasing the frequency of tentacle expansion (Porat and Chadwick-Furman 2004). Moreover, anemonefish excrete ammonia, which is derived from their zooplanktonic diet, thus enhancing both zooxanthellar and sea anemone growth (Porat and Chadwick-Furman 2005).

Probably the best-known example of a fish–fish symbiosis is that between cleaner fish and other, larger reef fish. While many non-symbiotic cleaner fish utilize designated cleaning stations, the symbiotic remoras—a number of species belonging to the family Echeneidae—remain attached to their hosts (often large sharks) for prolonged periods via a modified dorsal fin that forms a sucker. In return, the remora receives food and protection from predators.

5 Microbial, microalgal and planktonic reef life

5.1 Microbial reef life

Bacteria, archaea, viruses, fungi and protists (simple eukaryotic microbes) are the most diverse and abundant organisms on coral reefs. These microorganisms carry out a range of major ecological functions, including primary production, nitrogen fixation and the turnover, decomposition and cycling of most of the life which dies on the reef. Furthermore, microbes are an important food source for numerous reef organisms and, in some cases, are causative agents of disease. Microbial biomass is especially concentrated in the surface film of the benthos or in the shallowest layers of sediments (Ducklow 1990), although microbes are also abundant in the water column.

In addition to the enormous diversity of free-living planktonic and benthic microbes, many live in close association with corals (on their surfaces, in their tissues and in their skeletons), as well as form endosymbioses with other reef organisms such as sponges and ascidians, where they perform an array of metabolic and ecological roles (see Chapter 4).

5.1.1 Bacteria and archaea

Bacteria are abundant in both benthic and planktonic communities (see 'Molecular characterization of marine microbial communities'). The most numerous component is the picoplankton (0.2–2.0 μm), which is largely comprised of heterotrophic bacteria and cyanobacteria, and this exceeds the more familiar microplankton (20–200 μm) in terms of biomass and productivity. Nanoplankton (2–20 μm), which consists of the smallest protists, is also relatively abundant. The small cell size of bacteria, and hence their large surface-area-to-volume ratio, assists with nutrient scavenging in nutrient-poor reef waters. Bacteria constitute one of the most diverse components of coral reef communities (Rohwer et al. 2002; Blackall et al. 2015), although

The Biology of Coral Reefs. Second Edition. Charles Sheppard, Simon Davy, Graham Pilling, and Nicholas Graham, Oxford University Press (2018). © Charles Sheppard, Simon Davy, Graham Pilling, and Nicholas Graham (2018). DOI 10.1093/oso/9780198787341.001.0001

this diversity is nowhere close to being comprehensively quantified. Archaea are less well known than bacteria, but are common in marine environments and numerically dominate the picoplankton in meso- and bathypelagic zones of the ocean (Karner et al. 2001); they are also abundant and diverse on coral reefs (Tout et al. 2014; Frade et al. 2016).

Molecular characterization of marine microbial communities

Microbes within the marine system have long been recognized, but it is only relatively recently that we have begun to understand their exact roles. Traditional culture techniques have been limited because <1% of environmental microbes can be cultured (Amann et al. 1995). With the advent of molecular methods, our knowledge of these microbes and the roles they play in marine ecosystems have increased exponentially. These methods allow analysis of both free-living microbes and those associated with substrates and animal and plant tissues, and provide a range of specificity in the level of identification from that of individual species to functional groups, such as nitrogen fixers.

Characterization of members of the complex microbial communities present in the marine environment can be achieved using a range of methods that utilize the characteristics of the genetic material that all these organisms possess. The first step of the analysis requires the extraction of genetic material (DNA and/or RNA) from the environmental sample. DNA extracted from the environment can be used directly as the template for PCR, while RNA can be first reverse-transcribed into cDNA, which can then be amplified using PCR. Extraction protocols follow three basic steps that are adapted as required. These are (1) cell disruption, (2) isolation of the genetic material from other cellular components and (3) purification and recovery of genetic material. The purified genetic material can then be used to determine both community diversity and the identity and relative abundance of the community members.

PCR allows the amplification of specific fragments of DNA from the total genome (see Figure, Panel C). The technique uses cyclic enzymatic extensions of primers (short sequences of DNA, at opposite ends of the target sequence) and results in millions of copies of the target sequence. This enables the analysis of DNA, even when the DNA is present at very low levels. The primers used for the amplification of the target regions are designed using the microbial genetic sequence information of the regions of interest to identify areas of low variability, and can be designed to be specific for individual species (e.g. Kim et al. 2015) or can be general, targeting all members of a group such as Archaea (Gantner et al. 2011).

PCR can be used in conjunction with other methods, such as denaturing gradient gel electrophoresis (DGGE) (Muyzer et al. 1987). DGGE is used to create a 'community fingerprint' for the samples by separating out the PCR-amplified fragments, based on differences in their level of electrophoretic mobility when exposed to denaturants within the gel. As the fragments migrate through the gel by electrophoresis, those with identical sequences will stop at the same point, forming defined bands and creating the

Continued

Molecular characterization of marine microbial communities (*Continued*)

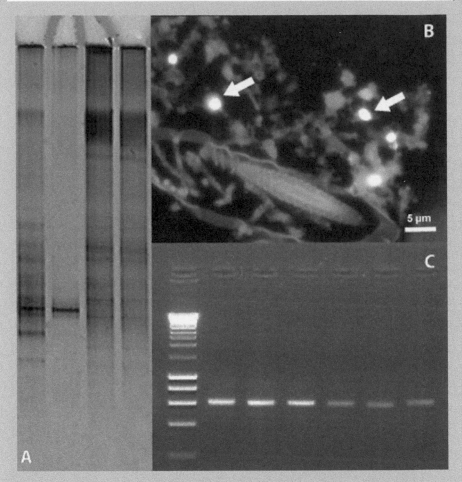

Figure. Molecular characterization of marine microbial communities. (A) Denaturing gradient gel electrophoresis profiles for bacterial communities associated with estuarine sediments and seagrass blades. (B) Fluorescence in situ hybridization image of coral tissues probed with a general bacterial probe. The labelled bacterial cells (marked by the arrows) are noticeable against the autofluorescence of the coral tissues. A nematocyst can be clearly seen. (C) PCR products amplified from six marine microbial mats, using bacterial-specific primers targeting a ~400 bp region of the sequence. The left lane contains a marker ladder with each band representing a different size fragment, allowing determination of the PCR product size. Band density reflects the concentration of PCR products.

fingerprint. Analysis of the fingerprints allows for the comparison of multiple samples, allowing rapid assessment of both similarities and differences between communities (see Figure, Panel A).

Continued

Molecular characterization of marine microbial communities (*Continued*)

PCR products can also be sequenced, whereby the genetic code of the amplified fragments of DNA is determined. Several different methods of DNA sequencing are available, including pyrosequencing and shotgun and chain termination sequencing. Before sequencing, individual sequence types can be isolated either by plasmid-based vector cloning and the production of clone libraries or by DGGE, where each band consists of DNA fragments with the same sequence. Comparison of the obtained sequences with online databases (e.g. the National Centre for Biotechnology Institute database) enables the identification of the species from which the DNA originated, or its nearest relative. Alternatively, next-generation sequencing technologies (also known as high-throughput sequencing technologies) can be used to sequence the PCR products from a mixed community. Analysis of the resulting amplicon sequence reads can then be carried out using open-source computer programmes such as QIIME (Caporaso et al. 2010), which can be used for the identification of a range of community parameters such as the relative abundance of individuals, levels of community richness and diversity, and the identification of core microbiomes. Knowledge gained from phylogenetic analysis can aid with the development of specific conditions required for the successful culturing of microbes of interest.

When characterizing microbial communities, it is important to know both the number of individuals present and their identity. It is difficult to distinguish between individual species by visual methods. Free-living microbes can be quantified using microscopy, by filtering them onto 0.02 µm filters and staining with a general nucleic acid stain such as SYBR®-Gold™ or DAPI (4′,6-diamidino-2-phenylindole). Stained microbes can then be easily viewed and counted using epifluorescence microscopy. To determine the biodiversity of an environmental sample, a range of molecular probes can be used in a process known as in situ hybridization (Gall and Pardue 1969; see Figure, Panel B). This enables identification of specific cells in heterogeneous populations and makes it possible to determine whether a gene is expressed at low levels in many cells or at high levels in only a few cells. Also known as hybridization histochemistry or cytological hybridization, this technique was originally used to identify the location of cellular DNA sequences but is now applied to the localization of viral DNA sequences, mRNA, rRNA, chromosomal regions and whole cells in thin tissue sections. It is therefore a very powerful research tool for looking at the interactions between microbes and their environment. These probes, like PCR primers, use the differences and similarities seen between species at the genetic level and can therefore be designed to target different phylogenetic levels. The growing number of published nucleotide sequences and whole genomes available has improved the specificity and enabled the development of probes to detect a wide range of targets. Probe visualization can be carried out using a variety of different labels. These include radioactive, fluorescent or coloured-precipitate labels, either incorporated into the probe during synthesis or conjugated to one end of the sequence post production. The availability of multiple different labels allows simultaneous staining and visualization of different sequence targets with different probes with distinct 'colours', allowing community composition to be determined quantitatively as well as qualitatively. The choice of marker molecule may be dictated by the chromatic and fluorescence properties of the tissues. For example, some tissues or phytoplankton caught on filtered samples may autofluoresce, obscuring the microbes, and therefore a fluorescent marker may not be suitable. There are, however, several methods available that allow the amplification of the signal, increasing the signal of the probe above that of the background.

Dr Olga Pantos, Institute of Environmental Science and Research, Porirua, New Zealand.

The composition of bacteria within sediments is affected by location and environmental conditions. Benthic bacteria in carbonate sediments were surveyed in four coral reefs on the Great Barrier Reef (GBR), two of which were inshore and hence subjected to enhanced terrestrial run-off, and two of which were offshore and considered to be pristine (Uthicke and McGuire 2007). They identified a highly diverse range of both heterotrophic and photosynthetic bacteria, with the most common groups being the γ-Proteobacteria (29.4% of the total diversity) and Cytophaga–Flavobacteria–Bacteroidetes (CFB) (20.4% of the total). They also noted a shift in the community composition from the nearshore to offshore reefs: Acidobacteriaceae and δ-Proteobacteria were most common in nearshore sediments, where anoxia is more likely, while cyanobacteria were more common in nutrient-poor offshore sites, where the better water quality promotes photosynthesis and where the capacity to fix nitrogen promotes a competitive advantage. Patchiness of sedimentary bacterial populations at a more localized scale was found at Heron Island (GBR), where it was speculated that abiotic environmental conditions such as wave energy and sediment depth led to spatial differences in the bacterial community composition (Hewson and Fuhrman 2006).

The abundance of bacteria in reef sediment is much higher than that in the overlying seawater, with a density of about 1×10^8 cm^{-3} of sediment being found at Heron Island. This density increased by up to 4.6 times after of a mass coral spawning event, as a result of spawned material sinking to the seabed, where it supported bacterial growth; it is also likely that some bacterial plankton (bacterioplankton) was transferred to the benthos on sinking material (Patten et al. 2008). In addition, nutrient enrichment can lead to blooms of benthic bacteria, most notably cyanobacteria such as *Lyngbia* spp. (Figure 5.1).

Planktonic bacterial communities do not mirror those in the benthos, even though some benthic microbes are stirred into the water column by events such as flood tides (Hewson et al. 2007). In fact, the sedimentary bacterial community on the GBR is more similar to that associated with Antarctic sediments than with overlying coral reef water (Uthicke and McGuire 2007). For instance, the proportion of α-Proteobacteria in seawater from the GBR (6.8% of the total diversity) may be similar to that in GBR sediment, but the seawater contains undetectable amounts of δ-Proteobacteria and CFB group members (Bourne and Munn 2005). Bacterial community composition in the seawater also varies spatially. For example, on the GBR, open water regions and lagoonal water overlying sand are dominated by oligotrophic microbes (i.e. those that favour low-nutrient environments), while water in close proximity to corals is dominated by more copiotrophic microbes (i.e. those that favour more nutrient-rich environments; Tout et al. 2014).

Figure 5.1   The benthic filamentous cyanobacterium, *Lyngbya majuscula*. Elevated levels of usually limiting nutrients are a major cause of blooms such as the one shown here, where the cyanobacterium is overgrowing corals. This bloom was near Great Keppel Island, Queensland, Australia (See Plate 8).

Source: Photo by S. Albert.

In terms of abundance, bacterioplankton density over reefs can vary considerably, but typical values are 2×10^5 to 9×10^5 mL^{-1}; higher concentrations ($>2 \times 10^6$ mL^{-1}) have been found in some lagoonal systems or where the coral reef is situated close to mangroves (Gast et al. 1998). Bacterioplankton abundances lower than 'normal' occur over reefs that are far removed from anthropogenic disturbance. This was illustrated in the Line Islands in the Central Pacific (Dinsdale et al. 2008), where increasing abundances of bacteria and archaea occurred along a gradient from uninhabited Kingman Reef, one of the world's most remote and pristine reefs, where abundance was 0.72×10^5 mL^{-1}, to the much larger Kiritimati, which is inhabited by ~5,500 people, where sewage is untreated and where concentrations were ten times greater, at 8.4×10^5 mL^{-1} (Figure 5.2). The abundance of planktonic bacteria also varies spatially and temporally on individual reefs. Spatial differences are particularly pronounced between overlying water and water within reef crevices, due to removal by benthic filter-feeders. For example, on the reefs of the Caribbean island of Curaçao, bacterial abundance was 4.5×10^5 mL^{-1} or less in reef crevices, under corals or in narrow spaces between corals, but up to 8×10^5 to 9×10^5 mL^{-1} in the overlying reef water 4–6 m above the reef, or in bottom water (the water layer from

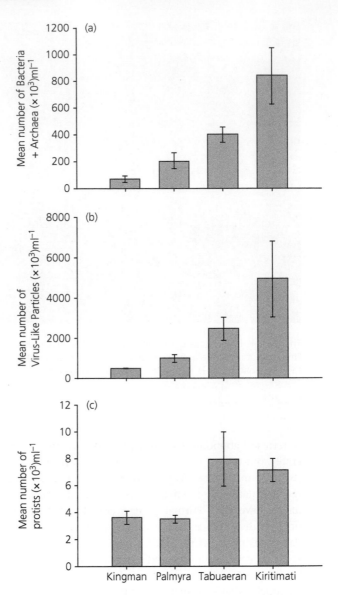

Figure 5.2 Mean abundance (± standard error) of (a) microbial cells (bacteria and archaea), (b) viruses and (c) protists on the four Northern Line Island atolls, Central Pacific. Kingman and Palmyra have little or no local anthropogenic impacts, with Kingman being uninhabited and Palmyra only being inhabited by about 20 people at a time, although, unlike Kingman, it does have a substantial population of seabirds. Tabuaeran and Kiritimati (both part of the Republic of Kiribati) have much larger land areas and both are inhabited, Tabuaeran by about 2,500 people, and Kiritimati by about 5,500 people; neither population treats its sewage.

Source: Dinsdale, E.A., Pantos, O., Smriga, S., Edwards, R.A., Angly, F., Wegley, L., et al. (2008). Microbial ecology of four coral atolls in the northern Line Islands. *PLOS ONE* 3:e1584.

the bottom to the top of the largest corals) (Gast et al. 1998); the overlying reef water had a bacterial abundance similar to that found in the surrounding open ocean. Temporally, bacterioplankton abundance may vary with season, as shown by a threefold increase in bacterial biomass between winter and summer at Lizard Island on the GBR (Moriarty, Pollard, Hunt, et al. 1985), and abundance also rises when bacteria attach to sinking organic matter such as coral spawn. Temporal changes also occur in response to the quality and size of oceanic water masses passing a reef or, in coastal regions, pulses of nutrients from freshwater run-off (Gast et al. 1998; Cox et al. 2006). However, bacteria and other members of the picoplankton tend not to respond to nutrient inputs as markedly as do larger members of the plankton because of their greater capacity to scavenge nutrients from oligotrophic water.

Within the bacterioplankton, much attention has focused on the cyanobacteria due to their roles in primary production and nitrogen fixation. Common planktonic genera include *Synechococcus* and *Prochlorococcus*, which together can comprise >90% of the pelagic cyanobacterial population. *Prochlorococcus* cells are about 0.5–0.8 µm in size. They are the smallest photosynthetic organisms of all, and probably the most abundant species on Earth. Importantly in the present context, they are almost universal in tropical and subtropical waters, and they dominate in the kinds of warm waters that are oligotrophic and which bathe coral reefs. Their several strains or species have been divided into low-light and high-light living groups (Kettler et al. 2007).

Synechococcus are a similar group of cyanobacteria, whose marine forms are only slightly larger at 0.6–1.5 µm. They occur more in waters which are not nutrient limited and, to date, two groups of strains have been identified: coastal and open ocean (Dufresne et al. 2008). Both of these genera form part of the picoplankton (<2 µm). In much of the oceans and in oligotrophic seas such as those that bathe coral reefs, their huge numbers and rapid turnover mean that they are responsible for two-thirds of primary productivity; this may rise to about 80% of the pelagic primary productivity.

Cyanobacteria typically contribute far less biomass to the picoplankton than do heterotrophic bacteria. For instance, at One Tree Island on the GBR, *Synechococcus* and *Prochlorococcus* each contribute 3%–6% of picoplankton biomass irrespective of season, while heterotrophic bacteria contribute 88%–90% (Pile et al. 2003). However, cyanobacteria dominate picoeukaryotic algae and often phytoplankton of all sizes. For example, on several French Polynesian atolls, *Synechococcus* and *Prochlorococcus* densities reach 210×10^3 to 370×10^3 cells mL^{-1} for each type, whereas picoeukaryotic algae peak at just 7.4×10^3 cells mL^{-1} (Charpy 2005). On the reefs of Okinawa in Japan, picophytoplankton contributes 45%–100% of total phytoplankton biomass (Ferrier-Pagès and Gattuso 1998).

Bacteria and archaea are extremely important for ecosystem function. Heterotrophic species are essential for the decomposition and recycling of organic matter (e.g. coral mucus and algal detritus) and the regeneration of nutrients; bacteria generally may consume one- to two-thirds of the primary production. Being small, planktonic species have a very low sink rate, so they generally remain higher up in the water column than larger organisms do. This means they are correspondingly more accessible to larger organisms and act as an important food source (Pile et al. 1996; Houlbrèque et al. 2004). Cyanobacteria and various other diazotrophic (nitrogen fixing) bacteria act as an important entry point for nitrogen into the coral reef ecosystem, while cyanobacteria are also important primary producers. Further details of microbial productivity, trophic links and the transfer of energy through the food web are given later in this chapter.

5.1.2 Viruses

Viruses are ubiquitous in the marine environment, reaching densities of 10^6–10^8 mL^{-1} of seawater, and probably infect all cellular organisms (Fuhrman 1999; Wommack and Colwell 2000). Viruses may not only be causative agents of disease, but they may also play a major role in a number of ecological processes, including the regulation of population dynamics, community structure and nutrient cycling in marine microbial communities (Fuhrman 1999). They also may be predated upon by sponges (Hadas et al. 2006). Despite their abundance and ecological importance, however, viruses are amongst the least well studied members of the coral reef ecosystem, and we are only now beginning to fully appreciate their diversity and functional significance on reefs (reviewed by Vega Thurber et al. 2017) (Figure 5.3).

Viral diversity is high not only on and within corals (see Chapter 4), but also in the surrounding seawater. For example, in a comparative study of viruses associated with coral mucus and seawater in Kane'ohe Bay, Hawaii, Lawrence, Wilkinson, et al. (2015) identified 26 different viral morphotypes via transmission electron microscopy, including abundant icosahedral or spherical viruses, as well as lemon-shaped, filamentous and rod-shaped viruses. There was considerable overlap between the morphotypes found on corals and in the overlying seawater at most reef sites, although not all. Furthermore, water quality (turbidity and chlorophyll *a* content) and temperature influenced virus consortium composition, with, for example, an increase in the abundance of large (>100 nm) lemon-shaped viruses increasing at more turbid sites; it was suggested that this increase could be related to an increased abundance of archaea—the potential hosts of these viruses—at turbid sites. In addition to morphological studies such as this, molecular studies, including metagenomic surveys, have revealed a wealth of single-stranded and double-stranded DNA(ssDNA, dsDNA) viruses in reef waters that infect a range of prokaryotes and eukaryotic protists and phytoplankton (Vega Thurber et al. 2017).

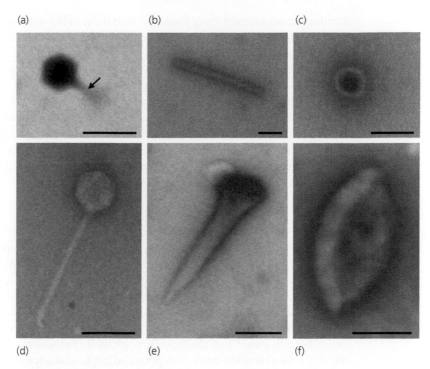

Figure 5.3   Coral reef viruses. A huge diversity of virus-like particles (VLPs; putative viruses whose identity has yet to be confirmed by appropriate molecular methods) exists in the water column, in sediment and on the surfaces of corals and other organisms. Panels (a)–(c) show examples of VLPs from the water column at Heron Island on the southern Great Barrier Reef, while Panels (d)–(f) show VLPs from the surface of corals at the same location. (a) Myovirus-like bacteriophage showing helical symmetry of the contractile tail (arrow). (b) Filamentous VLP. (c) Spherical VLP. (d) Siphovirus-like bacteriophage with a filamentous, non-contractile tail and isometric capsid. (e) Hook-like VLP. (f) Lemon-shaped VLP, similar to the fusellovirus SSV1, which is known to infect archaea from extreme environments. Scale bars = 100 nm.

Source: Photos by J. Davy and N. Patten.

The average concentration of viral plankton (viroplankton) measured over the GBR and the Florida Keys is 1×10^6 to $14 \times 10^6 \, \mathrm{mL^{-1}}$, which is up to seven times the concentration of bacterioplankton (Patten et al. 2008). The concentration is not uniform throughout the water column, but tends to be highest close to the benthos and the surface of corals. For example, at Magnetic Island on the GBR, the viroplankton concentration was about 0.5×10^6 to $1.0 \times 10^6 \, \mathrm{mL^{-1}}$ at >4 cm above the surface of corals, but peaked at about $1.5 \times 10^6 \, \mathrm{mL^{-1}}$ in the 4 cm closest to the surface (Seymour et al. 2005); diseased corals may have even higher densities close to their surface (Patten et al. 2006). Comparable densities of viroplankton were recorded at 10–12 m depth around the Line Islands, where there was an increase from the pristine Kingman Reef ($5.1 \times 10^5 \, \mathrm{mL^{-1}}$) to the anthropogenically disturbed Kiritimati ($4.9 \times 10^6 \, \mathrm{mL^{-1}}$; Dinsdale et al. 2008).

Benthic viruses are even more abundant than those in the water column, with viral densities in the carbonate sediments of coral reefs typically exceeding those in the overlying water by two orders of magnitude (Paul et al. 1993). At Heron Island (GBR), sediment contains 3×10^8 to 12×10^8 viruses cm^{-3}, compared with 1×10^6 to $5 \times 10^6 mL^{-1}$ in the seawater (Patten et al. 2008) while viral abundance on reefs near Key Largo, Florida is $\sim 5 \times 10^8 cm^{-3}$ compared with 1.5×10^6 to $1.8 \times 10^6 mL^{-1}$ in the water column (Paul et al. 1993). As with bacteria, however, viral abundance is not constant temporally and, for instance, may increase in the benthos as a result of organic matter such as coral spawn sedimenting out of the water column; this is accompanied by a concurrent decline in viroplankton abundance (Patten et al. 2008).

5.1.3 Fungi

Fungi on coral reefs have received very little attention, with the exception of *Aspergillus sydowii*, a pathogen of Caribbean gorgonian sea fans. Fungi are also found in hard corals and are frequently associated with the endolithic microbial communities in the coral skeleton (see Chapter 4), where they parasitize algal endoliths and the coral polyps. The bioeroding activity of these endolithic fungi contributes to the dissolution of coral skeletons, especially of dead corals; some fungi also bore through shells. Fungal diseases may also affect crustose coralline algae, with potential implications for reef consolidation and structure (Williams et al. 2014).

Free-living coral reef fungi are not at all well studied, yet they are both widespread and abundant. This was demonstrated in a study of fungi from Australian reefs (Morrison-Gardiner 2002), where a total of 56 fungal taxa were identified. Thirty-five of these were isolated from sediment, while others were isolated from a range of algae, sponges, cnidarians, bryozoans and vertebrates. The most common fungal genera were *Alternaria, Aspergillus, Cladosporium, Cochliobolus, Curvularia, Fusarium, Humicola* and *Penicillium*; 46% of all taxa identified belonged to one of these genera. Fungal diversity was greater in nearshore than offshore regions. We are still a long way, though, from a sound knowledge of fungal diversity and function on coral reefs.

5.1.4 Protozoa

Protozoa are heterotrophic protists. Protozoans provide a food source for benthic filter-feeders such as sponges and corals (Pile et al. 2003; Houlbrèque et al. 2004) and planktonic grazers such as copepods (Sakka et al. 2002), and act as degraders and consumers in the microbial food web. Protozoans can also be associated with disease in both invertebrates (e.g. corals) and vertebrates (e.g. fish) on reefs (Bernal et al. 2016; Sweet and Sere 2016).

Coral reef protozoans are no doubt diverse, with ciliates and flagellates, amongst others, frequenting both seawater and marine sediments in the tropics (Ekebom et al. 1996). Few have been characterized in detail but, of

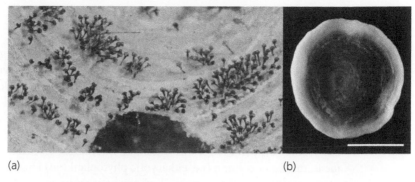

(a)   (b)

Figure 5.4   Coral reef Protozoa. (a) Clusters of the giant heterotrich ciliate *Maristentor dinoferus* on a blade of the brown seaweed *Padina* sp., on a reef in Guam. Each ciliate is about 1 mm tall and 300 μm across its cap, and contains about 500–800 zooxanthellae, which give a brownish hue to the cell. (b) The soritid foraminiferan *Marginopora vertebralis*, collected from Réunion Island in the Indian Ocean. Scale bar = 5 mm.

Source: Photos by T. Schils (copyright) and X. Pochon.

those that have, the benthic ciliates *Maristentor dinoferus* and *Euplotes uncinatus* harbour symbiotic dinoflagellate algae and so contribute to primary production on reefs (Lobban et al. 2002, 2005; Figure 5.4a).

Foraminifera is a phylum of amoeboid protists that range from less than 1 mm to more than 15 mm in size and live both as plankton and on the seabed. Some of these also form symbioses with various microalgae, including dinoflagellates. Importantly, most foraminiferans produce a calcium carbonate shell, or 'test' (Figure 5.4b) that may contribute substantially to coral reef sand deposits. Radiolarians, which comprise another phylum of amoeboid protists, also produce elaborate external skeletons, which are often very spiny and usually made of silica. These are generally buoyant animals that live in the plankton, and may also contain symbiotic dinoflagellates. This group's siliceous skeletons mean that they are not a particularly valuable component of the plankton for other plankton feeders but, as with the foraminiferans, their accumulated deposits on the seabed have proved to be a very useful tracer of past climatic changes.

Protists are not as abundant as bacteria, archaea and viruses, either in the plankton or in the benthos. In their survey of the Line Islands, Dinsdale et al. (2008) found that planktonic protist densities (including microalgae) ranged from about 3.5×10^3 mL^{-1} at the most pristine locations to about 8×10^3 mL^{-1} at the more impacted sites (see Figure 5.2); at the pristine Kingman Reef, 66% of these protists were exclusively heterotrophic (i.e. they did not contain chlorophyll and hence were not microalgae) while, around Kiritimati, only 22% of protists were strict heterotrophs. More specifically, densities of planktonic flagellates and ciliates on coral reefs have been reported as 180–193 mL^{-1} and 1.0–1.2 mL^{-1}, respectively (Moriarty, Pollard, Hunt, et al. 1985).

5.1.5 Microalgae

Microalgae are photosynthetic protists. The endosymbiotic microalgae of reef corals, and of numerous other reef organisms such as giant clams, sponges and sea anemones, are essential for the growth and survival of coral reefs in the shallow, nutrient-poor waters of the tropics. These microalgae are dinoflagellates of the genus *Symbiodinium*, which are discussed in detail in Chapter 4 in terms of their diversity, physiology and symbiosis with corals. However a diverse range of free-living microalgae are also of considerable importance on coral reefs. These include microalgae associated with sediment, the water column (i.e. eukaryotic phytoplankton) and the surfaces of corals, macroalgae, seagrasses and other reef organisms.

Free-living benthic microalgae, together with the benthic cyanobacteria, form the so-called microphytobenthos. On the GBR about 40% of the reef area is suitable for colonization by microphytobenthos (Uthicke and Klumpp 1998), although inter-reef areas also support microalgal primary production. Gottschalk et al. (2007) identified a total 209 different taxa of diatoms in the top 1 cm of sediment from inshore and offshore regions of the central and northern GBR, with the dominant taxa being pennate diatom genera such as *Diploneis, Nitzschia, Amphora* and *Navicula* (Figure 5.5). Similarly, on the southern GBR at Heron Island, Heil et al. (2004) recorded a range of pennate diatom genera as well as various dinoflagellate genera in the top 1 cm of reef sediment.

(a)   (b)   (c)

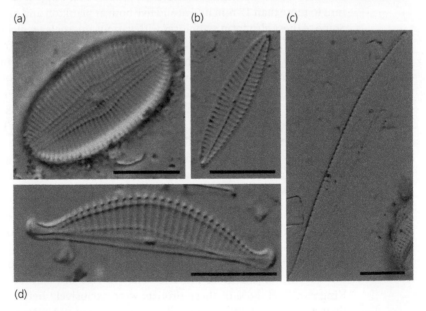

(d)

Figure 5.5   Benthic diatoms from sediment of the Great Barrier Reef. (a) *Diploneis* sp.; (b) *Navicula* sp.; (c) *Nitzschia* sp.; and (d) *Amphora* sp. Scale bars = 20 μm.

Source: Photos by S. Gottschalk, courtesy of the North Queensland Algal Identification/Culturing Facility (NQAIF), James Cook University, Australia.

The importance of benthic microalgae for coral reef biodiversity is also evident in the Florida Keys, where Miller et al. (1977) surveyed the diatom flora associated with corals, coral sand and the surface of nearby leaves of the seagrass *Thalassia testudinum*. They identified 331 species of diatoms on the surface of corals, 292 in the sand and 207 on the seagrass. As on the GBR, species of *Amphora* and *Diploneis* were commonly found in the sands of the Florida Keys, while the diatom communities associated with the corals were dominated by species of *Campylodiscus, Podocystis* and *Triceratium*; indeed, Miller et al. (1977) only found this latter genus on corals, indicating that some taxa may be very specific for live corals and coral skeletons. The seagrass leaves were dominated by different diatoms again, which belonged to the genus *Mastogloia*.

Not only are these benthic microalgae taxonomically diverse, but they occur at very high densities and biomasses, and are very productive. For example, on the central and northern GBR, average densities of 2.55×10^6 diatom cells mL^{-1} of sediment were recorded; densities were twice as high on inshore as on offshore reefs, most likely reflecting lower nutrient availability away from sources of terrestrial run-off (Gottschalk et al. 2007). Similarly, the micro-phytobenthos chlorophyll *a* concentration (an indicator of biomass) ranges from 8 mg m^{-2} to 995 mg m^{-2} at various sites around the world, with differences between sites reflecting factors such as nutrients, sediment type and hydrology, as well as the proximity of competing marine plants. Overall, the microphytobenthos may contribute 20%–30% of total primary production on coral reefs (Sorokin 1993). At Heron Island, this equates to a photosynthetic rate of up to 110 mg O_2 m^{-2} h^{-1} (Heil et al. 2004).

Coral reef phytoplankton includes a range of diatoms, dinoflagellates and other microalgae, which contribute to the nano- and microplankton (Sadally et al. 2014; Kurten et al. 2015). Dinoflagellates are a complicated group. About half are photosynthetic, possibly as a result of an ancestral ingestion and incorporation of algae, and some of these forms are the zooxanthellae found within corals and other benthic invertebrates. Others are free-living heterotrophic predators which feed on other protozoans. Free-living dinoflagellates are motile, using two flagella to provide motive force. Some dinoflagellates occasionally bloom to produce concentrations of millions of cells in each millilitre; these blooms may be recognized as 'red tides', slicks of which may extend for many kilometres. These blooms may become highly toxic due to production of neurotoxins and, even when they are not, their drawdown of oxygen alone can lead to substantial fish kills. In lower concentrations, they can be concentrated to toxic levels also within higher filter-feeding animals, especially shellfish, which are then eaten by people. Common toxic dinoflagellates include the genera *Gambierdiscus, Coolia* and *Ostreopsis*, all of which produce the toxin ciguatera; the best known species of ciguatera-producing dinoflagellate is *Gambierdiscus toxicus*. These various toxic dinoflagellates can be planktonic, benthic or epiphytic (live on the surfaces of macroalgae or seagrasses).

5.1.6 Microbial productivity and turnover

While densities of bacteria on the seabed are high, densities in the pelagic phase are still impressive, and may exceed one billion per litre even in clear blue oceanic water. Even so, in terms of volume, that litre is still 99.99999% water (Pomeroy et al. 2007) and hence remains very transparent. Many planktonic microbes adhere to suspended particles, many more may clump, but many are to a great extent 'free-living'. They include viruses, bacteria, archaea and protists, which together far exceed all macroscopic forms of life in almost all measures: biodiversity, biomass and metabolic activity. Microorganisms were, after all, the first forms of life, and they existed alone without any macroorganisms, for over half of the total span of life on Earth. That they are relatively unresearched is, of course, a function of the difficulties of examining them, so their importance has only been recognized for a relatively short time.

The density of phytoplankton is usually much less than that of the microphytobenthos. For example, at Heron Island, the chlorophyll *a* concentration of the microphytobenthos is about 100-fold greater than that of the water column (Heil et al. 2004). However, as with the microphytobenthos, phytoplankton abundance is influenced by nutrient concentration and, hence, proximity to coasts and river outlets. On the southern and central GBR, chlorophyll *a* concentrations in the water column decrease as distance from the coast increases (Brodie et al. 2007), and average concentrations on the far northern GBR ($0.23 \, \mu g \, L^{-1}$) were less than half those on southern and central regions of the GBR ($0.54 \, \mu g \, L^{-1}$), which adjoin particularly large freshwater catchments. Furthermore, chlorophyll concentration changed with season, with greater run-off and, hence, input of nutrients in the summer/wet season, causing a 50% increase from that occurring in the winter/dry season. In this case, the chlorophyll concentration represents contributions from both cyanobacteria and eukaryotic microalgae, and cyanobacteria rather than microalgae tend to be the dominant component of the phytoplankton on reefs.

Prokaryotic and eukaryotic microbes form the base of the coral reef food web, meaning that their productivity is an important consideration in ecosystem function. Yet, they are commonly overlooked in reef studies. Planktonic production and growth rates are in no way negligible, yet tropical water of the sort that bathes reefs is generally extremely clear, suggesting that there are commonly only low densities in the water column. This is misleading. The reason why reef water is typically so clear is partly due to the fact that most components of the plankton are very small, and because, although production may be significant, its consumption and hence turnover is very fast.

On the GBR, the average photosynthetic rate in the water column is 0.68 grams of carbon per square metre per day (= 1.57 µmol of carbon per litre per day), with values ranging from 0.1 to 1.5 g of carbon per square metre per day (Furnas et al. 2005) Primary production at inshore locations (<15 km from the coast) varies seasonally in response to rainfall and terrestrial run-off, which carries nutrients into reef waters but, irrespective of season, the highest productivity occurs at all times over the mid- and outer shelf due to greater light penetration and hence the greater depth at which primary production occurs. From this average productivity rate, Furnas et al. (2005) calculated that phytoplankton in GBR waters have a nitrogen requirement of 0.24 moles of nitrogen per litre per day, meaning that, in 50% of the survey sites, the phytoplankton potentially use up the available stocks of dissolved inorganic nitrogen (DIN) in just 8 hours or less; this highlights the need for the constant input of 'new' nitrogen or rapid recycling of existing nitrogen, and the potential for nitrogen to be limiting for phytoplankton growth. On the other hand, given a phosphorus demand of 0.015 moles of phosphorus per litre per day, stocks of dissolved inorganic phosphorus (DIP) would last over 24 hours at more than 80% of sites. This suggests that phosphorus is rarely limiting for phytoplankton.

The potential for nitrogen limitation is particularly a concern for large diatoms, dinoflagellates and other microalgae, which show signs of nitrogen limitation at DIN concentrations of <0.05 μM; such concentrations are at the lower end of concentrations found on the GBR. However, GBR waters are dominated by relatively small to medium-sized microalgae (e.g. the chain-forming diatom *Leptocylindrus danicus*, which has cells that are just 5–16 μm in diameter) and especially picoplanktonic cyanobacteria (which are <2 μm in diameter). These small microalgae and cyanobacteria have maximum population doubling rates of two to four and two to three doublings per day, respectively, although the actual doubling rates are usually closer to one per day because of limitation by the daily light cycle. Growth of these small phytoplankters is near maximal at DIN concentrations of as little as 0.02–0.05 μM, indicating that these organisms are rarely nitrogen limited because of their large surface-area-to-volume ratio and consequent capacity to scavenge nutrients.

Doubling rates recorded for heterotrophic bacteria in the water column range between <1 (0.0625) and 12 doublings per day (Gast et al. 1998), this considerable variability being a result of spatial and temporal differences in food availability or physical conditions. Seasonal variability is apparent at Lizard Island on the GBR, where bacterial productivity was measured as 4 μg of carbon per litre per day in winter, 37 μg of carbon per litre per day in spring and 56 μg of carbon per litre per day in summer; this seasonal increase is most likely a response to the greater availability of organic matter (e.g. algal exudates and fragments, coral mucus) in spring and summer rather than warmer temperatures (Moriarty, Pollard, Hunt, et al. 1985). These same authors also measured daily fluctuations in bacterial production, which was four times greater during the day than at night as a result of the release of photosynthetic products by algae and mucus by corals; higher bacterial growth rates over the reef flat than in lagoonal water and outside the reef were also thought to be related to the availability of these food sources.

The productivity of benthic bacteria is higher than that of the bacterioplankton, with production rates of 0.12–0.50 and 0.01–0.12 g of carbon per square metre per day, as measured in sand on the GBR during summer and winter, respectively (Moriarty, Pollard, Hunt, et al. 1985; Moriarty and Hansen 1990). These rates are about 30%–40% of primary production by benthic microalgae. Bacterial productivity also changes on a daily basis, with rates being four to five times greater during the day than at night. Increased bacterial productivity during both daytime and summer is related to the greater availability of food, especially algal exudates and debris, for bacterial growth; typical bacterial doubling times are 1–2 days in summer and 4–16 days in winter. On hard substrate (mainly dead coral), rates of bacterial production have been measured as 0.04–0.12 g of carbon per square metre per day in summer, which is 2–14 times lower than bacterial production in neighbouring sediments (Moriarty and Hansen 1990).

5.2 Trophic links

5.2.1 The microbial loop

Communities of microorganisms, notably bacteria, form a key part of the cycling of carbon and more complex organic compounds on a reef. The term 'microbial loop' or 'microbial food web' refers to this process, whereby detritus and dissolved organic matter (DOM) are incorporated into microorganisms which are then themselves consumed (Figure 5.6). Efficient recycling of matter is particularly important, given the nutrient-poor water in which coral reefs often live. Viruses also play a role in this recycling, as their 'predation' on both heterotrophic bacteria and cyanobacteria (which are infected by viruses known as bacteriophages and cyanophages, respectively) results in cellular lysis (the destruction of cells through membrane disruption) and the liberation of DOM to the seawater; viruses themselves ultimately become part of the DOM pool too (Rohwer and Kelley 2004; Vega Thurber et al. 2017). The microbial loop is an integral part of the more visible (to us) food web of herbivores, detritivores and carnivores. In many ways, the microorganismal trophic web can be viewed as being the main one, with that of large biota a relatively late-arriving side appendage of it.

Bacteria decompose animal and plant remains. This importantly includes the remains of species of organisms such as brown algae which may have large biomass but which are not palatable, or which cannot be consumed directly by many higher animals because the latter lack enzyme systems with which to do so. Another source of nutrition for these bacteria is dissolved

Marine microbial food web

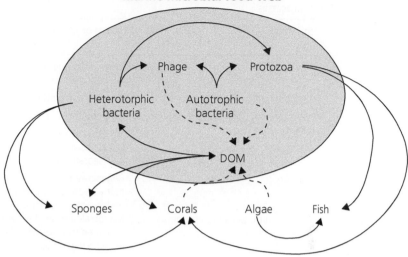

Figure 5.6 The marine microbial food web (microbial loop) and its trophic link to the 'wall of mouths'. The key part of the microbial loop is the uptake of dissolved organic matter (DOM) by heterotrophic bacteria, and their subsequent predation by the likes of Protozoa and viruses (phage). DOM arises from a variety of sources, including viral lysis of bacteria, photosynthetic exudates (of cyanobacteria, microalgae and seaweeds), and the mucus from organisms such as corals. Heterotrophic bacteria therefore play an essential role in the recycling of matter through the coral reef ecosystem. The consumption of bacteria, protozoans, viruses and DOM by larger reef organisms, including corals, fish, sponges and other benthic filter-feeders, which together comprise the 'wall of mouths', is substantial. Note that, for clarity, not all trophic links are shown; for instance, sponges are known to consume particles as small as viruses, while corals can consume bacteria.

Source: Figure modified from Rohwer, F. and Kelley, S. (2004). Culture-independent analyses of coral-associated microbes. In: E. Rosenberg and Y. Loya (eds), *Coral Health and Disease*. Springer, pp. 265–78; with kind permission of Springer Science + Business Media.

organic substances and minerals, derived from cell lysis, or from the leaking of such substances from living cells (e.g. photosynthetic exudates of algae and corals) into the water, and from faeces and other waste products of living animals. While it had long been known that bacteria in the sea have a key role in mineralization (which, is effectively, an end point of decomposition), it became clear in the 1970s that these microorganisms drive a quantitatively important cycle—the 'microbial loop' (Pomeroy 1974). This term particularly refers to the utilization and hence recycling of dissolved material by bacteria, although sometimes the decomposition aspects are referred to as well in the meaning of the term—indeed, the two are sometimes difficult to separate in an ecological context.

Rates of turnover vary depending on the substance involved, and may vary from seconds to many years (Table 5.1). Rapid microbial regeneration is key to much of this.

Table 5.1 Rates of turnover of carbon, nitrogen and phosphorus in several compartments of the reef, by different processes

Approximate Rate of Turnover	Substance	Nature of Turnover or Organisms
Seconds to minutes	Key nutrients	Dissolved nutrients, such as phosphorus, in water over reef
Hours	Carbon in microbial communities	Turnover in microbial communities
Days	Nitrogen and phosphorus in lagoon water column	Regeneration of phytoplankton
Weeks	Carbon in plankton	Life cycles of small benthic and pelagic invertebrates and benthic microalgae
Months	Carbon in larger organisms	Regeneration of macroalgae and many invertebrates, and turnover of metabolites in larger animals.
	Carbon, nitrogen and phosphorus in sediments	Microbial- and small-invertebrate-mediated turnover in sediments that are mixed and worked
Years	Carbon All organic materials	Turnover, grazing and natural mortality events in most large forms of biota Turnover in deeper or relatively immobile sediments

Source: Hatcher, B.G. (1997) Organic production and decomposition. In C. Birkeland (ed.), *Life and Death of Coral Reefs*. Springer, pp. 140–74.

5.2.2 Consumption of microorganisms

In the water column, protozoans (protozooplankton) are the major grazers of picoplanktonic microbes (heterotrophic bacteria and cyanobacteria). For example, in Tikehau lagoon in French Polynesia, phagotrophic nanoflagellates were found to be the major grazers of cyanobacteria, while ciliates and heterotrophic dinoflagellates grazed predominantly on both autotrophic and heterotrophic nanoplankton (e.g. small diatoms and flagellates) (Gonzalez et al. 1998). Similarly, at Miyako Island in Japan, Ferrier-Pagès and Gattuso (1998) measured that 30%–50% of cyanobacterial production was grazed by heterotrophic flagellates and ciliates, which in turn (50%–70% of their production) were consumed by higher trophic levels. These various proto-zooplankters are important for controlling populations of picoplankton and tend to prevent cyanobacterial bloom formation because their potential growth rate (one to three doublings per day) can match that of their prey.

In contrast, planktonic microalgae such as diatoms are principally grazed by micro-crustaceans, especially copepods (members of the metazooplankton), although other metazoans, including larvae of many reef organisms, and larger protozoans graze microalgae too. Copepods also graze on protozoans. Copepods are at least as diverse, and provide a greater biomass, than their more conspicuous and far better known larger crustacean relatives.

Copepods usually keep microalgal populations in check, although the relatively slow growth rate of these small crustaceans, which have a generation time of about 1 week, may be too slow to prevent microalgal blooms when nutrient levels are high (Furnas et al. 2005).

Grazing of planktonic primary producers by both protozoan and metazoan plankters is an important trophic link that facilitates carbon export to other parts of the coral reef ecosystem, through predation by fish and benthic filter-feeders; export also occurs via sinking of detritus, such as metazooplankton faeces. In their study of a lagoon at the French Polynesian atoll of Takapoto, Sakka et al. (2002; Figure 5.7) calculated that 70% of phytoplankton total net production (both particulate and dissolved matter) is lost through heterotrophic respiration within the plankton, leaving 30% for export. This export value is considered as being high and is related to the relatively low activity of planktonic heterotrophic bacteria at this site. The export of primary production was particularly aided by the strong grazing pressure of the protozooplankton, which consumed 41% of the phytoplankton particulate net production each day, and itself was largely eaten by metazooplankton. Indeed, the metazooplankton consumed the equivalent of the entire protozooplankton production every day.

Bacterioplankton, protozooplankton, phytoplankton and metazooplankton all act as food sources for larger reef organisms, especially fish and benthic filter-feeders, which together comprise the 'wall of mouths' (see 'Measurement of plankton consumption by corals'). Planktonic organisms are the main food source for most reef fish species while they are in their early larval stages, becoming increasingly important while they absorb their yolk reserves and change progressively to active feeding, and they remain the main or only source of food for many fish species as adults also.

Filter-feeding invertebrates include hard and soft corals (see Chapter 2, Figure 2.1), sponges, ascidians, polychaete worms, bryozoans, giant clams and crinoids. The great efficiency of sponges at filtering picoplankton from the water column is especially well known, with a retention efficiency of 65%–95% being estimated for coral reef sponges (Yahel et al. 2003; Lesser 2006; McMurray et al. 2016); sponges can even consume particles as small as viruses (Hadas et al. 2006). Corals also filter pico- and nanoplankton from the water column. In the corals *Stylophora pistillata*, *Galaxea fascicularis* and *Tubastraea aurea*, bacteria, cyanobacteria and picoflagellates were found to contribute 1%–7% of ingested pico- and nanoplanktonic carbon, although bacteria were the most frequently consumed prey items. In comparison, larger nanoflagellates contributed 84%–94% and 52%–85% of ingested carbon and nitrogen, respectively (Houlbrèque et al. 2004). The efficiency of such filter-feeding organisms means that plankton is removed very rapidly as water passes over the reef and that 'benthic-pelagic coupling'

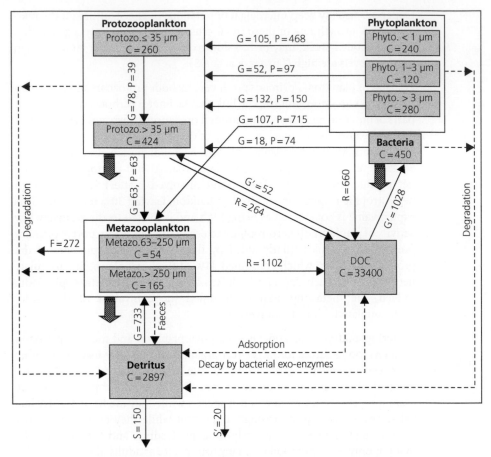

Figure 5.7   Carbon budget of the planktonic food web in the lagoon of Takapoto Atoll, French Polynesia. Standing stocks of carbon (in milligrams of carbon per square metre) are in boxes, and fluxes (in milligrams of carbon per square metre per day) are shown by arrows. Solid arrows are estimated fluxes and dashed arrows are non-estimated fluxes. As estimates of phytoplankton particulate production are net values, the budget does not include autotrophic respiration; heterotrophic respiration is represented by downwards-pointing block arrows. The microbial loop is shown by the uptake of dissolved organic carbon (DOC) by bacteria and their subsequent consumption by protozooplankton; these, in turn, are consumed by the larger metazooplankton (e.g. copepods); C = carbon stock; F = food-web transfer = production of metazooplankton; G = consumption of particulate organic carbon; G' = consumption of DOC; P = particulate organic carbon production; R = release of dissolved organic carbon; S = sinking of detritus; S' = sinking of organisms.

Source: Figure redrawn from Sakka, A., Legendre, L., Gosselin, M., Niquil, N. and Delesalle, B. (2002). Carbon budget of the planktonic food web in an atoll lagoon (Takapoto, French Polynesia). *Journal of Plankton Research* 24:301–20.

Measurement of plankton consumption by corals

Scleractinian corals are sessile benthic organisms that can feed on a wide range of food sources, from dissolved organic matter and bacteria to macrozooplankton (see Figure). Heterotrophy is a main source of essential nutrients such as nitrogen and phosphorus, and can sustain coral metabolism during stress events, such as bleaching. Feeding rates can be measured using different methodologies, either under controlled laboratory conditions or under in situ conditions, taking into account the whole reef community.

Figure. Polyp of the coral *Stylophora pistillata*, eating a brine shrimp (centre of the photo).
Source: Photo: E. Tambutté.

In laboratory experiments, predation on zooplankton is usually assessed using flow tanks or channels according to Levy et al. (2001), based on the procedures described by Vogel and LaBarbera (1978). These tanks can have different sizes to host one or more coral colonies. Since particle capture in corals is dependent on water flow, tanks have to be equipped with a motor-driven propeller with defined rotational speeds. The flow leading to maximal particle capture is itself dependent on coral species and food type. In such flow tanks, grazing rates are deduced from the disappearance of prey in a given amount of time. They are reported as the number of prey ingested (or the amount of carbon and nitrogen ingested) per hour and per polyp or skeletal surface area. At the same time, a

Continued

Measurement of plankton consumption by corals (*Continued*)

control tank with the same prey concentration but without grazers is needed to assess any change in prey concentration due to the experimental set-up. This control tank is especially important with small prey items such as pico- and nanoplankton, since internal grazing (within the microbial loop) can induce large changes independently of any grazer effect.

In situ, estimation of grazing rates at the individual level is more difficult than in laboratory conditions, because coral colonies are included in communities of benthic organisms, feeding more or less on the same plankton species. One technique consists of sampling polyps and probing each polyp with a dissecting needle and fine forceps under a dissecting microscope. All obvious prey items are removed from the coelenteron, which is then scraped out to expose any remaining prey. Zooplankton prey is identified, counted and preserved in alcohol. Since digestion in corals takes less than 2 h, prey found in the coelenteron is that ingested during the last hour before polyp sampling. This technique, however, does not allow measurements of small soft prey such as pico- and nanoplankton, which are too fragile to be observed using such a technique. Another method consists of enclosing in situ colonies in hemispherical Plexiglas incubation chambers, equipped with a pumping system for water renewal. In this case, plankton samples are taken at the beginning and at the end of the incubation, the difference corresponding to coral grazing.

Finally, plankton grazing rates by the whole reef community can be assessed in situ by doing transects above the reef, according to Yahel et al. (1998). A first set of transects, considered as a control, should be done on a reef portion formed only by rocks or dead corals. A second set of transects can be then performed above the reef flat or the reef slope. The per cent cover and abundances of each filter-feeding organism can be obtained using line-intercept transects on the reef. After measuring current flow and direction using a flow meter, plankton is sampled at different points along the transect (e.g. before, above and after the reef slope), either following the same water mass (called Lagrangian sampling) or not (using 'cross-shore transects'). The difference in plankton concentrations between upstream and downstream gives the amount of plankton grazed by the reef community during the time needed by the water mass to cross the reef. Zooplankton is usually collected in a 5 min haul using a plankton net and then preserved with formaldehyde. The nature and abundance of the planktonic prey are then determined using a dissecting microscope. Seawater samples for pico- and nanoplankton prey are sampled using Niskin bottles, preserved with formaldehyde, stained with DAPI and then counted using an epifluorescence microscope or a flow cytometer.

It should be noted that feeding rates are different from incorporation rates in tissue biomass, since a reasonably large fraction of the ingested prey can be lost through different processes (e.g. direct egestion of non-digested particles or via respiration or excretion). The most useful tools to assess the importance of heterotrophy to coral nutrition are

Continued

Coral microatolls and sea level *(Continued)*

radioactive or stable isotopes. Prey can be labelled/enriched with one or several of these isotopes (^{14}C, ^{13}C, ^{15}N; Benavides et al. 2016) and the signal can be traced within the coral tissue, allowing for a direct estimation of the proportion of heterotrophic nutrients allocated to coral growth (Tremblay et al. 2015). Stable isotopes can also be used in natural abundance to trace the importance of auto- and heterotrophy in a specific environment (Ferrier-Pagès et al. 2011; Nahon et al. 2013).

Overall, an accurate estimation of the feeding capacity and degree of heterotrophy of scleractinian corals requires a combination of these different techniques, each of which represents a different aspect of the trophic dynamics of corals.

Dr Christine Pagès, Centre Scientifique de Monaco, and Dr Fanny Houlbrèque, Laboratoire d'Excellence CORAIL, Paris, France, and UMR ENTROPIE, Nouméa, New Caledonia.

is a significant feature of this ecosystem. For example, in the Gulf of Aqaba (Red Sea), Yahel et al. (1998) observed that both phytoplankton (both eukaryotic and prokaryotic) abundance and chlorophyll *a* concentration are 15%–65% lower near the reef than in adjacent open waters, as a result of the water advecting across the benthic reef community; the decline in phytoplankton abundance was particularly pronounced within 1–3 m of the seabed. Through one 5 m section of reef, a depletion rate of ~20% per minute was recorded. In contrast, these same authors found no decline in phytoplankton abundance as water passed over a sandy-bottom site that lacked reef. From their measurements, Yahel et al. (1998) estimated a phytoplankton consumption rate of 719 g of carbon per square metre per year, which compares to a metazooplankton consumption rate by the reef of up to ~200 g carbon per square metre per year (Glynn 1973; Johannes and Gerber 1974). This highlights the potentially great importance of phytoplankton to reef nutrition.

Many filter-feeding benthic invertebrates belong to the 'cryptofauna', those species that live hidden on the reef, especially in cavities under coral overhangs and in voids in the reef framework. These cavities range in volume from several to hundreds of litres each, contribute 30%–75% of the total coral reef volume, and provide about 75% of the total surface area for colonization by benthic species (Ginsburg 1983; Scheffers et al. 2004). Reef cavities are therefore important sites for the removal of plankton from the overlying seawater, especially pico- and nanoplankton that are consumed by encrusting sponges. On the reefs of Curaçao in the Netherlands Antilles, Scheffers et al. (2004) closed off 70 L volume cavities and measured the removal of heterotrophic bacterioplankton from the water column over time. After just 30 minutes, 50%–60% of the bacteria had gone, which equated to a removal rate of 1.43×10^4 bacteria per millilitre per minute (0.62 mg of carbon per litre per day or 30.1 mg of carbon per square metre of cavity surface area per day). It was estimated that this removal rate met 60%–70% of the nitrogen demands of the cryptofaunal community, leaving just 30%–40% to be acquired from sources other than heterotrophic bacterioplankton. Bacterioplankton is therefore very important to the nutrition of benthic filter-feeders, not just in cavities, but on the coral reef as a whole.

Less is known about the consumption of benthic bacteria than about their planktonic counterparts; however, the considerable production of benthic bacteria presents an important source of food for detritivores and other organisms living in sediments. The consumption of benthic bacterial production by different reef organisms is shown in Figure 5.8, which for comparative purposes also shows consumption of bacteria in the water column. The most conspicuous detritivores on coral reefs are holothurians, which can

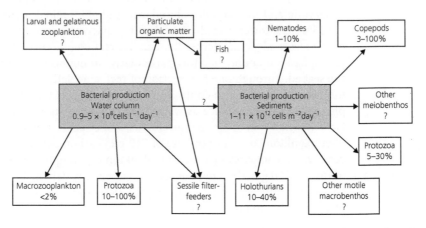

Figure 5.8   Major consumers of bacterial production in the water column and sediment of coral reefs. Values (percentage of bacterial production consumed in each system) estimated by Moriarty, Pollard, Alongi, et al. (Moriarty, D.J.W., Pollard, P.C., Alongi, D.M., Wilkinson, C.R. and Gray, J.S. (1985). Bacterial productivity and trophic relationships with consumers on a coral reef (Mecor I). *Proceedings of the 5th International Coral Reef Symposium* 3:457–62) from a range of published studies and their own findings on the Great Barrier Reef. Note that, while these authors did not calculate a value for consumption by sessile filter-feeders, removal of bacteria by organisms such as sponges and corals is substantial. Furthermore, these filter-feeding invertebrates, as well as fish, may consume particulate organic matter to which bacteria are attached; as much as 50% of bacteria in the water column may be attached to particulate organic matter.

Source: Modified from Moriarty, D.J.W., Pollard, P.C., Alongi, D.M., Wilkinson, C.R. and Gray, J.S. (1985). Bacterial productivity and trophic relationships with consumers on a coral reef (Mecor I). *Proceedings of the 5th International Coral Reef Symposium* 3:457–62.

consume 10%–40% of bacteria in reef sediments (Moriarty, Pollard, Alongi, et al. 1985; Moriarty, Pollard, Hunt, et al. 1985). Sea cucumbers frequently occupy backreef sandy areas, where they ingest large quantities of the sand, digest out the organic material which is, to a large extent, bacteria and then excrete the 'cleaned' sand. Other consumers of sedimentary bacteria include protozoans (5%–30%), harpacticoid copepods (3%–100%) and nematodes (1%–10%) (Moriarty, Pollard, Alongi, et al. 1985).

Benthic microalgae and other protists are consumed by a variety of grazers, with one of the key groups being the micro-molluscs, mainly gastropod snails. As with micro-crustaceans such as copepods, these tiny grazers are as at least diverse and have a greater biomass than their larger, more visible, relatives.

5.3 Zooplankton behaviour and ecology

5.3.1 Planktonic durations of larvae and planktonic dispersal

The larger zooplanktonic organisms are those which remain their entire lives in a pelagic phase, and larvae of sessile or benthic species. This pelagic phase, that is, the time that larvae of sessile and swimming species remain in the water column, has important consequences for the dispersal of that species. While many may swim and, for their minute size, may swim very vigorously, their movement is governed to a differing but usually great extent by water currents. Many regulate their vertical movements through the water column much better than they can regulate their horizontal movement, and may make use of the differing rates of current flows at different depths.

Durations of planktonic stages vary considerably. That shown by damselfish, for example, a large and important group of coral reef fish, varies from 12 to 39 days (Wellington and Vicor 1989). During this time, even a fairly gentle oceanic current may carry the larvae several hundred kilometres. Corals commonly show larval duration stages of 2–4 weeks. The importance of this for dispersal is considerable. Larval planktonic duration correlates with potential larval dispersal range, and this then correlates, roughly, with the total distribution range of the species.

The latter has important evolutionary consequences, because greater distribution equates approximately with greater gene flow and hence mixing. However, there are many cases of poor correlation between the planktonic stage and the geographical distribution of a species (Paulay and Meyer 2006). Studies on several reef fish have shown mixed results as, for many species, there is no good correlation between larval duration and geographic range. This is because several other factors are also important, including the ability to delay settlement, the ability to move vertically in the water column and the ability to sense favourable (or perhaps unfavourable) substrates prior to settlement and metamorphosis into adult form.

A few species have been shown to be able to arrest their settlement for very long periods when necessary, if they find themselves in open ocean, for example. This does not appear to affect the duration of their life following eventual settlement, but it does permit them to disperse further during their extended larval stage. Such larvae have been referred to as being 'virtually immortal' (Paulay and Meyer 2006), and some echinoderms have been shown to exhibit clonal reproduction too, further extending the stage of larval competency. The length of the larval stage may depend in part on the amount of energy reserves that the larvae contain, and the amount they will need, on settlement, to metamorphose into the adult form which can then feed efficiently. Those with large eggs may delay settlement for a year, or perhaps longer. Many larval species do, of course, feed. Lecithotrophic larvae hatch from large, yolky eggs upon which they depend for their initial nutritional needs, and this may permit the individual to reach a more advanced stage of development during their planktonic phase. Some of these appear not to need food at all while in their larval stages. In contrast, planktotrophic larvae hatch from small eggs with a limited, in-built nutrient supply, so that they then depend heavily on catching planktonic prey while still in the larval stage. Some of these may be able to abbreviate their larval stages if food is in short supply.

As well as duration of larval stage, equally important to dispersal is the behaviour of the larvae of each species. Many are geotactic, that is, they respond to gravity. Those which are positively geotactic quickly settle near their parents, and the apparent advantage of this is that settlement is more likely to occur onto favourable habitat—after all, the parent lived on it. Other larvae are negatively geotactic, and move upwards, which has the advantage of longer dispersal but obviously misses the advantage of a likely favourable habitat near the parent.

Other sensory mechanisms influence larval duration and settlement. Corals and soft corals settle preferentially onto limestone, and many, perhaps most, sense crustose coralline algae (CCA) in particular (Morse et al. 1996). The ability to do this spans corals across several genera, and is found in planulae of both Indo-Pacific and Caribbean regions. Detection of suitable reef substrate is controlled by chemosensory recognition of a chemical cue associated with the CCA (and perhaps microbes associated with these algae), as well as spectral cues, with planulae being attracted preferentially to red surfaces (Sneed et al. 2015; Foster and Gilmour 2016). This sensory perception of substrate declines with age of the larvae, thus favouring attachment and metamorphosis early on; but, because each species differs in timing and strength, this may avoid interspecific competition to some degree amongst the settling larvae by reducing the potential for post-settlement interactions.

Many CCA species, but not all, are attractive to coral larvae, although similar induction has been obtained with coral rubble and skeletons also (Heyward and Negri 1999). Such attraction, however, can be readily impaired by pollutants such as oil, to an even greater extent than inhibition of fertilization in the first place (Negri and Hayward 2000).

Other signals received and apparently used to remain near reefs, or to swim towards reefs, include sound. Sound on a reef is complex, and includes components from breaking waves to the sounds made by numerous organisms (Radford et al. 2014). Such auditory cues may enable larvae, for example of fish and corals, to be attracted to reasonably nearby reefs rather than to continue past and to disappear, perhaps, into open ocean (Parmentier et al. 2015; Lillis et al. 2016).

5.3.2 Diurnal cycles of demersal plankton

Much of the content in Section 5.3.1 concerned fish larvae, which is where a lot of the research has focused. But equally important is vertical migration of many kinds of zooplankton, both larvae and forms which are planktonic throughout their lives. The zooplankton on a reef is a mixture of demersal plankton, pelagic plankton swept onto the reef and larvae released on the reef. The demersal plankton is diurnal, and a major component. Demersal plankters live in the reef crevices by day, and migrate into the water column over the reef at night. They are mainly mysids, nematode and small polychaete worms, and a wide range of micro-crustaceans: amphipods, ostracods, isopods and copepods (Porter and Porter 1977). Their size is large in comparison with most members of the permanent zooplankton, and they are a major food source for planktivorous reef fish.

Zooplankton is important, because a large proportion of benthic reef animals depend on it. Many, such as the crinoids and basket stars, emerge only at night to take advantage of the zooplankton (Figure 5.9), remaining concealed themselves during daylight. Substantial vertical migration with a diurnal rhythm has long been known. For a long time, difficulties of detection and the inefficiencies of plankton traps for such diurnal plankton precluded any useful estimates of abundance, and therefore of any possible role in the ecosystem too. Madhupratap et al. (1991) sampled zooplankton in lagoonal atolls both by using the traditional way of plankton 'emergence traps' set over the sand and by counting those taken from the sandy substrates during the daytime. The core sampling revealed densities living in the sand in daytime that were 25 times greater than estimates made from the plankton traps. Numbers were many thousands of animals per square metre, and 80% of them migrated into the water column within an hour of sunset and returned close to dawn. Clearly, from the point of view of corals, this is also a good time to have expanded tentacles, whose purpose is to catch the zooplankton (Figure 2.1). During the night, the surface of the reef is covered with benthic

Figure 5.9   Filter-feeding. Top: A crinoid on the Great Barrier Reef, emerged at dusk, when it fishes for zooplankton. Crinoids comprise an ancient class of echinoderms. The crinoid's arms basically form filter-feeding nets which catch swimming zooplankton, which are abundant at night. Trapped plankton are then swept down into the mouth at the base. Bottom: A basket star. This is also an echinoderm, but is a greatly modified brittle star. Basket stars have arms which are repeatedly branching, forming an elaborately tangled mass. This is the golden-coloured Caribbean *Astrophyton muricatum*, which spends the day coiled tightly, in this case, in a branching soft coral. At night it uncoils, and orientates itself into the current so its web of arms can catch plankton (See Plate 9).

animals that are efficient at trapping zooplankton. Further, it is likely that substrates such as coral rubble house greater zooplankton densities than sand (Porter and Porter 1977). These numbers therefore provide an important contribution to the night-time zooplankton.

In the Gulf of Aqaba, vertical migrations of several major groups of emerging zooplankton exceed 25 m (Schmidt 1973). In the same location, the bottom 1 m of water becomes depleted relative to higher parts of the water column, where the abundance of large zooplankters doubled during the night, and returned to daytime levels around dawn in a clear diurnal cycle. Most zooplankton at night migrated to near-surface waters, creating a steep gradient in the concentration of zooplankton (Yahel, Yahel, Berman et al. 2005; Yahel, Yahel and Genin 2005). The lack of zooplankton near the surface of the reef is attributable to filter-feeding animals, and to larger numbers of planktivorous fish near the bottom.

Timing of the diurnal zooplankton in the Red Sea is very precise and consistent (Yahel, Yahel, Berman, et al. 2005). As detected by acoustic backscattering intensity, diurnal plankton emerge from the reef at sunset ±4 minutes, and the diurnal plankton disappear again 82 minutes ±5 minutes before sunrise in a way consistent throughout the seasons. Smaller zooplankters (500–700 µm) ascend first and the demersal zooplankton account for most of the increase seen shortly after sunset. Just after sunset, surprisingly, zooplanktivorous fish are usually still feeding, although at that time their feeding efficiency is declining, and most corals have not by that time extended their plankton-capturing tentacles, thus giving a window for the diurnal zooplankton to emerge and ascend to what appear to be safer levels higher in the water column above the reef.

It is assumed that predator avoidance is the main driver behind the diurnal behaviour of such a large quantity of tiny animals. A suggested contributing factor is that this behaviour also avoids depleted oxygen levels which may occur near the bottom during the night, although the water movement commonly seen on reefs, and the existence of sessile organisms, indicate that oxygen there should rarely be in short supply. Movement vertically up through the water column towards phytoplankton, or towards smaller zooplankton prey, is also a possible cause. Conversely, avoidance of becoming prey themselves to pelagic, planktivorous fish during daylight hours by living on the bottom, especially in interstitial spaces in sand and rubble for example, is another likely reason for many to have adopted the diurnal rhythm. Whatever the causes, their abundance is now thought to be a significant component of the trophic web of a coral reef. Surprisingly, perhaps, for nocturnal animals, they may be readily attracted to the light of a torch; if a torch beam is held beside an expanded, feeding coral at night, numerous zooplankton are readily attracted towards the light. On touching a coral, their capture and immediate immobilization on the coral tentacles takes place, at a speed which illustrates the potency and efficiency of plankton capture by corals (Figure 5.10).

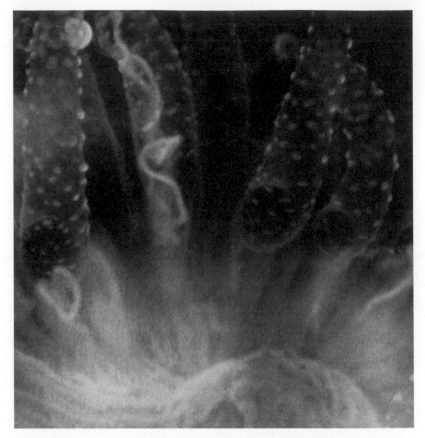

Figure 5.10   Close-up of tentacles of a coral polyp, taken on the reef at night. The white dots on the tentacles are groups of nematocysts. On some tentacles, a few tiny worms, probably nematodes or polychaetes, have been trapped. The mouth of the polyp is at the top of the cone, which can be seen at the bottom of the photo, and is about 2 mm wide.

5.3.3 Reef connectivity

Plankton, therefore, is far from being composed of passive particles swept by currents. Old views of swimming organisms (nekton) and passively swept organisms (plankton) no longer hold. It has long been presumed that species which cannot disperse well are more likely to become extinct, because unpredictable, large impacts occur periodically to eliminate local populations. Examples of such impacts include hurricanes, freshwater deluges or large changes in sea level. Species which can disperse, in contrast, are more likely to have populations in locations which are sufficiently distant from the impact to be able to survive and, possibly, recolonize the depleted area.

In the case of reefs, an additional complication arises. Over most of the tropical world, reefs contain concentrated densities of species, separated by hundreds or thousands of kilometres of hostile habitat in the form of open ocean. This is the case with atoll chains, for example. Even apparently continuous reefs, such as the GBR, or the extensive Meso-American reef system, are rarely continuous for very far but are comprised of many smaller units, again with greater areas of 'space' between them than there is area of reef substrate. Spacing may be closer in such cases, but the idea of patches of conducive habitat separated by gaps of hostile habitat is the same in principle, if not scale.

As noted in Section 5.3.1, most larvae show a varying and sometimes marked ability to swim by themselves. Some surgeonfish larvae have been shown to swim continuously for nearly 200 hours, such that distances covered by some of them can be as much as 8–60 km. Such distances would seem to be more than adequate for good dispersal but, in many cases, experiments have shown that patch reef components in a system of reefs are mainly 'self-seeded', meaning that there is commonly a marked preference for settlement relatively close to the parents, at distances much less than the potential maximum. However, using genetic techniques, it has been found that some fish species have a high degree of genetic connectivity between different areas, demonstrating a high degree of larval exchange between them. This has been shown to apply also to elkhorn coral in the same general region (in Mexico). With both corals and fish, factors influencing connectivity between reefs include not only dispersal of course, but also larval buoyancy and other ecological factors such as planktonic and post-settlement survival.

In this subject, more than most, the problems of scale are central. Coral reef systems depend for continued survival on connectivity which links reproductive populations and, for this to occur, several important criteria are involved: successful reproduction, dispersal, settlement and then post-settlement survival. Throughout all these, nutrition is obviously essential too.

Early studies of reef fish assumed that fish were in equilibrium, that stable coexistence was the underlying characteristic, and that the high diversity seen on any reef was a function of the many niche spaces that existed. But, as noted by Sale (2002) in relation to reef fish, reefs are open systems filled with local populations that are replenished by the settlement of larvae which have come from varying and sometimes great distances. This applies equally to corals and other benthic groups as well. The larval mode of reproduction should ensure that species are well mixed but, as noted, various chemosensory and swimming behaviours may ensure that any one species makes much less use of its maximum potential for dispersal than might appear to be possible. The patchy distribution of reefs, and of precisely suitable habitat on any one reef, leads to subdivision of populations on many spatial scales, and many fish do not move far from their initial settlement site once they have metamorphosed into adults. There are fish species which migrate to other habitats, such as seagrasses and mangroves, in order to breed or which travel long distances to spawning sites, but these are a minority. Thus, a population of adults on a reef, of fish, corals or other benthic species, may be locally restricted but may receive and export larvae from and to other populations, and the latter may be closely adjacent or many kilometres distant. It is becoming clearer, for reef fish at least, that their dispersion is greatly influenced by reef distribution and by currents that connect them. Larvae are more competent than previously thought in this respect, and are far from being passively swept particles, relying on chance to end up near suitable habitat.

6 Reef fish

Evolution, diversity and function

While scuba diving on or snorkelling over coral reefs, you will be struck by the huge diversity in the fish surrounding you. These fish display a very wide range of shapes, colours and sizes, from small reef-associated damselfish and diversely coloured parrotfish to massive groupers. The diversity is so great that, despite covering less than 0.1% of the ocean, coral reefs are home to about a third of the marine fish species that have been identified (Helfman et al. 1997). Indeed, they are the most speciose vertebrate communities on Earth. Clearly, there must be strong drivers leading to the development of these diverse forms and functions amongst fish species. In this chapter, the evolution and biogeography of reef fish will be explored. Age and growth will be discussed, followed by larval fish ecology. Drivers of the exceptional diversity of reef fish will be outlined, along with the utility of the diversity of colours. The science around the abundance, biomass and trophic structure of reef fish assemblages will be examined. The range of fish feeding habits will be detailed and functional roles of fish explored. Finally, the factors and issues that can affect reef fish assemblages will be examined. It is not possible within a single chapter to do full justice to all issues regarding coral reef fish biology and ecology; for further comprehensive information, see Mora's (2015) book.

6.1 Evolution and biogeography

From 230 to 90 Ma, pycnodonts, which were an early form of bony fish but are now extinct, showed some morphological characteristics that may link them to reef fish; however the stem lineages of many modern reef fish families arose 90–66 Ma (Bellwood et al. 2015). It was during the period when dinosaurs were lost, 66–34 Ma, that the taxonomy, phylogeny and functions of reef fish really started to come together to form the foundation of modern

The Biology of Coral Reefs. Second Edition. Charles Sheppard, Simon Davy, Graham Pilling, and Nicholas Graham, Oxford University Press (2018). © Charles Sheppard, Simon Davy, Graham Pilling, and Nicholas Graham (2018). DOI 10.1093/oso/9780198787341.001.0001

reef fish families. The origins of different families occurred during distinct periods; for example, the common ancestors of wrasses (Labridae) arose 66–50 Ma, while the butterflyfishes (Chaetodontidae) did not appear until around 32 Ma. This initial phase of reef fish diversification occurred in the West Tethys Sea, in the area now occupied by Europe. Fish species became more specialized in feeding, for example developing diets focused on corals, detritus and turfing algae, from 34 to 5 Ma. A huge amount of speciation and the species of fish seen on today's reefs occurred in the past 5 million years, although there were few additional broad feeding groups beyond those that developed from 34 to 5 Ma (Bellwood et al. 2015).

This evolutionary history has been associated with 'hopping hotspots' of marine biodiversity, with the centre of biodiversity shifting from the Tethys Sea region to South East Asia and western Melanesia, where the highest concentration of species is found today (Remena et al. 2008). Over more recent evolutionary periods, large biogeographic patterns have gradually separated species pools. One key event occurred 12–18 Ma, when contact between the African and Eurasian continental plates formed a land bridge in the Middle East, separating the Indian and Atlantic Oceans (Figure 6.1). Another major barrier developed 3.1–3.5 Ma, with the elevation of the Isthmus of Panama (see Chapter 1), leading to the separation of the Eastern Pacific from the Caribbean. These events effectively isolated different regions; however, barriers to species transfer do not need to be physical. In modern times, the East Pacific Barrier, which divides the Indo- and East Pacific regions, limits transfer purely through the presence of a large expanse of deep open ocean, approximately 5,000 km wide, which forms an effective barrier to many species of fish.

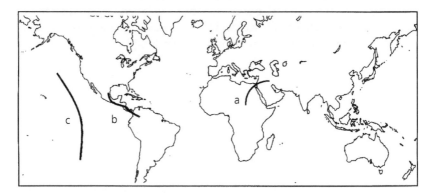

Figure 6.1   The major biogeographic boundaries for reef fish taxa: (a) closing of the Tethys Sea by the Red Sea land bridge; (b) the isthmus of Panama; (c) the East Pacific Barrier.
Source: Modified after Blum, S.D. (1989). Biogeography of the Chaetodontidae: An analysis of allopatry among closely related species. *Environmental Biology of Fishes* 25:9–31; and Bellwood, D.R. and Wainwright, P.W. (2002). The history and biogeography of fishes on coral reefs. In: P.F. Sale (ed.), *Coral Reef Fishes: Dynamics and Diversity in a Complex Ecosystem.* Academic Press, pp. 5–32.

Following macroscale separations of species pools, drivers and natural selection within regions led to differences in species communities (Rocha and Bowen 2008). For example, fish species on Caribbean reefs are less numerous than those in the Indo-Pacific. This may result from periods of reduced sea levels that may have led to species losses. Within any one region, diversity may be further limited by local currents, linked to biological factors such as the swimming capacity and length of the larval phase of species. Short larval phases and unfavourable currents will lead to reduced dispersal and hence reduced diversity in particular areas. Therefore, as the geographic range of species is reduced, biological and physical factors interact greatly on resulting biodiversity and species evolution. A result of all these factors is the wide variety of geographic ranges shown by reef fish species. Even closely related species can have vastly different ranges; for example, within the wrasses, *Thalassoma purpureum* has a range across the entire Indo-Pacific, from the east coast of Africa to the west coast of Central America, while the entire range of *T. robertsoni* is restricted to a single 6 km-long Pacific atoll (Ruttenberg and Lester 2015).

6.2 Age and growth

The principal way to age a fish is through the ear bones, or otoliths. Much like a tree, otoliths lay down distinct growth rings, which can be used to assess age and are linked to growth rates and key life history stages, such as age at maturity and lifespan. In juvenile and short-lived species, daily growth rings can be discerned for some species, whereas annual growth rings are easier to see in longer-lived species. While scales and fin spines also lay down distinct growth bands in some species, there are problems with these methods, particularly for tropical species, and otoliths have proven more useful.

Reef fish have huge variation in maximum ages. Indeed, the coral reef pygmy goby (*Eviota sigillata*) has the shortest recorded vertebrate lifespan—just 8 weeks (Figure 6.2; Depcznski and Bellwood 2005). Three of those weeks are spent as a larval fish in the plankton, 1–2 weeks maturing and then just 3.5 weeks as an adult. At the other end of the scale, some reef fish can live for more than 40 years. There tends to be some similarities within families; for example, parrotfish tend to have shorter lifespans (mostly <20 years) than surgeonfish (many of which live for >20 years) do. However, there is also considerable variability within families; for example, the maximum age of parrotfish species can range from ~3 years to ~25 years, and that of surgeonfish from 10 years to 40 years (Choat and Robertson 2002).

In terms of growth, temperate fish tend to show quite indiscriminate growth, where growth can continue throughout life. While some reef fish, such as some species of groupers, show similar growth patterns, many reef fish show a more

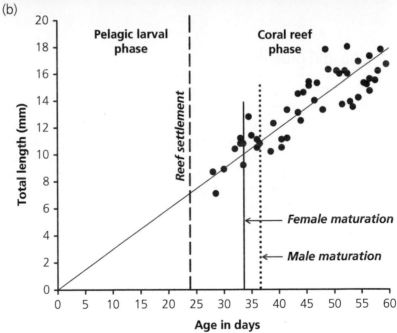

Figure 6.2   (A) The vertebrate with the shortest recorded lifespan, the cryptic coral reef pygmy goby (*Eviota sigillata*). (B) The size of *E. sigillata* at different life stages (*n* = 50): the pelagic larval stage (~24 days); settlement onto reefs; sexual maturation (Days 34–38); and death, which occurs within 59 days.

Source: (A) Photo: J.E. Randall. (B) Depczynski, M. and Bellwood, D.R. (2005). Shortest recorded vertebrate lifespan found in a coral reef fish. *Current Biology* 15: R288–9. Figure 1 in <http://www.cell.com/current-biology/pdf/S0960-9822(05)00,387–8.pdf>.

square growth pattern. For example, for species of surgeonfish and snapper, young fish grow rapidly until they reach maturity and then abruptly reduce somatic growth, such that their length remains stable as they continue to age (Choat and Robertson 2002). The size or age at which somatic growth slows can vary substantially amongst species within a family, and even within a species amongst different locations. For example, the growth rate of the surgeonfish *Ctenochaetus striatus* flattens out at 100 mm in parts of Papua New Guinea, is just under 140 mm on the outer Great Barrier Reef (GBR) and is closer to 180 mm on sheltered reefs closer to shore on the GBR (Choat and Robertson 2002).

6.3 Larval fish ecology

Some species of coral reef fish from families such as the damselfishes and the triggerfishes stick their eggs to the substrate and guard them much like a nest. Others still, such as cardinalfish, are mouthbrooders, whereby the male fish keep the fertilized eggs in his mouth, and even maintains the larval fish in his mouth for some time post hatching. Another example of paternal care is found in the seahorses, where the males have a brood pouch, into which the female deposits her eggs to be fertilized and protected. The male eventually gives birth to fully formed larval seahorses. Despite these diverse examples of parental care, many other species of coral reef fish spawn into the water column. For an estimated 120 species of fish, this takes the form of dramatic spawning aggregations, where individuals of a species come together in a particular location to breed, often having travelled substantial distances. These aggregations may involve tens of individuals for some species, but for many it is hundreds or even thousands of individual fish. Fertilized eggs of spawning fish drift off into the plankton, representing a bipartite life history, where the larval phase is spent in the pelagic realm, and the adult phase is spent on the reef.

The pelagic larval duration of fish is the length of time larvae of a given species spend in the plankton before settlement. On average, this may be 30 days, but there is great variability amongst species, with some only spending about a week in the plankton, whereas larvae of other species may spend as long as 100 days in the plankton. Reef fish larvae are far from passive particles. They have impressive swimming capabilities, which can be sustained for long periods, enabling them to modify their dispersal patterns beyond where the dominant currents may be flowing. These abilities increase as the larval fish develop. For example, larvae of the damselfish *Pomacentrus amboinensis* can swim at 3.5 cm s^{-1} for 0.11 hours on hatching from its egg but, by the time they are ready to settle on the reef, they can swim at 30.3 cm s^{-1}, and sustain swimming for 90 hours (Fisher et al. 2000). This translates to the larvae being able to swim 40 km by Day 20.

Larval fish also have considerable sensory abilities which can assist in locating reefs and suitable habitats to settle (Atema et al. 2015). Many species have highly developed olfactory abilities, enabling them to seek out appropriate settlement locations. For example, the anemone fish *Amphiprion melanopus* can recognize its obligate anemone host through olfaction. Using tank experiments where larvae have a choice of different tunnels to swim up, species have been shown to prefer water with the odours of conspecifics, corals and even common leaves from coastal plants. Many species also have highly developed auditory abilities. Reefs are noisy environments, from the crashing of waves, to the many animals clicking, snapping, grunting and scraping. Experiments have demonstrated that nearly four times as many larval fish will settle onto experimental reefs if sounds made by fish and shrimps are played next to the reefs, compared to controls with no sound (Simpson et al. 2005). Vision also plays a role for some species, enabling larvae to choose groups of conspecifics, or certain habitat types to settle into. The settlement process onto a reef is somewhat of a gauntlet, with many predators lying in wait to feed on the naive new arrivals.

These swimming abilities and sensory capabilities give larval fish substantial capacity to influence their eventual settlement site. Indeed, despite some larval fish travelling long distances before settling onto a reef, a surprising proportion of many reef fish larvae settle close to their parents. By marking the ear bones in embryos of the damselfish *Pomacentrus ambionensis* and then capturing returning larvae at the same reef, Jones et al. (1999) calculated that as many as 15%–60% of the new recruits were returning to their natal reef. Work in this area has moved on to using parentage analysis to genetically link larval fish back to their actual parents. This work has shown that even larger reef fish, such as coral trout (*Plectropomus* sp.), have a high proportion of larvae that settle back to natal reefs, but with a large range of dispersal distances displayed. The shortest distance from the parents recorded at settlement was <200 m, while one larval fish was found on a reef 250 km away from its parents (Williamson et al. 2016).

6.4 Reef fish diversity

There are an estimated 6,000–8,000 coral reef fish species. This diversity is unevenly distributed globally, with 500–700 in the Caribbean, and 4,000–5,000 in the Indo-Pacific region. Large-scale diversity patterns mirror those of corals, with the greatest diversity in parts of South East Asia and western Melanesia, and progressively fewer species on reefs the further you travel from this centre of diversity. At small scales, area of reef available and isolation become important. For example, the GBR in Australia hosts an estimated 1,500 species of fish, while the geographically nearby, but smaller, New Caledonia Barrier Reef contains around 1,000 species of fish.

On individual reefs, coral reef fish distribution and diversity can be affected by hydrodynamic conditions, as well as reef zonation, and depth. Different species inhabit the rough conditions on outer reef areas compared to the more sheltered reef flat or lagoon areas. Across reef zones, distinct assemblages of fish exist, often highest in diversity in the reef crest and slope areas.

While some species are adapted to have broad depth ranges, others inhabit more specific depths. This can lead to particular fish species compositions on the deep reef slope (e.g. larger snappers such as predatory *Pristipomoides* spp.) and different ones in shallower waters (e.g. some plankton-feeding damselfish such as *Chromis margaretifer*) (Jankowski et al. 2015).

Hone in further still, and the processes driving diversity over small scales can be examined. Early work suggested a lottery hypothesis, whereby a fish species got to inhabit an area of reef on a first-come, first-served basis; this hypothesis assumed that species were generalist in habitat use and competitive abilities (Sale 1977). However, much subsequent work has shown that interspecific competition amongst fish can lead to distinct resource partitioning and to fish inhabiting distinct niches. Niches can be related to essential properties, such as food or habitat space. Indeed, a great deal of work has demonstrated that fish diversity is strongly controlled by the three-dimensional structure of the reef and the diverse niches it provides (Graham and Nash 2013). Diversity of the fish assemblage is greater where the rugosity of coral reefs is greater. This is linked to the abundance of hard corals, substrate topography and the number and range of gaps within the coral that provide refuges and homes for different species (Nash et al. 2016). Greater rugosity moderates the level of predation, providing more areas for shelter, and also competitive interactions, both within and between species. Disturbances that reduce live-coral cover, such as crown-of-thorns starfish outbreaks (see 'The crown-of-thorns starfish'), can lead to a reduction of rugosity and, ultimately, reef fish diversity. The processes shaping reef fish diversity and enabling such a high diversity of species to coexist still constitute an area of active research in coral reef science.

The crown-of-thorns starfish

Crown-of-thorns starfish are generally referred to as *Acanthaster planci* (Linnaeus, 1758), based on the original description by Plancus and Gualtieri (Vine 1973). However, molecular sampling has revealed that there are at least four differentiated and geographically separated species, located in (i) the Red Sea (*Acanthaster* sp.), (ii) the Northern Indian Ocean (*Acanthaster planci*), (iii) the Southern Indian Ocean (*Acanthaster mauritiensis*), and (iv) the Pacific Ocean (*Acanthaster* cf. *solaris*) (Vogler et al. 2008; Haszprunar and Spies 2014). Formal species descriptions are still being prepared and specific differences in their biology and behaviour are yet to be explored, but all four species inhabit coral reef environments and feed almost exclusively on scleractinian corals.

Crown-of-thorns starfish (*Acanthaster* spp.) are world renowned for their capacity to devastate coral reef ecosystems (Pratchett et al. 2014) during population outbreaks (Figure 1), when local densities can increase >100-fold within 2–3 years (e.g. Chesher 1969). *Acanthaster* spp. are also one of the largest and most efficient predators on scleractinian corals.

Continued

The crown-of-thorns starfish (*Continued*)

Whereas most other coral-feeding organisms (e.g. *Chaetodon* butterflyfish and *Drupella* snails) cause only localized injuries or tissue loss (Cole et al. 2008), large crown-of-thorns starfish readily kill entire corals, including very large colonies. High densities of large crown-of-thorns starfish can therefore cause rapid and extensive coral depletion. Around Moorea in French Polynesia, for example, recent outbreaks of crown-of-thorns starfish killed >96% of corals (Kayal et al. 2012), with significant impact on the biodiversity and productivity of the reef ecosystem.

Figure 1. Outbreak of crown-of-thorns starfish (*Acanthaster* cf. *solaris*) on the Great Barrier Reef in 2014 (See Plate 11).
Source: Photo: Ciemon Caballes.

Outbreaks of *Acanthaster* spp. are often attributed to anthropogenic degradation of marine and coastal environments. The foremost hypotheses to account for outbreaks of crown-of-thorns starfish are the 'predator-removal hypothesis' (Endean 1977), which suggests that overfishing of putative predators (e.g. the giant triton and/or predatory reef fish) allows increased numbers of adult starfish to survive and reproduce, and the 'larval starvation and terrestrial run-off hypotheses' (Birkeland 1982), which argue that phytoplankton blooms induced by terrestrially derived nutrient input enable rapid development and higher survivorship of larval crown-of-thorns starfish. Explicit tests of these hypotheses yield varying results (e.g. Wolfe et al. 2015; Cowan et al. 2017), and much more research is required to understand the inherent complexities in the population dynamics of crown-of-thorns starfish.

Major population fluctuations of crown-of-thorns starfish may be entirely expected, given their life history characteristics (Uthicke et al. 2009). Most notably, *Acanthaster* spp. are

Continued

The crown-of-thorns starfish (*Continued*)

amongst the most fecund of all the starfishes (Figure 2): a single female may release >100 million eggs in a single spawning (Babcock et al. 2016). The reproductive potential of crown-of-thorns starfish is, however, conditional upon their recent feeding history (Caballes et al. 2016), highlighting an important feedback loop that may contribute to coupled oscillations in the abundance of crown-of-thorns starfish and that of the local cover of their preferred coral prey (mainly *Acropora* spp.).

Figure 2. Large volume of ovarian tissue in crown-of-thorns starfish (*Acanthaster* cf. *solaris*) on the Great Barrier Reef in November 2014.
Source: Photo: Ciemon Caballes

Regardless of the cause(s) of outbreaks of crown-of-thorns starfish, preventing such outbreaks is considered to be one of the most direct and effective mechanisms for minimizing the ongoing coral loss across the Indo-Pacific (e.g. De'ath et al. 2012). The most efficient method for culling crown-of-thorns is to inject adults and larger juveniles in situ with bile salts (Rivera-Posada et al. 2014) or an alternative and more readily available acid (e.g. Buck et al. 2016). Even so, the comprehensive removal of starfish (and effective protection of corals) following the emergence of outbreaks requires considerable and sustained control effort. The best way to prevent outbreaks is to concentrate control effort in areas where outbreaks are known to initiate (see Wooldridge and Brodie 2015), although this will require unequivocal evidence of an impending outbreak, together with timely and appropriate investment in control activities.

Morgan Pratchett and Ciemon Caballes, James Cook University, Townsville, Australia.

6.5 Reef fish: Colourful for a reason

The bright colourations of many reef fish must provide some evolutionary or ecological advantage. Research into this issue has shown that a range of reef fish can see in colour, and this suggests that their markings and colouration are likely to provide signals for others (e.g. Siebeck et al. 2008). The appearance of fish colourations is affected by the wavelengths of ambient light. Light in water varies in quality as well as quantity as depth increases, with longer wavelengths—red, orange and yellow—being absorbed more rapidly than short wavelengths of blues and greens. The colour patterns of some coral reef fish contain a UV-reflecting component that is invisible to the human eye. Some reef fish have four photoreceptors in their eyes, rather than the three in humans, with the fourth often making them sensitive to UV light. Such species may use UV wavelengths and markings for communication (Siebeck 2004). However, life is never that easy, and some fish show UV patterning without the ability to see UV wavelengths (Siebeck and Marshall 2001). UV sensitivity may also be useful for detecting zooplanktonic prey, with the zooplankton appearing dark on a bright background. Therefore, fish may be able to see UV wavelengths without displaying those colours themselves.

Fish colouration can play a range of different roles within the coral reef system. These include mate recognition and display of reproductive potential, such as with some parrotfish which become more brightly coloured as they mature (Figure 6.3). Many fish appear to use the principle of disruptive colouration, using bold bands, stripes or spots to confuse, and disruptive camouflage patterns used by such fish must be understood in this context. This may serve a number of functions, making the fish look larger than it really is, giving the impression, perhaps, that the fish is facing the opposite direction to what is really the case, with possible escape advantages, or providing a false target for predators to protect the head region, increasing chances of survival. Three uses of colouration are discussed below.

6.5.1 Poisonous and venomous fish

Many venomous (toxic to touch) and poisonous (toxic to eat) fish tend to be brightly or distinctly coloured. A classic example of a colourful, venomous fish is the lionfish (*Pterois volitans*), with its distinctive mane of fins and red-, white- and black-striped body colouration. These fish have a poison gland at the base of their dorsal spines. In the face of potential predators, these spines point forwards as protection. If the spine is pushed inwards, the pressure on the venom gland triggers toxin to shoot up the hollow spine to its tip, injecting the attacker. This fish corners its prey of small fish using its fins as sweepers, and then can jab the prey before gulping it. This Indo-Pacific fish has recently been introduced into the Caribbean, probably from aquarium releases, where it is expanding its numbers

Figure 6.3   Top: Initial phase (female) parrotfish tend to have quite drab colouration. Bottom: In contrast, terminal phase (male) parrotfish have very bright colourations, often with vivid greens, blues, yellows and reds. Different species are shown in this example.
Source: Photos by Nick Graham.

and range (Schofield 2009). It is proving a highly successful predator to naive prey fish in the Caribbean. Similarly, lionfish are an unfamiliar prey to predatory grouper in the Caribbean; there is some evidence of grouper predating on lionfish, with possibly fatal consequences for the grouper.

6.5.2 Camouflage

Camouflage can be a key mechanism for survival on coral reefs, hiding fish from predators, as well as from potential prey. The bright and varied colouration of fish may act as a direct adaptation to camouflage them against the colourful backdrop of corals around which they reside. Patterns can also be used for camouflage; good examples of this are provided by rabbitfish, which can change colour patterns, in a chameleon-like fashion, to match their background for protection, and lizardfish, which can camouflage themselves to hide in the background to assist their feeding on prey. More direct attempts at camouflage can be seen in the stonefish (*Synanceia* sp.), a carnivorous fish with highly venomous spines. This is considered the world's most venomous fish. It lives on reef bottoms, its mottled greenish-brown colour allowing it to camouflage itself amongst the rocks of tropical reefs and hence earning it its common name. Algae can even grow on its skin to increase the concealment.

6.5.3 Mimicry

Mimicry occurs when a species (the mimic) resembles a different species (the model) by duplicating colour, morphology or behaviour, to confuse other species. Batesian mimics are harmless, gaining protection from predators by evolving colours and body patterns that resemble a species that is undesirable to eat. Aggressive mimics are predators that get close to and attack prey by looking like a species that their prey perceives as harmless. A combined example is the blenny *Aspidontus taeniatus*. This fish obtains food (pieces of fins and epidermis of larger fish) by looking and behaving like the harmless cleaner wrasse *Labroides dimidiatus* (Randall and Randall 1960). The approach also reduces predation because the cleaner wrasse itself does not get predated due to the cleaning activities it performs. Studies have identified up to 60 different species as being mimics (Moland et al. 2005).

6.6 Abundance, biomass and trophic structure

Abundance and biomass are two common metrics of reef fish assemblages that give you complementary yet different information. Abundance, often captured as density, is the number of individuals in a given area, while biomass is the mass of those individuals. Biomass for reef fish is commonly calculated using established length–weight relationships, which differ amongst species based on their growth forms. Larger fish obviously contribute disproportionately to biomass, such that few large fish may give a much higher biomass value than abundant small fish. At small spatial scales on reefs, the abundance and biomass of fish is driven by zonation, with more fish on reef slopes than reef flats, and by structural complexity, with more fish

Plate 1 The richest part of any coral reef lies below the depths where waves break, where light is abundant for photosynthesis, sedimentation is low, the salinity is near that of the open ocean (about 31–34 ppt) and the temperature is about 20–29 °C. (See Preface to 2nd edition).

Plate 2 Caribbean reefs have a relatively low number of reef-building coral species, yet the reefs and limestone banks they have produced are enormous (British Virgin Islands) (See page 17).

Plate 3   An assemblage of many species of typical Caribbean soft corals. These do not help to build a reef, although they may support a rich diversity. In such areas, stony corals may struggle to survive (See page 52).

Plate 4   Representatives of three major invertebrate groups with high diversity on coral reefs: echinoderms, represented by the brilliantly coloured starfish *Protoreaster lincki*; crustaceans, represented by a mantis shrimp (Stomatopoda); and molluscs, represented here by the predatory cone shell *Conus geographus*. This is a piscivorous cone and so has a very strong and fast-acting toxin (See page 57).

Plate 5   In marginal areas, corals live amongst macroalgae. In this case, *Acropora* tables are seen in southern Oman amongst the macroalgae *Sargassum* at the top and (foreground) the more kelp-like *Nizamuddinia zanardinii*, which is a regional endemic (See page 86).

Source: Photo by Rob Baldwin.

Plate 6   Coral bleaching. The usual brown colouration of this *Acropora* coral (left) is lost under environmental stress when the zooxanthellae (inset) are expelled from its tissue, leaving the white coral skeleton visible through the translucent tissues (right) (See page 122).

Source: Courtesy of O. Hoegh-Guldberg.

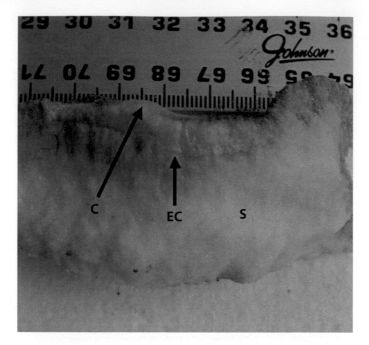

Plate 7   Cross section through the skeleton of an Indo-Pacific coral (*Porites* sp.), showing the endolithic community just below the surface. Such communities may contain a variety of microorganisms, including filamentous green algae, red algae, fungi, heterotrophic bacteria and cyanobacteria. The endolithic community forms distinct horizontal bands within the skeleton. In this specimen, the endolithic band is especially pronounced midway across the image, where the overlying coral tissue has been damaged, so allowing more light through to photosynthetic members of the endolithic community (e.g. the filamentous green alga *Ostreobium quekettii*); EC, endolithic community; S, skeleton; C, coral tissue on surface. Scale is in centimetres (See page 127).

Plate 8   The benthic filamentous cyanobacterium, *Lyngbya majuscula*. Elevated levels of usually limiting nutrients are a major cause of blooms such as the one shown here, where the cyanobacterium is overgrowing corals. This bloom was near Great Keppel Island, Queensland, Australia (See page 139).

Plate 9   Filter-feeding. Left: A crinoid on the Great Barrier Reef, emerged at dusk, when it fishes for zooplankton. Crinoids comprise an ancient class of echinoderms. The crinoid's arms basically form filter-feeding nets which catch swimming zooplankton, which are abundant at night. Trapped plankton are then swept down into the mouth at the base. Right: A basket star. This is also an echinoderm, but is a greatly modified brittle star. Basket stars have arms which are repeatedly branching, forming an elaborately tangled mass. This is the golden-coloured Caribbean *Astrophyton muricatum*, which spends the day coiled tightly, in this case, in a branching soft coral. At night it uncoils, and orientates itself into the current so its web of arms can catch plankton (See page 162).

Plate 10   Feeding groups of coral reef fish. Top left: Plankton-feeding fusiliers. Top right: A facultative coral-feeding butterflyfish. Bottom left: An invertebrate-feeding wrasse. Bottom right: A piscivorous reef shark (See page 185).

Source: Photos by Nick Graham.

Plate 11   Outbreak of crown-of-thorns starfish (*Acanthaster* cf. *solaris*) on the Great Barrier Reef in 2014 (See page 174).

Source: Photo: Ciemon Caballes.

Plate 12   Indo-Pacific representatives of the four primary herbivorous fish functional groups. (A) The excavating parrotfish *Bolbometopon muricatum*. (B) The scraping parrotfish *Scarus rubroviolaceus*. (C) The cropping rabbitfish *Siganus puellus*. (D) The macroalgal-browsing *Naso unicornis* (See page 184)

Source: Photos by Andrew Hoey.

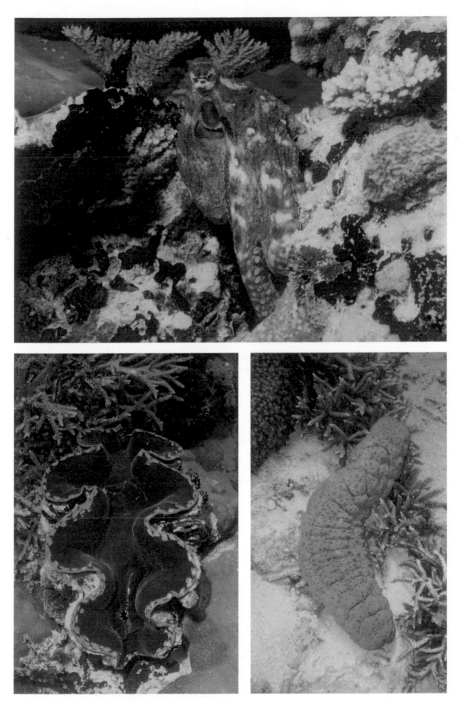

Plate 13 Important invertebrate reef resources: Top: Octopus. Bottom left: The large bivalve *Tridacna*. Bottom right: A sea cucumber (holothurian). (See page 202).

Plate 14   Diseased corals. Left: A recently killed elkhorn coral (*Acropora palmata*), (British Virgin Islands). Right: A *Diploria* brain coral with black band disease. The band is visible slanting diagonally across the coral, the top half is dead and the band is creeping downwards, destroying the living polyps as it goes (Bermuda) (See page 245).

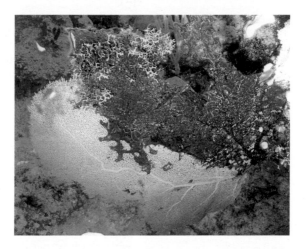

Plate 15   The very abundant Caribbean sea fan *Gorgonia ventalina*. The bottom part is living, but the dark part of the fan is dead, decaying and covered with a red fungus (See page 248).

Plate 16   Terminal at Yanbu, Saudi Arabia, crossing the Red Sea fringing reef. Shortly after the terminal was constructed (in the late 1970s and early 1980s), coral diversity and cover were still very high in this area, in part due to the huge quantity of healthy corals on the otherwise untouched Red Sea reef (See page 286).

supported by a structurally complex reef. At larger spatial scales, the productivity of the environment can come into play (Nadon et al. 2012). However, humans are also major drivers of reef fish abundance and biomass, particularly through fishing, as discussed in Section 6.8 and Chapter 7.

One way of examining abundance and biomass patterns in reef fish is through the distribution amongst trophic levels. Trophic levels reflect the position in the food web that an organism feeds; for example, algal-feeding fish are at the bottom trophic level, as they feed on primary produced algae, whereas predatory fish such as sharks are at the top trophic level, as they feed on other fish, which have in turn fed on organisms and algae further down the food web. The distribution of abundance or biomass amongst trophic levels can be useful in capturing the stocks of energy in a food web. The classic expectation of such a distribution is in the form of a pyramid, with a large base of low-trophic-level organisms, fewer organisms at each level as the trophic level increases and the fewest organisms at the top (Figure 6.4). This expected pyramid distribution is logical, as available energy in the system is lost through metabolism and other processes through each trophic step. Coral reef food chains were thought to follow this pyramid structure; however, early work was based on studies of reefs affected by various anthropogenic factors, such as overfishing and pollution.

More recently, studies on remote coral reefs in the Pacific have suggested that unfished coral reefs may have a substantially greater proportion of top-level predators than anticipated. On Kingman Reef in the Line Islands (Central Pacific), predators such as sharks accounted for a massive 85% of the total fish biomass, while on other protected Pacific reefs, large apex predators such as groupers and snappers represented 56% of the biomass, compared to less than 10% on fished reefs (Stevenson et al. 2007). This high proportion of top predators was comparable to that found at the remote and uninhabited Northwestern Hawaiian Islands (Friedlander and DeMartini 2002). These

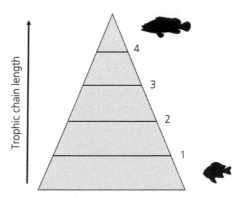

Figure 6.4   The coral reef fish trophic pyramid.

patterns suggest inverted trophic pyramids, being 'wider' at the top than the bottom. An inverted food web is difficult to explain based upon energetics. However, rates of turnover and not just biomass must be considered; a more rapid turnover of species at lower trophic levels, and longer lifespans of top predators accumulating biomass could help to support these patterns. However, the sampling of fish in a discrete area by divers is likely biasing the findings substantially, because top predators often forage over large areas, sourcing food and energy from non-reef sources, and numbers of large mobile predators can be inflated through attraction to divers in remote locations (Bradley et al. 2017).

Studies in the Pacific and Indian Oceans have also shown a high herbivore biomass, driven by abundant large-bodied parrotfish, suggesting that, despite high numbers of top predators, herbivores can still flourish (Graham et al. 2017). This could be due to low direct predation rates by top predators when herbivores are large, to prey release where apex predators actually feed on those species that consume herbivores directly or merely to reduced exploitation at all trophic levels. There is a marked contrast on reefs that are fished, with a substantial reduction in biomass of most trophic levels, and a near loss of upper-trophic-level fish. Pyramids become more bottom-heavy in structure, until very heavy fishing, where the bottom of the pyramid can be depleted due to the relatively large herbivorous surgeonfish and parrotfish being removed. In these situations, herbivorous sea urchins often increase in abundance to replace the algal feeding role previously played by the fish (Graham et al. 2017).

6.7 Feeding and ecosystem function

Reef fish have a wide diversity of feeding modes, reflecting the multiple energy pathways available in these complex food webs. Fish are a key component of this coral reef food web, ensuring the transfer of energy through the system. At the top of the chain, fish feed on other fish, influencing community structure and abundance of smaller species. Further down the food chain, fish graze on available algae, controlling the growth of macroalgae and preventing the smothering of coral, while others can have direct impacts on the structure and function of corals themselves. In Section 6.7.1, an overview of the common feeding guilds will be given, followed by an examination of the functions reef fish play in the ecosystem and how scientists are capturing this information.

6.7.1 Detritivores

A particularly notable component of the benthos is algal turf, which is a matrix that grows on most surfaces that are not occupied by larger organisms such as corals. The quantity of detritus, settling organic material and fish faeces in this surface turf can be between 10% and 78% of the total quantity of the organic matter within the turf, providing a greater nutritional value than the algae itself. At least 26 fish species from five families focus on detritus for nutrition, with a further 30 or so where detritus is an important part of their diet. These fish account for at least 20% of individuals and 40% of the biomass of those fish feeding on the algal matrix on the GBR, including species of the Scaridae and the Acanthuridae. Many of these species selectively feed on detritus within this matrix, particularly small, organic-rich particles <125 μm, as these can provide a nutritional value equal to or greater than that of filamentous algae. Species such as the bristletooth surgeonfish from the genus *Ctenochaetus* have a specialized, long feeding apparatus that combs the algal turf, removing the detrital matter (Figure 6.5; Bellwood et al. 2014). The ingestion and assimilation of detritus by these fish species represents a significant and important pathway for transferring energy from organic matter within sediments and from rocky reef substrate to secondary consumers. Given the importance of detritus for carbon and nitrogen production within the generally impoverished waters surrounding coral reefs, the role of detritivorous fish is seen to be a critically important component of coral reef trophodynamics (Wilson et al. 2003).

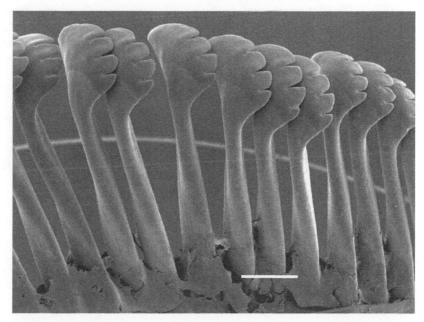

Figure 6.5   Long, comb-like teeth of the surgeonfish *Ctenochaetus striatus*. Scale bar represents 500 μm.
Source: Bellwood, D.R., Hoey, A.S., Bellwood, O. and Goatley, C.H.R. (2014). Evolution of long-toothed fishes and the changing nature of fish-benthos interactions on coral reefs. *Nature Communications* 5:3144. Figure 2a from <http://www.nature.com/articles/ncomms4144>.

6.7.2 Herbivores

Herbivorous fish can be highly prolific, and contribute considerable biomass to the reef system. They are commonly found in shallower waters, under the brighter conditions required for their algal food to grow. Adapted to their feeding habits, they are commonly narrow-bodied, with mobile fins that allow them to manoeuvre accurately in order to obtain their food. Herbivorous fish can feed on a variety of food sources, from turfing algae to fleshy macroalgae that may be more common on inshore reefs. They range from generalists, which can feed on a variety of algal forms, to specialists adapted to specific algal types (see 'Functional group approaches to herbivory on coral reefs'). This specialism is reflected in their morphology. Some herbivores such as parrotfish may have beak-like mouthparts which scrape at turfing algae. Their strong mouthparts and fused teeth (which form the beak) adapt some of the parrotfishes, termed excavators, to a highly calcareous diet (Choat et al. 2004). In these, up to 75% of the stomach contents consists of inorganic calcareous material which is subsequently excreted; adult parrotfish can excrete over 1 tonne of this material a year, adding considerably to the coral reef fine sediments. Surgeonfish, in contrast, have rows of teeth than can bite at algae; for this food source, they do not need such strong jaws. Specialism may also be reflected in behaviour. Some herbivorous damselfish vigorously maintain and defend areas for grazing specific algae, repelling from these claimed territories all competitors (including diving humans) even when they are very much larger than the damselfish itself. This activity modifies the benthic community. Palatable algae have been shown to increase in abundance within territories defended by damselfish, compared to quantities outside those territories (Ceccarelli 2007).

Functional group approaches to herbivory on coral reefs

The ongoing degradation of many of the world's ecosystems, coupled with widespread reductions in biodiversity, has led to an increased focus on the relationships between biodiversity and ecosystem function and, importantly, on identifying and understanding the ecological roles of individual species. Under such scenarios, species are not viewed in terms of their taxonomic labels or phylogenetic relationships, but rather in terms of their 'function' within an ecosystem (Bellwood et al. 2004). An ecosystem function may be defined as the movement or storage of energy through trophic and/or bioconstructional pathways, and a functional group as a collection of organisms that primarily shape or modify the extent of a given function (following Done et al. 1996). Functional approaches to ecology allow not only the contribution of a species (or group of species) to an

Continued

Functional group approaches to herbivory on coral reefs (*Continued*)

ecological process/es but also the implications of population reductions or loss on ecosystem functioning to be quantified (Bellwood et al. 2012).

It is widely accepted that large, mobile (or roving), herbivorous fish perform a key ecosystem function in coral reef ecosystems: the removal of algal biomass. Indeed, reductions in herbivorous fish through fishing or experimental exclusion (i.e. cages) and the subsequent expansion of macroalgal biomass highlight the importance of the group to reef processes (Hughes et al. 2007; Graham et al. 2015). There is, however, considerable variation in the morphology, diet and feeding behaviour amongst herbivorous taxa and, hence, in their contribution to reef processes. The standing algal biomass on reefs is, therefore, dependent on the overall abundance and the functional composition of herbivorous fish communities (Rasher et al. 2013). It is important to note that feeding by any organism is driven not by their contribution to ecosystem processes, but rather their need to acquire sufficient nutrients for maintenance, growth and reproduction. Despite this, several generalities can be drawn based on where and how species feed.

Herbivorous fish may be broadly classified into two groups: grazers and (macroalgal) browsers, based on the algal surfaces from which they feed. The terminology used to describe these groups has been adopted from terrestrial ecosystems and, although useful, does not direct translate to marine systems. The term 'grazing fish' refers to species that feed on reef surfaces covered by algal turfs or an epilithic algal matrix (a conglomerate of filamentous algae, macroalgal propagules, crustose coralline algae, detritus, microbes and sediment; Wilson et al. 2003); such fish are analogous to terrestrial grazers that feed on grass. In contrast, the term 'browsing fish', although often used to refer to species that feed on any type of macroalgae, should be reserved for those species that feed on tall, leathery macroalgae such as *Sargassum* spp. and *Turbinaria* spp. This distinction is important, as numerous recent studies have shown that many grazing fish will feed on smaller branching and foliose macroalgae but have limited capacity to consume larger, leathery macroalgae (Hoey and Bellwood 2009).

Grazing fish may be further divided into croppers, scrapers and excavators, based on the amount of the underlying substratum that is removed during feeding (see Figure). Algal croppers (many acanthurids, siganids, non-browsing kyphosids and some sparisomatine parrotfishes) remove the upper portions of the algae and associated epiphytic material when feeding, leaving the holdfast and basal portions intact, whereas scrapers and excavators remove parts of the underlying reef substratum together with the algae and associated material. Scraping parrotfish (*Scarus, Hipposcarus* and some *Sparisoma* spp.) take relatively shallow bites (<1 mm), thereby clearing space for the settlement of benthic organisms. Excavating parrotfish (*Bolbometopon, Cetoscarus, Chlorurus* and *Sparisoma viride/ampulum*) take deeper bites, remove a greater volume of carbonates and are major contributors to the external erosion of the reef framework (Bellwood et al. 2012).

Continued

Functional group approaches to herbivory on coral reefs (*Continued*)

Figure. Indo-Pacific representatives of the four primary herbivorous fish functional groups. (A) The excavating parrotfish *Bolbometopon muricatum*. (B) The scraping parrotfish *Scarus rubroviolaceus*. (C) The cropping rabbitfish *Siganus puellus*. (D) The macroalgal-browsing *Naso unicornis* (See Plate 12). *Source:* Photos by Andrew Hoey.

These broad functional groupings, while far from comprehensive, have provided a useful construct for examining various aspects of herbivory; however, like all reductionist approaches, they are far from comprehensive. Considerable inter- and intraspecific variation is evident within these groups. For example, it is estimated that a single bolbometopon can remove over 5.5 tonnes of reef carbonates per year, while other excavating parrotfish can remove 0.02–1.02 tonnes (Hoey and Bellwood 2008). Further distinctions are evident based on ontogeny, body size (Lokrantz et al. 2008), the macroalgal species or components targeted (Streit et al. 2015) and/or the microhabitat in which species feed (Fox and Bellwood 2013).

Andrew S. Hoey, ARC Centre of Excellence for Coral Reef Studies, James Cook University, Townsville, Australia

6.7.3 Planktivores

Reef structures lead to interactions with currents that result in hydrological patterns that concentrate nutrients, food particulates and crustacean plankton above the reef (Wolanski and Hamner 1988). This attracts many planktivorous species, from whale sharks (*Rhincodon typus*) and large manta rays, whose size allows them to feed during the day without fear of predation, to planktivorous coral reef fish, such as caesionids (fusiliers) (Figure 6.6, top left), which may school for protection from predators, and abundant damselfish of the genus *Chromis*. Some planktivorous fish species adopt a nocturnal strategy, such as the soldierfishes, whose large eyes help them identify suitable prey, which consists of relatively large zooplankton. While the nocturnal feeders may achieve a lower rate of feeding success, the nocturnal release of gametes and larvae by fish and corals, the occurrence of larval recruitment to the reef during the night and the nocturnal ascent of demersal zooplankton species into the water column (Chapter 5) help to compensate. The switch from feeding to non-feeding behaviour in day- or night-feeding planktivores appears to occur at species-specific light intensities, and represents a careful trade-off between the ability to identify and successfully feed on prey, and the risk of falling victim to predation. Most major coral-reef-related families of fish contain one or a few species that specialize in planktivorous feeding.

6.7.4 Corallivores

Corallivores, as the name suggests, feed on live coral. There are an estimated 128 corallivorous fish species from 11 families, with 68 of these species belonging to the butterflyfishes (family Chaetodontidae; Figure 6.6, top right;

Figure 6.6   Feeding groups of coral reef fish. Top left: Plankton-feeding fusiliers. Top right: A facultative coral-feeding butterflyfish. Bottom left: An invertebrate-feeding wrasse. Bottom right: A piscivorous reef shark (See Plate 10).
Source: Photos by Nick Graham.

Cole et al. 2008). Corallivores can be facultative or obligate feeders. Facultative (opportunistic) corallivores have diets including corals, but also including a range of other items, such as polychaetes, sponges and algae. Obligate corallivores feed almost exclusively on coral (representing over 80% of their diet), feeding on coral mucus and coral polyps. While obligate corallivores are found in the Indo-Pacific, none have been found in the Caribbean (Cole et al. 2008). Obligate corallivores can have very broad diets, feeding on many species of corals, or be highly specialized and only feed on several species of (often *Acropora*) corals. The loss of live corals can have a severe impact on corallivorous species, in particular on those fish species with specialized obligate diets (e.g. *Chaetodon trifascialis*, which concentrates on *Acropora hyacinthus*).

6.7.5 Invertebrate feeders

Fish can feed on a variety of invertebrates, from those found in sand at the edge of reefs, to those feeding on corals themselves (Figure 6.6). The invertebrates found in sand are a rich source of food for many very differently shaped fish, some of which scrape from the surface, while others are able to extract prey more deeply buried. Fish that feed on buried prey may have large mouths, using suction to draw in large volumes of water and, hopefully, their prey. Other species, such as goatfish (family Mullidae), use barbels beneath their chin to detect prey by touch. Many invertebrate prey inhabit the complex reef matrix, particularly dead reef structures. Fish need to have the ability to precisely control their swimming movements to reach in and pull prey from inaccessible areas. This leads to body characteristics including good close-up vision and, often, beak-like mouths, strong jaws and pharyngeal teeth which are located in the throat and which are used to grind up prey. The searching predatory habit of invertebrate feeders means there are advantages to foraging alone. However, a lack of schooling conspecifics means they need alternative means of protection from predators. A range of protective mechanisms have therefore evolved. For example, porcupinefish (Diodontidae), as the common name implies, are covered with protective spikes. Adaptations may also be behavioural. For example, fish that feed on invertebrates off the reef, and hence away from the protection that the reef topography provides, do so at night, to reduce the risk of predation.

6.7.6 Piscivores

Higher up the food chain are those fish that feed on other fish. These species can range from small species of hawkfish that ambush juvenile fish, to large groupers and sharks (Figure 6.6). Reef piscivores also include species such as trevally and barracuda, which are capable of feeding in open water, but are found in high numbers close to coral reefs. Piscivores can feed on all types of fish down the food web, meaning coral reef fish food webs are not simple linear chains, but complex convoluted systems of energy exchange (Graham et al. 2017). Different piscivores feed during the day or night, dependent upon the species. During the day, identification of prey often requires binocular vision to assist in range assessment of their food items. Piscivores therefore often have larger eyes than herbivores, likely to be a function of the increased demands on the visual system necessary to their mode of feeding. Feeding techniques can vary widely, from chasing down prey to sit-and-wait or stalking-and-ambush strategies. Larger groupers, for example, ambush prey. They wait for passing prey, swallowing them whole using the powerful suction that is generated by rapid expansion of their mouth and gills. To succeed in the ambush approach, it helps to be well camouflaged. Stalking predators manoeuvre slowly into position, waiting for prey to approach close enough to strike. Their bodies are often thin and long, reducing the area visible to their prey, while their tails are relatively large, allowing rapid acceleration for the strike. For example, trumpetfish (*Aulostomus maculatus*) orient themselves vertically amongst corals, gaining a level of camouflage, and strike out at passing prey, expanding their jaws to suck in fish.

While this section might appear to pigeonhole species into particular feeding niches, in many cases a species does not conform to a single trophic preference. For example, many piscivores supplement their diet by feeding on invertebrates. It is also of note that taxonomy is not often a good indication of diet. For example, different species of the Labridae (wrasses and parrotfish) can feed on other fish, plankton, invertebrates, algae, coral and many other things besides, and adults range in body weight by four orders of magnitude. Fish size is also not always a good guide to diet, with piscivores ranging in length from less than 10 cm to over 2 m.

6.7.7 Ecosystem function

Coral reef fish feeding groups can be linked to key ecosystem functions that regulate the wider ecosystem. For example, the orange-lined triggerfish (*Balistapus undulatus*) feeds on sea urchins. If the abundance of the triggerfish falls too low on East African reefs, the sea urchins can increase their population densities to very high numbers, with detrimental impacts through bioerosion of reef structure (McClanahan 2000). Herbivorous fish play a critical role in controlling algal growth state and abundance, and enabling corals to outcompete algae on reefs. Some herbivorous fish, such as many species of parrotfish and surgeonfish, maintain cropped turf algae enabling coral settlement and growth. A different suite of herbivores, such as species of rabbitfish, and some surgeonfishes of the genus *Naso*, feed on mature macroalgae and are essential in helping to reduce it once it has outcompeted corals. The overfishing of herbivores (see Chapter 7) can therefore lead to increases in algal cover, causing shifts towards states where corals are less abundant (e.g. Mumby, Dahlgren, et al. 2006).

In reality, things are even more complex than this. Most reef fish are likely to play some role in the ecosystem, and grouping fish by broad feeding groups does not reflect how fish interact to have different roles in the system. For example, seemingly similar species of parrotfish may influence the reef differently by feeding in different locations of the reef, at different times of the day or in different foraging patterns. As such, species are likely to complement each other in delivering ecosystem functions, rather than replicating each other.

The recognition that species of fish may have quite distinct and important roles has led to attempts to characterize species by their traits. Traits can reflect things like diet (as outlined in this section), behaviour or morphology, which all reflect different aspects of the function performed by a species. Body size, for example, is tightly linked to foraging and home-range size in fish, capturing the scale over which a species performs a function in the reef (Nash et al. 2015). Body size can also reflect the magnitude of impact through diet. For example, as parrotfish grow, they take larger and larger bites, such that the area of coral reef grazed increases exponentially with body size in parrotfish. Similarly, gape size in groupers increases with body size, directly influencing the size of prey they can target.

Morphology is another trait that can be informative for ecosystem function. Mouth structure, size and shape are often tightly linked to the specific diet fish have. For example, some rabbitfish have very specialized extended mouths that enable them to feed on algae growing in reef cracks and crevices that many other reef fish cannot access. Fin shape can indicate swimming speed and the ability of fish to inhabit certain flow regimes. Species with tapered pectoral fins are able to attain faster swimming speeds and cope with higher energy environments (Fulton et al. 2005).

Using a combination of various traits of reef fish, ecologists are now able to address questions of how diversity in reef fish links to ecosystem function in more nuanced ways. These approaches capture the functional diversity of the fish assemblages. Recent research has shown how this functional diversity can influence recovery rates on reefs following disturbances such as coral bleaching. More concerning, these approaches have highlighted the high number of potentially unique ecosystem functions that are performed by only one species of fish (Mouillot et al. 2014). This suggests that maintaining a high overall diversity of fish on reefs, along with sufficient biomass, is likely to be critical to maintaining ecosystem functioning and enabling reefs to cope and rebound from disturbance.

6.8 Disturbances and coral reef fish

Disturbances influencing coral reef fish assemblages can be broadly categorized as bottom–up or top–down. Bottom–up impacts influence reef fish through benthic change; that is a change in reef fish habitat. A wide range of disturbances, including storms, crown-of-thorns starfish outbreaks, nutrient run-off and coral bleaching events, cause changes in the reef benthos, with ramifications for reef fish. Top–down disturbance occurs primarily through fishing, which is a pervasive impact on most coral reefs. In this section, we first examine the different pathways by which bottom–up disturbances influence reef fish, and the timescales on which the effects of the disturbances take place. We then discuss fishing impacts on reef fish assemblages. Finally, we evaluate the combined influence of bottom–up and top–down disturbances.

Bottom–up disturbances can be termed biological, or physical. Biological disturbances, such as coral bleaching events or crown-of-thorns starfish outbreaks, cause the loss of live-coral cover, but the physical, three-dimensional matrix of the coral reef remains intact for some time following the disturbance. Physical disturbances, such as tropical storms, cause both a loss in live coral and a reduction in the three-dimensional structure of the reef. The responses of reef fish assemblages to these different categories of disturbances are quite different.

In the short term, biological disturbances that lead to the loss of live-coral cover principally influence reef fish that are specialized in their use of coral, through diet, habitat use or settlement preferences. Around 10% of fish may feed directly on live coral, although the level of dependence can vary greatly, with some species being only facultative feeders, while others will only eat coral. Following disturbances that cause a loss of live coral, the physiological condition of these fish can decline, and substantial declines in abundance have been reported, typically most severe for more specialized feeders (Pratchett et al. 2006; Graham 2007; see Figure 6.7). Live-coral habitat dwellers are similarly vulnerable to coral loss. Various species dwell in live coral, with perhaps the gobies being a classic group. Some species of the coral gobies (*Gobiodon* spp.)

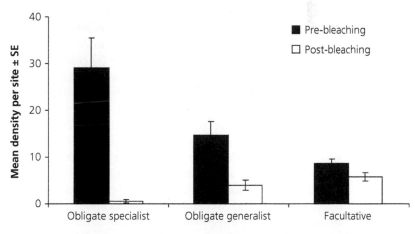

Figure 6.7   Coral-feeding fish declines following mass coral mortality in the Seychelles. The facultative coral-feeding fish showed small declines, whereas obligate feeders declined in much larger numbers, especially for those specialized on only several species of coral.
Source: Graham, N.A.J. (2007). Ecological versatility and the decline of coral feeding fishes following climate driven coral mortality. *Marine Biology* 153:119–27. Figure 5 (bottom panel) from <http://link.springer.com/article/10.1007/s00227-007-0786-x>.

live, feed and reproduce within live corals, and often remain within a single coral host for the duration of their lives, such that the abundance of that coral defines the population size (Pratchett et al. 2008). Again, species vary in their level of specialization, reflected in the diversity of corals they are observed inhabiting. Following live-coral loss, declines in the abundance of these live-coral-dwelling fish is greatest for species that inhabit only a few species of coral (Munday 2004). Another type of specialization that renders fish species vulnerable to live-coral loss occurs at settlement from the plankton. Larval fish use various auditory and olfactory ques to locate suitable settlement substrate, and a study by Jones et al. (2004) demonstrated that 65% of reef fish in Papua New Guinea preferentially settle into live-coral substrates. This pattern correlated to changes in abundance of these species following major coral loss from crown-of-thorns starfish and sedimentation. Some other groups, most notably larger-bodied herbivores such as parrotfish, can increase in abundance and biomass following coral loss, presumably due to increased algal resources on which they feed, promoting faster growth and survivorship.

In the long term, if live-coral cover does not recover, the three-dimensional structure of the dead reef matrix begins to erode. This results in effects that are similar to those documented following physical disturbances such as storms which cause a loss of live coral and reef structure. A synthesis of studies that documented changes in the abundance of fish species following coral loss showed that, on average, over 60% of fish species declined in abundance following a loss of 10% or more in coral cover (Wilson et al. 2006). However, these declines were far greater when the physical matrix of the reef was damaged in conjunction with a loss in live coral.

Once the physical structure of the reef erodes, the many niches and refuge holes it provides are lost. This structure is incredibly important for mediating competition amongst fish, and enabling prey to avoid predators. As such, the impacts on the reef fish community become far greater once the structure is lost, with declines in a wide range of species. These declines are most acute for smaller bodied reef fish species, dependent on the reef to avoid predation. However, many species of larger reef fish rely on the reef matrix as juveniles. This results in the size structure of the reef fish assemblage being fundamentally altered, with a decline in the abundance of smaller fish, including smaller cohorts of the larger fish (Graham et al. 2007). The implications of this are substantial, and suggest a lag effect in the full impacts of habitat loss on reef fish. As the long-lived adult fish die of natural mortality, or are caught by fishers, there are fewer and fewer smaller cohorts to grow up and replace them, meaning the ecosystem functions and fisheries potential of those fish species will decline. Further, the decline in smaller fish leads to a reduction in the condition and abundance of piscivores, as the types and availability of prey fish changes (Wilson et al. 2008).

The temporal dynamics over which these changes occur have been investigated in some detail (Figure 6.8; Pratchett et al. 2008). Live-coral-dependent

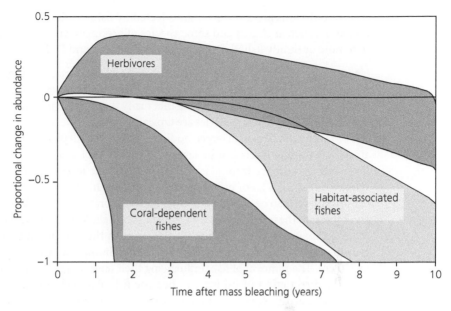

Figure 6.8   Diagrammatical representation of changes in communities of coral reef fish species following mass coral mortality.
Source: Adapted from Pratchett, M.S., Munda, P.L., Wilson, S.K., Graham, N.A.J., Cinner, J.E., Bellwood, D.R., et al. (2008). Effects of climate-induced coral bleaching on coral-reef fishes: Ecological and economic consequences. *Oceanography and Marine Biology* 46:251–96.

fish can begin to decline in the first year following coral loss, with more and more species declining depending on their level of specialization. The reef structure typically begins to erode several years after a biological disturbance, and so reef habitat or structure associated fish begin to decline in abundance after this. Finally, species of herbivores or piscivores that may have benefitted from a food bounty soon after a disturbance, take the longest to respond, associated with declines in smaller cohorts within species not surviving to adulthood, or the eventual decline in prey for the piscivores. Of course, should the reef benthos recover, and live corals flourish again building three-dimensional structure, the reef fish assemblages will respond. Such a dynamic occurred on some reefs of the inner Seychelles, where severe bleaching caused the loss of ~90% live-coral cover in 1998. Reef fish assemblages changed dramatically following this live-coral loss, but on half of these reefs where coral cover rebounded 10–15 years later, the fish assemblages began to reassemble to the composition that was present prior to the disturbance (Graham et al. 2015).

Top–down disturbance through fishing influences the fish community in quite a different way. As will be discussed in more detail in Chapter 7, fishing tends to impact top predators first, with greatest impacts on larger-bodied species. These larger-bodied fish, which are often higher up the food chain, tend to produce fewer young, be slower-growing and mature later in life to attain greater maximum ages. Consequently, the sizes of their populations are more strongly affected by fishing. As such, large reef predators, such as sharks and some species of groupers, are only abundant in remote or lightly fished reef systems (Friedlander and DeMartini 2002).

Under more heavy fishing, an increasing array of fish is targeted, by gears such as spears, hooks, traps and nets. Indeed, fishing on coral reefs is typically multigear and multispecies, meaning that many species are targeted (McClanahan et al. 2008). These fisheries are typically size selective, rather than species or taxon selective, with larger-bodied species being depleted first. As fish at various trophic levels can be large-bodied, the typical fishing down the food web from upper trophic levels to lower trophic levels is not as strong (Graham et al. 2017). For example, many species of herbivores at the bottom of the food web can be as large as species of invertebrate-feeding fish, or piscivores towards the top of the food web. Examples include the herbivorous parrotfishes, many of which grow to 50–80 cm, with the bumphead parrotfish growing to 130 cm, and the invertebrate-feeding humphead wrasse growing to 230 cm. The removal of fish, influencing their abundance, can be thought of as the direct effect of fishing but, because fish play such important roles in coral reef ecosystems, their removal can also have indirect consequences.

Studies and simulations which look along gradients of fishing have shown considerable changes in the structure of coral reef systems (see 'Coral reef models: Hasty conservation action beats procrastination to enhance reef resilience'). For example, Dulvy, Freckleton, et al. (2004) noted a decline in predatory fish by 61%, and increases in crown-of-thorns starfish densities by three orders of magnitude on Fijian reefs across a range of fishing pressure. This was mirrored by declines in reef-building corals and coralline algae by 35%, as a result of starfish predation, with corals subsequently being replaced by filamentous algae. A replacement of herbivorous fish as the dominant grazers by echinoderms occurred in the Caribbean, where removal of herbivores through fishing allowed *Diadema* sea urchins to become the dominant grazer. *Diadema* densities became so great that a disease took hold, causing mass mortality and population crashes. With few fish herbivores remaining to compensate, a number of reefs transitioned from coral cover to fleshy algal cover (Hughes 1994).

Coral reef models: Hasty conservation action beats procrastination to enhance reef resilience

Models of coral reef ecosystems tell us something about the need for urgency in reef management. The model illustrated here uses a spatial simulation to represent the demographic and ecological processes influencing corals and algae on a Caribbean forereef (Mumby et al. 2007). This model can be used to explore whether coral cover will tend to decrease, increase (recover) or remain steady between major disturbance events such as hurricanes.

Continued

Coral reef models: Hasty conservation action beats procrastination to enhance reef resilience (*Continued*)

Clearly, managers wish to avoid a situation where the natural trajectory of the reef is one of coral decline, so the model can help identify the circumstances that avoid this.

To start with, a wide range of starting conditions, in which coral cover and the amount of grazing are altered, are used. Grazing can be carried out by the urchin *Diadema antillarum* and/or parrotfish (Figure 1). Reefs are then simulated for 50 years without any acute disturbance, although chronic sources of coral mortality such as corallivory do take place. While this is an artificial situation, such simulations allow us to match the current state of a reef to the expected direction of coral trajectory (Figure 2). The open squares in Figure 2 represent unstable equilibria that join upper and lower stable equilibria. Reefs above and to the right of an unstable equilibrium follow a trajectory towards a stable equilibrium at high coral cover, whereas those below and left of the line decline to a stable, coral-depauperate state, which is usually dominated by macroalgae. Multiple equilibria occur because of ecological feedbacks (Mumby and Steneck 2008): for example, a decline in coral cover liberates new space for algal colonization. Once maximum levels of grazing have been reached, further increases in grazable area, such as occur during mass coral mortality, reduce the mean intensity of grazing and increase the probability that a patch of macroalgae will be established, ungrazed, from the algal turf. The resulting rise in macroalgal cover reduces the availability of coral settlement space and increases the frequency and intensity of coral–algal interactions, thereby reducing coral recruitment, reducing the growth rate of corals and causing limited mortality (Nugues and Bak 2006). The resulting increase in coral mortality further reduces the intensity of grazing, thereby reinforcing the increase in macroalgae.

Figure 1. The parrotfish *Sparisoma viride*, one of the most important consumers of macroalgae on Caribbean forereefs.
Source: Photo credit: Bob Steneck.

Continued

Coral reef models: Hasty conservation action beats procrastination to enhance reef resilience (*Continued*)

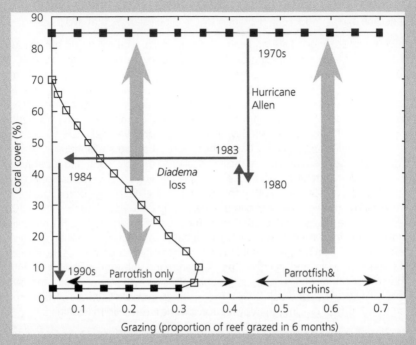

Figure 2. Stable and unstable equilibria for Caribbean forereefs with modest algal growth. Stable and unstable equilibria are denoted (■) and (□), respectively. Underlying trajectories of reefs towards either stable equilibrium are illustrated using thick, light grey arrows. A time series of events in Jamaica is illustrated by the thin, dark grey arrows.

The interpretation of such plots is easily illustrated for the decline in health of some Jamaican reefs between 1981 and 1993 (Figure 2). By 1979, forereefs had not experienced a severe hurricane for 36 years and coral cover was high at ~75% (Hughes 1994). In 1980, a combination of coral disease and Hurricane Allen reduced coral cover to around 38% but, since urchins were present, the reef began to recover. When the urchins died out in 1983, grazing levels were decimated, in part because long-term overfishing had removed larger parrotfish. With a coral cover of approximately 44% and a grazing intensity of only 0.05–0.1, meaning that parrotfish maintained only 5%–10% of the reef in a grazed state every 6 months, the reef began a negative trajectory towards algal domination; this trend was then exacerbated by further acute disturbance. By 1993, coral cover had fallen to less than 5%. A key feature of this graph is that reversing reef decline becomes ever more difficult as the cover of coral declines; as coral cover drops, the level of grazing needed to

Continued

**Coral reef models: Hasty conservation action beats procrastination
to enhance reef resilience (*Continued*)**

place the reef on the reverse trajectory (to the right of the unstable equilibrium) increases. Thus, based on the model used here, conservation action in the mid-1990s would have required grazing levels to have been elevated at least fourfold, to the maximum observed levels for fish in the Caribbean. In contrast, if action had been taken a decade earlier, when coral cover was still around 30%, target grazing levels would have been more achievable, requiring only a two- to threefold increase.

An appealing aspect of this approach is that the model's output helps define management objectives by explicitly integrating the impacts of disturbance with the effects of conservation action. The locations of thresholds (unstable equilibria) and bifurcation points reflect the underlying ecosystem dynamics and are influenced by processes such as primary production (Steneck and Dethier 1994), coral growth rate (Bozec and Mumby 2015) and eutrophication impacts on algal growth rates (Mumby, Harborne, et al. 2006). Acute disturbance phenomena such as coral bleaching (Hoegh-Guldberg 1999) cause sudden coral mortality and shift the state of a reef down the y-axis, whereas periods of recovery allow the reef to move back up the y-axis. Changes in grazing on the x-axis represent the effects of fishing herbivores (shifts from right to left) and active herbivore management such as marine reserve implementation or the banning of fish traps (shifts in the opposite direction).

Models of this type have been used to consider where and by how much simple management interventions, such as greater regulation or even closure of herbivore fisheries, would elevate coral reef resilience under climate change. For example, the closure in 2009 of the parrotfish fishery in Belize was predicted to increase the resilience of reefs by fivefold (Mumby et al. 2014). Similarly, coupling ecosystem models with those of parrotfish dynamics suggested that Caribbean parrotfish fisheries should cap harvest rates at a maximum of 10% and impose 30 cm minimum body sizes so that the fisheries would have only modest impacts on future reef health (Bozec et al. 2016).

Prof. Peter J. Mumby, Marine Spatial Ecology Lab, University of Queensland, Brisbane, Australia.

Of course, disturbances rarely occur in isolation, and many coral reefs are influenced by multiple disturbances at any one time. For example, fishing occurs on most reefs around the world, and climate-induced coral bleaching is an increasingly common phenomenon impacting almost all reef locations globally. It is therefore sensible to understand how reef fish differ in their response to bottom–up versus top–down disturbances if they were to co-occur. Graham et al. (2011) did just this, assessing the vulnerability of ~130 fish species to habitat change versus fishing (Figure 6.9). Interestingly, the species of fish vulnerable to habitat disturbances are not particularly

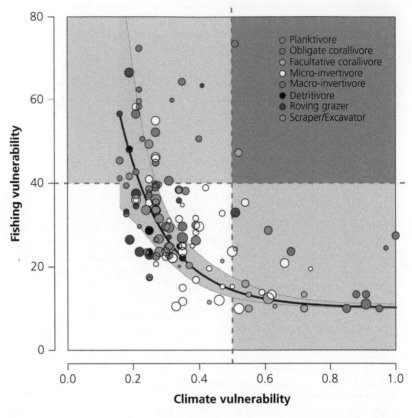

Figure 6.9 Relationship between vulnerability of coral reef fish species to climate change disturbance (i.e. bottom–up habitat degradation) and fisheries (i.e. top–down exploitation). Each dot is a different species of fish, with the colour corresponding to its feeding group. Shading in the boxes is hypothetical stress levels.

Source: Figure 3 from Graham, N.A.J., Chabanet, P., Evans, R.D., Jennings, S., Letourneur, Y., MacNeil, M.A., et al. (2011). Extinction vulnerability of coral reef fishes. *Ecology Letters* 14:341–8; <http://onlinelibrary.wiley.com/doi/10.1111/j.1461–0248.2011.01592.x/full>.

vulnerable to fishing, and vice versa. Of particular note, the species of fish that are known to exert very strong influences on the rest of the ecosystem are more vulnerable to fishing rather than to habitat change. This includes herbivore species that can control algae and enable coral recovery, species of invertebrate feeders that control echinoderm outbreaks, and piscivores. Given that fishing is a local/regional problem, and a range of options exist for fisheries management (Chapter 7), this is a positive finding, suggesting that good local management may enable reefs to continue to function and bounce back in the face of disturbances such as climate change, at least in the short term; reducing carbon emissions will dictate the ultimate future of coral reef ecosystems (Chapter 8).

7 Reef fisheries and reef aquaculture

Where there are reefs, there tend to be fisheries exploiting the resources on them. Reef fisheries represent only 2%–5% of the total global estimated catches. However, their importance lies not only in the level of fish removals or their associated monetary value (itself estimated at more than US$5 billion each year). Reef fisheries, particularly those adjacent to or surrounding tropical small island states, play an important role in the provision of protein to islanders and coastal populations that have few alternative food sources, including some of the world's poorest people (Moberg and Folke 1999). To provide context, an estimated six million people fish on reefs in 99 reef countries and territories around the world (Teh et al. 2013). Over 50% of protein consumed in the Maldives comes from fish resources, while marine resources represent 50%–94% of animal protein in the diet of coastal and urban communities in Pacific Island countries and territories. The role of reef fisheries as a source of dietary protein cannot be overstated, particularly for those island nations with a narrow resource base due to limited land for agriculture and grazing, often poor soil, crops that are being increasingly affected by extreme climatic events such as typhoons and droughts, and an increasing risk of atoll soil salination due to sea level rise (Pilling et al. 2015).

In addition to providing protein, reef fisheries have considerable historical and traditional significance. They offer livelihood opportunities, for example through fishing-related tourism and recreation, and provide employment for an increasing number of people, particularly where fish are marketed at regional and global scales. Reef fisheries also offer a fallback position when economic or social problems occur (Sadovy 2005), and help combat negative national food trade balances, which in island states can be exacerbated by increases in global oil prices that affect the cost of both local food production and the transport of imported food.

The considerable benefits and services arising from reef fisheries are under threat from many different sources. In this chapter, the variety of reef fisheries that have developed around the world are examined, with a focus on the

The Biology of Coral Reefs. Second Edition. Charles Sheppard, Simon Davy, Graham Pilling, and Nicholas Graham, Oxford University Press (2018). © Charles Sheppard, Simon Davy, Graham Pilling, and Nicholas Graham (2018). DOI 10.1093/oso/9780198787341.001.0001

direct exploitation of both invertebrate and vertebrate resources, their international trade and the more recent development of aquaculture. In the face of increasing pressures on and threats to reef resources, the resultant impacts on those resources are discussed, and the potential ways in which these pressures may be mitigated are examined.

7.1 Fisheries resources on reefs

The vast range of species congregating around and on reefs mean that reef fisheries are often highly heterogeneous. Worldwide catches cover over 200 species of reef fish, molluscs and crustaceans, and involve a variety of fishing methods. Here, the discussion is divided into two sections: vertebrate resources and invertebrate resources. In each section, key species are discussed, and the methods that can be used to catch them are detailed.

7.1.1 Vertebrate resources

Finfish resources around coral reefs can be diverse and plentiful. The inverted food chain, with a relatively high concentration of larger top predators in unexploited reef systems, makes these areas ideal for fishing.

Think of coral reef fisheries and you may think of groupers (Serranidae) and snappers (Lutjanidae). These are commonly the species that holidaymakers will see on menus in hotels and restaurants while on tropical holidays. Species from these families can reach an impressive size and offer consumers an excellent meal. Indeed, local fishers may sell prize specimens direct to hotels and restaurants for considerable amounts of money. As a result, the most desirable reef fish species for fishermen tend to be predators higher in the trophic scale. In particular, fishers target the larger (generally mature) individuals of these species, the price for which tends to be greater. The long-lived predatory nature of those species makes them particularly vulnerable to fishing. Further, the behaviour of these species may also increase their vulnerability. For example, particular fish such as many grouper species aggregate at specific sites to spawn. Where stocks are healthy, these aggregations can be huge; grouper aggregations may number up to 30,000 strong, although numbers in the hundreds are more common. Once located, these spawning aggregations are ideal for fishers, allowing large catches with much less effort than would be required at other times of the year. However, the impacts on stocks can be considerable, as rapid depletion of adult numbers can lead to serious overall population declines.

Fisheries for reef fish species vary widely in intensity and scale. The structural complexity of coral reefs does not generally allow the use of heavy fishing gear such as trawls. As a result, most coral reef fisheries are dominated by semi-industrial subsistence and artisanal fisheries (Johannes 1978).

Handlines, which are long vertical lines baited with a number of hooks that are lowered and raised over the reef manually from small boats or directly on the reef, have been frequently used to exploit reef fish resources, particularly in artisanal fisheries. They can also be used in a semi-industrial manner, as in the case of Mauritian mothership–dory ventures that target demersal resources on reefs and banks in the Indian Ocean to the north of Mauritius (Figure 7.1). These fisheries are based on motherships with blast-freezer storage facilities. They deploy up to 20 dories (up to 8 m in length), each crewed by three fishermen employing handlines rigged with three to five baited hooks. Those dories return to the mothership daily or twice daily to offload their catch for cleaning and freezing. Fishing is generally conducted in shallow waters (less than 50 m depth), targeting Lethrinidae and shallow-water groupers. However, these handlines can be used at greater depth, changing the fish assemblage targeted. Fishing at depth catches deep water snappers and groupers (Mees et al. 1999). A challenge with deep water species is not only the skill required to locate and maintain fishing position in order to exploit them, but the fact that these species are frequently characterized by extended longevity, slow growth rates, late maturity and low rates of natural mortality, making them typically vulnerable to over-exploitation. In the tropical and subtropical regions of the Pacific Ocean, commercial fishing for deep water stocks began in the 1970s, mostly supplying local markets.

Figure 7.1   Mauritian mothership–dory venture operating on the banks to the north of the island. Individual pirogues, with three fishermen each, are lowered over the side of the mothership to fish within a ~10 nm radius, before returning their catch for blast freezing.
Source: Photo by Graham Pilling.

Only where numerous habitats are sufficiently localized have higher catch rates been supported, as fishers move between fishing grounds (e.g. Fiji and Tonga). However, there is continued interest in the potential to renew these fisheries, given that shallow reef and lagoon fisheries have limited potential for expansion. Based upon past observations of localized depletion in the Pacific Islands region, fishery managers are approaching such opportunities with caution (Williams et al. 2013).

Given the time, and considerable effort, involved in manually raising lines with large heavy fish on the end by hand from 100 m depth, semi-industrial fisheries may use automated lines (electric reels), with a number of baited hooks. These lines are lowered and raised mechanically onto the reef and reef slope, and represent a considerable advance on the manual hand-lining techniques. While particularly useful when targeting deep water reef species, their use has not always been cost efficient, and they require further skill to maintain and deploy.

Longlines consist (as the name suggests) of a long main line from which branch a number of smaller lines ('snoods') with baited hooks. The approach is more semi-industrial in nature, requiring specialized machinery and skilled personnel to deploy this gear. The line may be used to fish in the water column or on the seabed, but around reefs their use is generally pelagic, due to the challenges of deploying gear amongst uneven reef structures. In this way, they can be used to target large pelagic species, including swordfish, marlin and tuna, above reefs.

Traps are a common method of fishing for reef fish and invertebrates. These are used throughout the tropics, being particularly common in the Middle East, Caribbean and Asia. Their construction and design varies according to the location and species targeted. They can be made from wire or bamboo or, when fishing for invertebrates like octopus, can form pots into which the octopus moves for shelter. Traps are usually set on the shelf, on or near reefs, but can be set in deeper waters on the reef slope or deeper banks. Baits used within pots vary depending upon the target species (Figure 7.2), while the size of fish caught can be adjusted through the size of mesh used in trap construction.

Nets can also be used to capture reef fish. They are frequently set just off the reef, to avoid entanglement, and the fish driven off the reef into the net. As for traps, the size of fish to be caught can be adjusted dependent on the mesh size used.

Another fishing approach is the use of spears, either on the surface (which takes considerable skill) or by using spear guns under water, particularly when targeting relatively large predatory fish. These methods tend to be more artisanal, with significant cultural and social importance, or recreational. While a selective fishing approach, it can still have substantial negative effects on target fish populations, particularly where considerable numbers of recreational fishers are active.

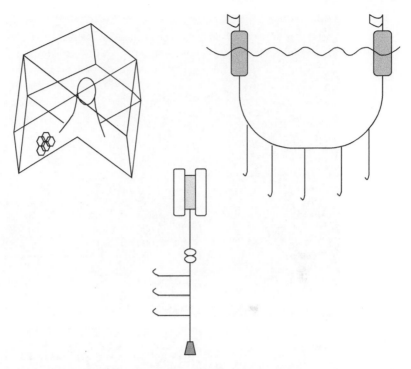

Figure 7.2   Diagrammatical representation of alternative fishing gears: trap (top left), longline (top right) and handline (bottom centre).

Indeed, the use and impact of recreational fishing gear on reef fish resources cannot be ignored. In the United States, the large number of recreational fishers can have a significant effect, even though they may only catch a few fish per fishing trip. For example, estimates of recreational fish landings from southeast Florida coral reefs in the 1990s were approximately 1,500 tonnes per annum. Grouper species around Florida are a target for recreational fishers, who can use dayboats to fish offshore. While 'bag limits' for these species can be applied to limit the number of fish retained in each recreational fishing trip, and minimum size limits are used, the mortality of those fish returned to the sea that are too small to retain needs to be considered, as fish brought up from significant depths can suffer higher mortality than those from shallower waters. This begins to highlight the complexities in estimating the overall impacts of fisheries, the fish population sizes at sea and, hence, sustainable levels of exploitation.

7.1.2 Invertebrate resources

Invertebrates are an important component of reef fisheries, and have played key roles in the livelihoods and welfare of island states for many years. Compared to reef fish, invertebrate species are more sessile (although some can

Figure 7.3   Important invertebrate reef resources: Top: Octopus. Bottom left: The large bivalve *Tridacna*. Bottom right: A sea cucumber (holothurian) (See Plate 13).

travel long distances over time). The productive waters around reefs allow these organisms to concentrate and multiply to levels that can support significant extraction. As a result, fisheries have developed for a wide range of reef-associated invertebrate species. These range from the spiny lobsters of the Caribbean, to the sausage-like holothurians called bêche-de-mer, to octopus species, to the massive giant clams of the Pacific (Figure 7.3). Species such as octopus may form a staple of the diet of local communities (see 'Sustainable reef octopus fisheries'). In contrast, due to their high value in Asian markets, catches of bêche-de-mer, trochus and giant clams are usually fished for export rather than local markets. For example, trochus shell forms a key export of the South Pacific islands of Wallis and Futuna (Adams and Dalzell 1995). These invertebrates provide local communities with notable levels of income, and countries receive considerable foreign exchange. In turn, the resource may be eaten during periods of hardship. For example, in Fiji a particular bêche-de-mer species (*Holothuria scabra*; sandfish) can form an important food source and income after cyclones.

Sustainable reef octopus fisheries

The increasing international demand for marine products fuels the over-expansion of fisheries in many tropical and subtropical coastal countries. Declines in reef fish catches and the low price paid for the catches have encouraged fishermen to look towards invertebrate stocks as an alternative source of revenue.

Octopuses are ubiquitous across the world's reefs and support a global fisheries industry that has grown significantly in recent decades. Octopuses occupy a broad habitat range, and are well suited for exploitation because of their short lifespan and rapid growth. Madagascar's fishing industry is relatively undeveloped compared to those of many other countries in the region, and this country is one of the few African nations that are still increasing their octopus fishery output.

Artisanal fishing for octopus is the primary economic and subsistence activity for indigenous Vezo coastal communities in the region of Andavadoaka, in the remote southwest of the country (see Figure). Traditionally, octopus was dried for sale in inland markets, but the arrival in 2003 of commercial fisheries collection companies brought a readily available and higher paying market, dramatically increasing fishing intensity and raising concerns of direct reef damage and unsustainable biomass removal. Data collected since 2003 show a decrease in the average weight of octopus, indicating over-exploitation.

The decline of the octopus fishery in south-west Madagascar is of grave concern. The arid climate prevents large-scale agriculture, and the few alternatives available to Vezo fishers often involve more ecologically damaging techniques, such as gill netting for sharks and small-mesh nets dragged over coral reefs and seagrass for fish. Declines in invertebrate

Continued

Sustainable reef octopus fisheries (*Continued*)

Figure. Octopus caught on shallow reefs in south-west Madagascar.

Source: Photo by Al Harris.

catches may also encourage over-exploitation of reef fish stocks. Fisheries monitoring has shown that reef fishing increases significantly during periods outside the octopus fishing time. In the absence of an octopus fishery, other marine environments—such as seagrass beds and coral reef communities—would be even more heavily exploited, primarily as a source of reef and pelagic fin fish.

In an effort to restore the local octopus population, an entire reef flat in Andavadoaka was declared a temporary octopus no take zone (NTZ) for 7 months from 1 November. The reef closed was the nearby sand cay of Nosy Fasy, whose reef flat—a favoured octopus fishing ground—is approximately 200 ha in area. The closure of the NTZ followed meetings with local stakeholders who supported the proposal so that fishers could have access to larger (and more valuable) octopus once the closed area was reopened.

The NTZ showed that the closure produced a highly significant increase in both the average weight of octopus and the catch per unit effort for fishers. During the closure, the average weight of octopus more than doubled from 0.5 to 1.1 kg. However, the high intensity of fishing on the days following the reopening reduced the potential longer-term

Continued

Sustainable reef octopus fisheries (*Continued*)

benefits of the NTZ closure. Subsequent management trials showed that a reduction in fishing pressure after a reserve's reopening can lead to longer-lasting benefits.

These studies demonstrated that seasonal, temporary closures of fishing sites are potentially a strong management tool in maintaining the sustainability of traditional octopus fisheries. The active participation and support of the local fishers is seen as key to the success in the implementation and respect of closed areas.

The success of these pilot NTZ trials led to the expansion of efforts to conserve octopus fisheries regionally. By 2007, 3 years later, a network of community-managed marine and coastal protected areas had been created, involving 23 villages. This unprecedented community-based support for developing marine conservation strategies came about as a direct result of the octopus NTZ trials, which, notwithstanding their limitations, served to demonstrate the potential economic benefits of effective fisheries management.

Dr Alasdair Harris, Blue Ventures Conservation, London, UK.

The wide range of invertebrate species caught for both local food and international markets means that only a few groups will be concentrated upon here. The aim is to provide information on particular species and their associated fisheries, rather than a comprehensive survey of exploited invertebrates.

Spiny lobster (*Panulirus argus, Jasus edwardsii* and related species) is the focus of reef fisheries around the world, including in the Caribbean, the Americas and Australia. These fisheries have considerable importance in many developing countries, both for protein and for export. The relatively simple fishing methods used can exert considerable pressure on the resource. For example, in Western Australia, the west coast rock lobster (*Panulirus cygnus*) fishery is one of that region's most valuable fisheries. The 6,000 tonnes of commercial catch taken in 2014, using baited traps, was worth almost AUD 360 million. The catch is exported either live or frozen, to Asia, the United States and Europe as well as servicing a small domestic market. It is also the focus of a recreational fishery. Managed through total catch limits, the fishery has been certified as ecologically sustainable by the Marine Stewardship Council.

While perhaps not as obvious as a valuable resource, bêche-de-mer, the dried or smoked product from sea cucumbers (Phylum Echinodermata, commonly holothurians) (Preston 1997), is a Chinese delicacy, sought for its homeopathic medicinal and aphrodisiac properties. Holothurian fisheries have a long history and have been traditionally located in the Indo-Pacific, although modern fisheries have spread to other regions as demand from the expanding Chinese market has increased (see 'Coral reef sea cucumbers: Exploitation and trade'). Tropical fisheries in the Pacific Islands, Indian Ocean and western central Pacific countries tend to be multispecific, with around 300 species found in relatively shallow waters (Conand 1997). For the Pacific region, bêche-de-mer has represented the second-largest fishery economically, second only to the tuna fishery, artisnal fisheries providing up to USD 50 million per year to coastal communities. Commercial fishing along the coast is also economically important, contributing over USD 165 million to communities each year. However, that commercial value has led to depleted abundance and availability.

Coral reef sea cucumbers: Exploitation and trade

Sea cucumbers (Echinodermata: Holothuroidea) are large invertebrates common in reef communities. Amongst the different orders of this class of echinoderms, aspidochirotids are dominant in tropical seas. They live on different types of substrates, with some buried in sand, and they ingest large quantities of sediment. They take their food from the microfauna, microflora and bacteria, and some species can rework huge quantities of sediments each year. They are very important in recycling nutrients.

The left panel of the figure shows a specimen of *Stichopus chloronotus*, or greenfish, in a seagrass bed; this species is common in the reefs in the Pacific and Indian Oceans where, as well as undergoing sexual reproduction, it can reproduce by transverse fission, dividing itself into two smaller individuals. Note, in the top right of the panel, the greenfish's coils of faeces, which consist mainly of sand.

Figure. Left: *Stichopus chloronotus* (commercial name, greenfish) in a seagrass bed. Right: Variety of bêche-de-mer species (dried sea cucumbers) awaiting collection, in Madagascar
Source: Left: photo by P. Frouin, La Réunion University; right: photo by C. Conand.

In the tropical Indo-Pacific, some 60 species are harvested in artisanal multispecific fisheries by hand collecting via diving using scuba or hookah (surface air supply), and dredging. These tropical fisheries were initiated by the Chinese, who consume bêche-de-mer (or hoi som) traditionally, and have a long history dating back, in some countries, to the seventeenth century. The sea cucumbers are then processed during several phases of cleaning, boiling and drying, to prepare a dried product called bêche-de-mer or trepang. The processing can be done by the fishers or by local collectors who then send the product to

Continued

Coral reef sea cucumbers: Exploitation and trade (*Continued*)

the main towns or larger dealers. The dried product is generally not consumed locally, but exported to Asians who consume it as a delicacy or who use it as medicine. Export is not direct to the consumer country, but goes first to a few important intermediate markets in Hong Kong, the United Arab Emirates or Singapore. Indonesia is the major world exporter, and Hong Kong is the most important importer but, because some reciprocal exchanges of products occur between these markets, many details of the trade are unclear.

The fishery has recently expanded due to increasing demand, and resources have become depleted, even in distant remote fishing grounds. Sea cucumber stocks now require effective conservation measures. In countries with reefs, sea cucumber fisheries provide an important contribution to the economies and livelihoods of coastal communities. These small fisheries share a few characteristics: (i) the demand is fluctuating, but presently very important; (ii) new species, formerly considered to have low market value, are now fished, as the higher-valued species become scarcer; (iii) fishermen have to dive deeper and go further from their traditional fishing grounds (this can raise problems of illegal fisheries and conflicts between countries); (iv) the fishers' catch per unit effort has therefore decreased; and (v) the size of the collected specimens decreases, making the stocks vulnerable if individuals are collected before the onset of sexual maturity.

The right side of the figure shows a collector's store house in Madagascar, with several different sea cucumber species. But the decrease of the stocks has meant that alternatives have to be found for fishers; one alternative is aquaculture of another species, the 'sandfish' (*Holothuria scabra*). The aquaculture venture is being developed as a co-management project by villagers, who get juveniles from a hatchery and grow them to commercial size.

Effective management plans for these reef fisheries are rare. Difficulties come from the lack of regular data on catches; these data are difficult to obtain, given the variety of species exploited, with some species not yet being described! Some management measures have been adopted to regulate fishing pressure, including closed seasons, minimum sizes, total allowable catches, gear restrictions, spatial and temporal closures and the establishment of marine protected areas. However, lack of enforcement in many small islands has posed considerable constraints on the effectiveness of such management measures. The trade data are complex and often not accurately reported. The international export and import statistics are important for quantifying the catches, using a conversion factor, which is often around 1/10, between the fresh catch and the processed product, but this value is variable with species and treatment uses.

In some locations, cucumber populations have collapsed, with no or very slow recovery of overfished stocks. Species are under heavy fishing pressure, or even depleted throughout the Indo-Pacific. High prices and increasing demand have seen the expansion of the range of collection areas and an ongoing search for new species. In many regions, the socio-economic dependency on bêche-de-mer is so important that fishers continue to collect sea cucumbers despite falling catches, further affecting the stock's capacity to reproduce

Continued

> **Coral reef sea cucumbers: Exploitation and trade (*Continued*)**
>
> and repopulate the fishing grounds. Generally, when one commercial species is depleted or becomes 'economically extinct', traders will encourage fishers to search for new species, or fish deeper or further afield.
>
> Many additional threats have been identified for sea cucumber populations worldwide, including global warming, habitat destruction, unsustainable fishing practices (e.g. blasting), development of fisheries with little or no information on the species, and lack of natural recovery after over-exploitation. The present trend of overfishing, and examples of local economic extinctions, require actions for conserving stocks' biodiversity and ecosystem functioning, and therefore sustaining the ecological, social and economic benefits of these natural resources.
>
> Prof. Chantal Conand, University of La Réunion, Saint-Denis, Réunion.

Another invertebrate that forms the focus of an export fishery is *Trochus niloticus*. Trochus is a marine gastropod found in the Eastern Indian and Western Pacific Oceans. Fisheries for trochus in the Pacific began in the first decade of the twentieth century. The shell is used primarily for the manufacture of mother-of-pearl buttons, but other uses include jewellery, handicrafts and polishing agents. Additionally, subsistence fishers harvest trochus for its meat. Following a decline in demand from the mid-1950s, primarily due to the replacement of natural shell by plastics in button manufacture, the world market and fishery then revived in the late 1970s when an increased demand resumed for trochus shell buttons. Factories in Pacific island countries process trochus shells for export, handling around 100 to 200 tonnes annually, with markets in Italy, Korea and Japan. Like bêche-de-mer, its value has led to heavy fishing pressure, and translocation of wild stocks has been used as a management option.

A fishery that falls between that of export and local consumption is that for giant clam (family Tridacnidae). Members of this family have species-specific habitat preferences, in that all species in this group are depth limited by their symbiotic zooxanthellae which live within their brightly coloured mantle tissue and provide nutrients to the clam, complementing direct feeding on phytoplankton (see Figure 7.3, bottom left). The target of giant clam fisheries is generally the large adductor muscle, which closes the clam's massive shell. There is a lucrative market for this muscle in countries such as Taiwan and China. To service both this international demand and the growing demand from the expanding human populations of local islands, fishing for such clams in the South Pacific islands has expanded rapidly. Declining population sizes have meant that giant clams of particular species are categorized as endangered in many areas of the Pacific.

As invertebrates are relatively sedentary but less abundant than fish, fisheries for invertebrates tend to be manual and targeted. Traditionally, these species were generally gathered through hand collection at low tide (e.g. holothurians) or freediving (in particular for trochus, which is found in waters shallower than 8 m). However, the increasing use of scuba gear has expanded the pressure on these resources by allowing fishers to remain underwater for extended periods, rather than be limited to the time they are able to hold their breath. Scuba gear has also allowed access to new, previously unexploited resources in deeper waters. Traps can also be used, for example to catch lobster and octopus.

7.1.3 Practical issues with reef fisheries

As just noted, apparently low levels of fishing by gears can still exert considerable effort and have serious impacts on resources (Figure 7.4). While low numbers of fishers may be observed fishing each day, this can still add up to very high levels over the course of a year, particularly if the reef area is small. For example, Craig et al. (2008) found that, although the average number of fishermen observed fishing at any one time was only 2.7, this continued level of fishing added up to 20,282 hours per year on small reefs. In

Figure 7.4   Example of the potential fishing pressure resulting from 'traditional' methods: traps on a Korean fishing vessel.

Source: Photo by Dr Stephen Lockwood, Pusan Harbour, Korea, 1995.

turn, increasing use of outboard motors, larger vessels and GPS systems can increase the pressure on specific resources over time.

The assessment and management of fisheries for both vertebrates and invertebrates is complicated by their very nature. Collecting information on fisheries that may target a huge variety of fish, using many different methods, is complex enough, particularly where data may be grouped at the level of a family (e.g. 'groupers') rather than at the species level. It becomes even more complicated when small-scale subsistence fisheries operate continuously throughout the year, or land catches all along a shoreline, or sell direct to retailers (e.g. hotels and restaurants) without going through markets (Craig et al. 2008). Information is also difficult to collect from recreational fisheries, as a result of the large number of individuals that may be fishing at different times of the year. In Florida, information on the recreational fishery for snappers and groupers is obtained through a variety of approaches, including logbooks, interviews on return to shore and, in particular, telephone interviews from a cross section of fishers. The potential bias in data that results, particularly given the decline in landline telephones in some regions, can be significant. Overall, the difficulties in obtaining data for stock assessment means that reef fisheries represent a critical, but often understudied, resource. Indeed, accurate data collection is often beyond the capabilities and funding of many fishery departments, whose attention may necessarily be focused on fisheries for highly valuable large pelagic resources. However, the development of new mobile-phone-based apps for artisanal and recreational data collection holds promise.

7.1.4 Fishing with dynamite and poisons

Fishing by explosives or poisons has been a relatively common approach in parts of southeast Asia and parts of Africa. Explosives may be either homemade recipes, or stolen from road building programmes or mines, for example. The effects of such activities are obvious in that bombed patches of reef are totally destroyed over a radius of several metres, with blast effects sufficient to kill and injure fish with swim bladders (almost all except the shark family) over a greater radius still. Where the practice occurs, it may be widespread, and the damaged patches soon merge. Where the practice continues relentlessly, there is little chance of recovery over time. Estimates of the impact of this are mostly anecdotal, although in Bolinao in the Philippines, for example, losses of coral cover were determined to be approximately 1.4% per year from blasting, with a further 0.4% per year lost to fishing with cyanide (McManus et al. 1997), reducing the potential recovery rate by a third. In Borneo, over 35% of the coral reefs (some estimates say over 50%) have been blasted flat and can no longer provide fish, and have lost most of their rich diversity.

Figure 7.5   Top: 'Ghost fishing' by abandoned fish trap on a reef, where fish continue to be caught and perish for years until the rope or wire mesh rots or rusts. Bottom: Lost nets on a reef in Oman. These directly blanket and kill the coral.

Often the practice begins when traditional fishing with nets, lines and hooks is no longer capable of producing sufficient food for coastal populations. This is exacerbated when traders from other countries, mainly east Asian, arrive in ships with freezing capacity to purchase fish for cash. This results in new pressures and incentives to catch more than is required for local consumption. When each reef is destroyed, the fishermen simply move on, although this activity itself becomes less sustainable as the numbers of healthy reefs progressively diminish. However, profits remain tempting despite the terribly high injury rates to fishermen that come from using explosives and poisons. Some dynamite fishermen in Indonesia can earn three times that of a local university professor, for example, which obviously encourages the activity. In almost all countries the practice is illegal but, while the practice has been successfully reduced in some areas, it may continue due to absent policing and management, or the practical challenges in monitoring large coastal fishing areas.

'Ghost fishing' is the term given when fishing gear continues to catch fish and impact reefs after being lost (Figure 7.5). There appear to be no estimates of quantities of resources destroyed in this way, other than many anecdotal reports suggesting that mortality can be substantial. This is a wasteful practice, especially when non-degradable materials are used. Loss of pots can be substantially reduced by the simple expedient of attaching two lines to each pot so that, if one breaks, the second can be used to retrieve it. Alternatively, biodegradable panels can be used, which 'release' fish caught within traps after a period. This reduces the impact on fish, and hence on the reef itself. In the case of nets, several (but far from all) countries have now banned their use on or around coral reefs, to reduce impact on coral, for example. However, the issue is not confined to fishing gear. Marine pollution is a growing issue, with almost 700 marine animal species having been recorded as encountering marine debris, including entanglement in gear or ingestion of plastic debris.

7.2 Live reef fish trade

The live reef fish trade comprises two parts. First, it supplies high-value fish for sale as luxury items for restaurants, such as those in China, with fish being transported live to the market and chosen from tanks in restaurants by customers just prior to cooking and serving. Second, it supplies ornamental fish and invertebrates for the aquarium trade. The trade for restaurants will be concentrated upon here, and the aquarium trade is described in 'The marine aquarium fish trade'.

The marine aquarium fish trade

Marine aquarium fish fisheries are based on the hand collection of small, colourful marine fish that are exported live to markets throughout the world to be kept in private and public aquariums. While there may be some overlap with food fisheries, the majority of species collected are not utilized for food.

Between 1.5 and 2 million people worldwide are believed to keep marine aquaria. The trade which supplies this hobby with live marine animals is a global multimillion dollar industry and operates throughout the tropics. The marine aquarium fish trade began as long ago as the 1930s, and became firmly established on a commercial scale in the 1950s. Marine fish species are collected and transported mainly from South East Asia but also increasingly from several island nations in the Caribbean, Indian and Pacific Oceans, to consumers mainly in the United States, which is followed by the European Union, Japan and, increasingly, China.

Estimates suggest that, globally, around 1,800 species of fish from around 50 families are captured for this fishery, with an estimated 25 million specimens traded annually. Relatively low-priced fish, such as common and widely occurring species of damselfish (Pomacentridae), account for the majority of the trade, with species of angelfish (Pomacanthidae), surgeonfish (Acanthuridae), wrasses (Labridae), gobies (Gobiidae) and butterflyfish (Chaetodontidae) making up most of the remainder. Due to advances over the last two decades, mainly in lighting technology, hobbyists can now keep, grow and propagate corals, and live coral reef aquariums, rather than fish-only aquariums, have grown steadily in popularity. They are, in fact, so popular that the demand for fish species has dramatically changed in accordance with this trend. Coral reef aquariums today usually contain fish that perform an important function, such as cleaning the sand, eating algae (e.g. Acanthuridae), or eating parasites (e.g. Labridae)—fish that do not eat or damage coral.

Marine aquarium fisheries are extremely selective, with fish typically collected by hand, snorkelling or using an underwater breathing apparatus such as scuba gear. Sustainable fish collection revolves around the use of small-mesh nets that do not damage the target fish or non-target species, or the habitat. In Indonesia, the Philippines and possibly other countries in South East Asia, sodium cyanide is sometimes used for collection. Cyanide is a toxic substance, and its use is universally outlawed for the capture of both aquarium and food fish, as it has detrimental effects on target and non-target species, including corals. It continues to be used because it is easy to obtain, is inexpensive and makes fish catching easier. While enforcement is difficult, it is critical to ban the use of chemicals, including cyanide, and to promote the sustainable collection of fish using nets.

After collection, fish are held in land-based holding systems or in cages in the ocean. When the fish are ready for export they are packed—one individual per plastic bag, which is approximately one-third filled with seawater. Air is squeezed from the bag and replaced with oxygen. The final step is to seal the bag with a rubber band or mechanical clip before placing it in an insulated box. All steps from collection to export must be executed with extreme care.

Continued

The marine aquarium fish trade (*Continued*)

Marine aquariums can help to inform people about reef biodiversity and conservation, and the trade provides jobs and economic benefits to local fishing communities in supplying countries where opportunities for livelihoods are often limited. However, some have raised conservation concerns associated with the trade. First, aquarium suitability and trade data indicate that species known to be poorly suited to aquarium conditions are traded in high numbers. In the hands of experienced hobbyists, such species may thrive, but premature mortality is a more likely outcome, leading to wastage of resources. Therefore, it is the trade of such species in high volume that is of concern, rather than their presence in trade statistics. Second, over 95% of trade is in wild-caught fish specimens. The aquarium trade mostly targets species characterized by high turnover rates and high abundance (e.g. *Chromis* spp.). It also targets juvenile fish of longer-lived and less-abundant species, with collection removing individuals that are subject to naturally higher mortality rates, primarily from predation, than adults. For these species, leaving the bigger adults on the reef also provides protection for spawners, particularly large ones, who produce greater numbers of eggs that are of higher quality, and contribute disproportionately more to population replenishment and stability. In both instances, impact on stocks will be reduced. However, at a local level, if effort is high and targeted at specific species, over-collection may still be of concern. In addition to not being ecologically sustainable, over-collection is also economically 'unwise', as it will undermine the longevity of any business. As a note, many species that are considered rare in a given area are, in fact, highly abundant at other locations. For fish that do have an exceptionally limited geographical range (i.e. globally rare), are found at low densities and are of high interest in the aquarium trade, collection should be carefully monitored and regulated.

Significant advances and breakthroughs have been made in understanding the development from larvae to adult of a number of key fish species, as well as the technology and nutrition requirements required to achieve success in their culture. However, to date, costs of culturing fish remain largely prohibitive and there is no immediate prospect of significantly reducing reliance on specimens taken directly from the reef. Moreover, while, in some instances, sourcing of species from cultured facilities may reduce stress on the ecosystem, given that such facilities tend to be situated in developed countries, doing so can put the livelihoods of those in source countries at risk.

A few countries have effective aquarium fishery management plans in operation, but others have poorly enforced regulations or no regulations at all. As with any fishery, sustainability is enhanced through the implementation and enforcement (e.g. by random checks and fines) of a suite of management tools often described and mandated in a management plan to regulate and monitor the fishery. The following are examples of recommended procedures:

- issuance of permits for aquarium fishing and/or export
- ensuring that all collectors be professionally certified if they use any type of underwater breathing apparatus

Continued

The marine aquarium fish trade (*Continued*)

- limited entry into the fishery through control on the number of collectors and/or exporters
- size restrictions on fish collected, by species
- establishment of protected or no take areas (i.e. areas that are off limits to aquarium fishermen, often placed to reduce user conflict, such as a popular dive site; this is often the simplest means to maintain stocks)
- submission of reports to the management authority on species and number caught by fishing area
- submission of reports to the management authority on species and number exported and price per piece
- in-water surveys and/or risk assessments to determine which species may be vulnerable to over-exploitation
- species-specific quotas or bag limits, including zero quotas for species considered to be at risk or unsuitable for the trade due to high mortality in captivity
- a ban on the use of chemicals and other natural or artificial substances used as 'anaesthetics'
- a ban on any damage to coral during collection
- regulations relating to fishing gear
- a set of minimum requirements for fish-holding facilities
- a set of recognized best-practice standards (i.e. a code of conduct for responsible fisheries)

Dr Colette Wabnitz, Institute for the Oceans and Fisheries, University of British Columbia, Vancouver, Canada, and

Dr Liz Wood, Marine Conservation Society, Ross-on-Wye, UK

While fish have been transported live for consumption for hundreds of years in Asia, the trade has developed considerably in recent years to the extent that it has become a serious issue for key reef resources. Estimates of the trade in South East Asia have reached around 30,000 tonnes per year (Sadovy et al. 2003), with 20,000–25,000 tonnes trading through Hong Kong annually, and 40%–50% shipped to the increasingly affluent mainland China market. Fish are commonly sourced from South East Asia, the Western Pacific, and the east coast of the United States (the last mainly for restaurants within the United States). It can represent an important income to coastal communities; the annual export value of the grouper trade from Palawan, Philippines, is USD 25 million, both supporting the local economy and providing livelihoods for many thousands of people.

The overall value of the live fish trade is difficult to identify, but numbers of around USD 800 million per annum have been suggested (Sadovy et al. 2003). Individual fish can fetch an extremely high price, with a sale value of USD 200 per kilogram or more when demand is high (e.g. during festivals such as Chinese New Year), while restaurant prices are 100%–200% higher than reported wholesale prices. This price difference fuels the supply chain. It also increases the impact on fish stocks, since rarity of a species results in higher value, and hence leads to targeting of species with lower rates of growth and reproduction. As a result, the trade focuses on the capture of a small number of species. The initial focus of the Hong Kong market was the subtropical Hong Kong grouper (*Epinephelus akaara*), but this expanded to more tropical species, including the leopard coral grouper (*Plectropomus leopardus*), the humphead wrasse (*Cheilinus undulatus*) and the humpback grouper (*Cromileptis altivelis*). The focus on these species is due to their fine-textured flesh and, for species such as *P. leopardus*, due to their red colour, as red is the luckiest colour in China and signals wealth and prosperity. These species are generally vulnerable to even low levels of fishing due to their life history, being relatively rare, long-lived and only fully maturing at older ages. Target species may change sex during their lifespan (e.g. from females to males, being protogynous hermaphrodites), and, as noted earlier, they may form spawning aggregations. Both features increase the impact of any fishing on stock size and long-term viability.

The focus on rarity also means that economics works differently from other fisheries. Unlike capture fisheries, where stock rarity means extra effort to catch species so that fishing becomes uneconomical as a result, the focus on rarity means that targets of the live reef fish trade become more attractive as they decline in numbers. For example, the Chinese bahaba (the snapper *Lutjanus fulviflamma*) is valued for the medicinal properties of its swim bladder. As this species became almost extinct through its limited geographical range, prices rose enormously, such that a large swim bladder has fetched USD 64,000 per kilogram (Sadovy and Cheung 2003).

The large economic incentive to fish has led to a boom-and-bust pattern of exploitation. Depletion of the resource to very low levels has led to fishers moving to new areas to meet market demand. This has led to serial depletion of localized resources, which in turn has led to a rapid geographical expansion of the fishery following depletion of reefs near Hong Kong, so that now fishing to supply the Asian markets encompasses the entire Indo-Pacific region.

Live fish are caught using many different methods. These include traditional approaches such as traps, nets and hand-lining with the use of barbless hooks. More destructive practices have involved the use of chemicals such as bleach, formalin and cyanide, combined with use of scuba or hookah (surface-supplied air) by fishers. All approaches can lead to undesirable fish mortality, either directly at the point of capture or during transit to markets, while chemicals in particular can result in mortality of non-target species and damage to habitats. Chemicals are squirted at target fish, which can then easily be collected in their stunned state. However, recent management controls have resulted in an increasing use of non-destructive fishing practices (Sadovy and Vincent 2002).

Wild-caught fish can be held briefly before being exported by sea or air, depending on the fishery and transport links. As wild fish stocks decline, subadult fish may be retained and grown in captivity to market size in net cages, although this mariculture is not possible for all species. This expansion of the target sizes to subadult fish further increases pressure on the target resource. Finally, specimens can be reared directly from eggs to market size in aquaculture facilities (see Section 7.3). Cultured fish raise a lower price in markets, due to consumer preferences, but values still outweigh the costs involved.

The live reef fish trade has often led to social conflicts. Conflicts can arise between operators and local fishing communities over several issues: price, how the wealth is distributed amongst individuals and villages, the use of particular local fishing areas and the impact of destructive fishing methods used to obtain target species and which affect other species in the area. Furthermore, once such fisheries move on, local fishers are left with degradation of habitats and reduced fish resources.

This does not mean that the live reef fish trade cannot be controlled, however. Control can be applied at several steps along the market chain, from fishers (e.g. controlling numbers of fishers or boats, closing seasons or areas, limiting equipment used), to traders (e.g. controls on species sold, export controls), to the market itself (by influencing consumers). In Australia, for example, effort in the fishery has been managed, leading to a sustainable and lucrative fishery. However, the effectiveness of these methods is highly variable, due to the difficulties and costs involved in their implementation and monitoring. Given the impacts of these fisheries, however, controls are needed to prevent the continued over-exploitation of resources, the associated degradation of the reef habitats, indirect impacts on the abundance of other species and impacts on the livelihoods of those that rely on them. Enhancing the local capacity of fishing communities in sustainable fishing practices and improving governance will be key.

7.3 Aquaculture on reefs

Mariculture (aquaculture of marine species) is an alternative to wild capture and is expanding rapidly worldwide. The process aims to increase the natural production rate of a species through the optimal feeding of fish and

Figure 7.6   Aerial view of fish farms along the Red Sea coast of Saudi Arabia.

reduction of mortality (Figure 7.6). For coral reef species, activities tend to concentrate on those species with high market value. However, aquaculture is not a recent phenomenon. Both South East Asia and the Pacific have a long history of mariculture activities, for example, although the level of success has varied widely. In turn, pond aquaculture of species such as Nile tilapia (*Oreochromis niloticus*), ranging from subsistence culture in small household ponds to intensive industrial farms, has been pursued to increase supply of fish for food.

In the Pacific, pearl farming has proved a valuable activity. In French Polynesia, the value of cultured black pearls from the black-lipped pearl oyster (*Pinctada margaritifera*) was USD 173 million in 2007, representing 66% of the region's fishery and aquaculture production at that time, and employing 5,000 people. The viability of this approach is high, as pearls are available from harvesting wild shells, while oyster spat can either be grown in hatcheries or collected from the wild for culture. As a result, pearl production has spread to other Pacific island nations.

The value of giant clams and their vulnerability to exploitation has led to considerable efforts to culture them in many Pacific Islands. This has focused both on aquaculture to sell adult specimens, with the combined exports of cultured giant clams totalling 75,000 pieces in 2007, and on rearing of sprat and juveniles to reseed reefs in over-exploited areas. However, larval survival rates are low, and hence considerable investment is required before sufficient larvae can be produced to recover stocks. On a smaller scale, communities in the Solomon Islands have sustainably operated clam 'gardens'. In these areas, small *Tridacna* are nurtured on the outer reefs before being brought closer to home for live storage and later consumption (Hviding, 1998). Levels of transfer and consumption are maintained at rates that ensure the stock remains viable.

Few reef fish species can currently economically be grown (from larvae to exploitable adults) within hatcheries. A notable success is the hatchery production of milkfish (*Chanos chanos*). These activities have provided livelihoods and cheap protein for millions of people (Sadovy 2005). However, this success is being overshadowed by the greater monetary value of species involved in the live reef fish trade in South East Asia. As a result, milkfish hatcheries are being turned over to the rearing of more valuable grouper species that give faster and greater returns on investments. Often, these are 'grow-out' operations, where small wild-caught fish are grown in cages to a size suitable for the market. Given the profitability of these enterprises, in comparison to milkfish production, demand for wild broodstock is increasing, with growing impacts on wild populations. However, more hardy grouper species such as tiger grouper (*Epinephelus fuscoguttatus*) and green grouper (*Epinephelus coioides*) have been successfully bred and reared in captivity since 2000. In turn, advances in aquaculture technology have allowed higher-value species such as humpback grouper (*Cromileptes altivelis*) and coral groupers (*Plectropomus* spp.) to be raised from hatchery-reared seed stock in Taiwan. Nevertheless, large amounts of wild-caught fish are used to feed cultured individuals to market size—often five to ten times the biomass that is recouped, due to the slow growth of the target species. However, considerable research is being focused on techniques for the hatchery production of very young groupers and the development of pellet diets, to reduce the impact on wild fish populations.

The problems of aquaculture need to be overcome, since current rates of expansion cannot be maintained without a sustainable supply of juvenile fish (be they caught wild or hatchery reared), feed (be it fish caught as bycatch, targeted higher-trophic-level fish or synthetic feed with its associated high costs) and reduced environmental impacts (e.g. increased nutrient loads in surrounding waters, spread of disease, potential implications for local population genetics of escaped reared individuals). In turn, coastal aquaculture can be vulnerable to extreme weather events, and the increasing effects of climate change. The solutions to these problems, while by no means straightforward, could contribute to aquaculture's sustainable development and the movement away from destructive fishing practices (Sadovy et al. 2003). Although the range of species reared in aquaculture is increasing, the issues that need to be overcome and the fact that not all species can be effectively reared mean that this approach will not solve all the problems in reef fisheries. However, if the culture of fish is combined with the involvement and benefit of local coastal communities, considerable socioeconomic benefits can result.

7.4 Impacts of fishing

Increases in local economies, the arrival of external market drivers, improvements in technology and, in particular, growing coastal populations have massively increased the pressure on renewable reef resources. In turn, the trade in reef fish, in both the luxury food market and the general international trade, have increased the monetary value of reef fish catches, further increasing demand. Reef fisheries have therefore often moved from supporting local protein requirements to marketing fresh fish for catch and export, while cheaper imported fish or other protein sources are supplied to the local tables. In some cases, this can have implications for local nutrition and health, with increasing risk of chronic diseases such as type 2 diabetes, hypertension, anaemia and goitre.

As a result of these changes, overfishing has become one of the best examined drivers of reef fish composition changes, with many examples of resultant trophic and ecological implications (see Chapter 6). Fishing is seen as one of the principal threats to the diversity, function and resilience of coral reefs, with the potential to lead to economic hardship for local communities (Bellwood et al. 2004). Over half of island coral reef fisheries examined by Newton et al. (2007) were found to be unsustainable, with total catches 64% higher than can be sustained. They note that 'consequently, the area of coral reef appropriated by fisheries exceeds the available effective area by 75,000 km², or 3.7 times the area of Australia's Great Barrier Reef'.

Overfishing of a resource can occur in a number of ways. The removal of larger fish, leaving behind individuals too small to maximize the yield that could theoretically be obtained, is termed 'growth overfishing'. The excessive removal of too many adult fish, thereby reducing recruitment and stock productivity, is termed 'recruitment overfishing'. These types of overfishing can occur separately, or in combination. Finally, the removal of particular species within the ecosystem can have serious consequences for the ecosystem as a whole. This is termed 'ecosystem overfishing'. All these effects have been seen in reef fisheries.

Over the past decade or more, increases in fishing pressure in tropical reef ecosystems have affected both species composition and habitat structure (Dulvy, Freckleton, et al. 2004; Dulvy, Polunin, et al. 2004) and have led to declines in species (Munday et al. 2008), resulting in the depletion of higher-trophic-level carnivorous species and the consequent dominance of fish from lower trophic levels (Jennings and Polunin 1997). Changes generally result because, as already noted, fishing tends to target predators higher up in the food chain, focusing on the larger individuals of both a particular species and the fish assemblage of the reef. These larger species are generally relatively long-lived with slow growth rates, mature at older ages and have limited distribution (Jennings et al. 1999; Sadovy and Domeier 2005). As a result, fished coral reef systems can become dominated by smaller individuals and species.

Mature fish have all but disappeared from many reef areas that are associated with large human populations, due to direct fishing pressure as well as indirect impacts on coral reef structure. The impact can be seen by comparing the species and size composition of remote (and hence less heavily exploited) reefs with those closer to population centres. For example, megafauna like sharks and, in the Indo-Pacific, humphead wrasse (*Cheilinus undulatus*) and bumphead parrotfish (*Bolbometopon muricatum*) dominate remote, lightly fished reefs. Over half the fish biomass on such reefs can be in apex predators like sharks (and giant trevally (*Caranx ignobilis*) in Hawaii). In the larger populated Hawaiian Islands at the south-east end of the chain, fish communities are almost entirely composed of small colourful fish, including many herbivores, but large predators are very rare. On the reefs around the remote, unpopulated Northwestern Hawaiian Islands (now part of the Papahānaumokuākea Marine National Monument), apex predators comprise the majority of the biomass.

Vulnerability (and catchability) of species is also affected by their habits. In the face of powerful spearguns, scuba equipment and waterproof flashlights, bumphead parrotfish that sleep in schools on the open sand or in shallow holes become easy targets, such that a skiff is easily filled with them in a single night. In the Solomon Islands, such technology made it possible to fill the fish markets with bumpheads, and the vulnerability of this species to fishing has led to local management action to prevent the consequences seen in Fiji, where bumphead parrotfish rapidly became rare and now are locally extinct on some islands.

Worldwide extinctions of coral reef fish have been identified in semi-industrial and artisanal and subsistence fisheries (Dulvy et al. 2003). As examples, the rainbow parrotfish (*Scarus guacamaia*) has been driven extinct locally through exploitation in the Caribbean, while exploitation of giant clams has removed them from some Pacific Islands. The great value of invertebrate species and their relatively sedentary habit means that intense fishing pressure can be brought to bear and often leads to a boom-and-bust cycle. For example, in the Maldives, an export fishery for bêche-de-mer started in 1985 and quickly resulted in the sea cucumber resource being heavily overfished (Adam et al. 1997), and similar patterns have been seen in the Pacific region. In Sri Lanka, gleaning for sea cucumbers was a traditional activity that had lasted for generations, but the introduction of scuba gear to extend and increase catches led to almost immediate depletion (Kumara et al. 2005). Trochus used to exist in huge numbers in the pristine coral reef flats of King Sound in Western Australia, but trochus populations have declined over the past 20 years due to overharvesting and, to a lesser extent, poaching by illegal foreign fishers.

Koslow et al. (1988) suggest that the complexity of coral reef fish communities may make them less stable and more vulnerable to overfishing. With declines in catch rates of larger fish, fisheries tend to subsequently focus on smaller species lower down the food chain, for example omnivores and herbivores, the abundance of which may have increased due to 'prey release' as their predators have been fished out. Their removal reduces the food available for top predators, limiting the ability of this group to recover in size and number. Even low exploitation levels are capable of affecting reef resources and the trophic structure of reef fisheries (Dulvy, Freckleton, et al. 2004; Dulvy, Polunin, et al. 2004). The result to human populations is ultimately a lack of fish to catch, and the result on the reef is a grossly distorted ecosystem that, because of positive feedbacks, rapidly becomes unable to support continued coral growth and reef development. Indeed, removal of levels of the food chain can cause significant ecological shifts on reefs (Campbell and Pardede 2006). Increased algae cover may be one result when grazing herbivores are depleted and nutrient levels are increased by land run-off and sewage. Conversely, where grazing urchins are kept in check by predators, overfishing of those predators can lead to explosions of urchin numbers, and increased bioerosion. None of these effects act in isolation. Fishing in the Bahamas removed large parrotfish as well as large predators and reduced total fish herbivory. Increases in numbers of the urchin *Diadema*, due to reduced competition, increased grazing sufficiently to compensate. That species was virtually eliminated by disease in the 1980s, at which point the balance on the reef became massively disturbed in favour of algae. As a result, the removal of herbivores in the Caribbean unexpectedly contributed to a

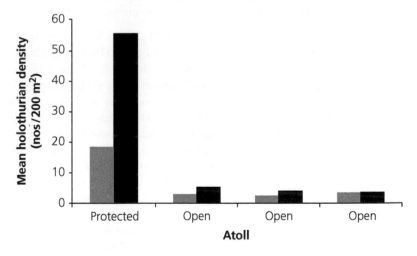

Figure 7.7   Effects of poaching on mean abundance of the much-sought-after holothurians. Densities on the atoll Diego Garcia, which is out of bounds for military reasons, versus abundances on three more 'open' atolls, which are only lightly guarded.

Data from Price, A.R.G., Harris, A., McGowan, A., Venkatachalam, A. and Sheppard, C.R.C. (2010). Chagos feels the pinch: Assessment of holothurian (sea cucumber) abundance, illegal harvesting and conservation prospects in British Indian Ocean Territory. *Aquatic Conservation: Marine and Freshwater Ecosystems* 20:117–26; determined from visual censuses along seventy-two 100 m × 2 m (200 m²) transects; shown for all transects (grey) and for transects with suitable habitat (black).

phase shift from coral- to algal-dominated ecosystems when disturbances such as hurricanes or coral disease killed corals. Decades later, many algal-dominated reefs there have not recovered (Mumby et al. 2006).

Finally, poaching is another important issue, and one that is becoming greater as human populations and issues of food security increase, and the reef resources to support them diminish under increasing fishing pressure. Illegal catches, which may be substantial, tend not to be recorded in official statistics, and generally remain unquantified. The amount of carnivore or herbivore biomass that is actually extracted from some reefs, therefore, may remain unknown. In places where even legal fishing records are poorly kept, in addition to the data collection challenges already described, this adds a further unknown quantity into the fisheries equation. Poaching may be substantial and, because poachers usually have no incentive to fish sustainably and, indeed, tend to maximize their catch in the minimum possible time, the problem is exacerbated. This is illustrated in the Indian Ocean in the group of atolls comprising the Chagos Archipelago, which has been designated as a marine reserve. One atoll is strictly protected by virtue of supporting a large military base, while several adjacent atolls are afforded lighter monitoring and protection. Holothurian poaching has taken place on the latter, with clearly measurable effects on holothurian population numbers (Figure 7.7), although other factors are also implicated (Price et al. 2013).

7.5 Approaches to managing coral reef fisheries

There is a range of methods that have been used, or suggested, to manage coral reef fisheries. Initially, management approaches were based upon those traditionally used in large industrial fisheries in temperate waters. However, it became clear that the nature of coral reef fisheries, being frequently artisanal and concentrated in a relatively small area, meant that new approaches were required.

Catch quota systems, as frequently used in temperate fisheries management, have been applied to reef fisheries in an attempt to limit resource exploitation. However, these approaches run the risk of causing a rush to harvest within a short time period. This was found in Aitutaki, Cook Islands, where total allowable catches were put in place for trochus within individual regions. In the ensuing rush to fish, villagers could not select the best shells for removal and export, and there was a drive to ignore size limits (see below). The approach was therefore modified, and individual transferable quotas were put in place, where catch quotas of trochus could be transferred between households. This involved the communities affected by management in the decision-making process, and the management approach led to nearly optimal long-term yields (Adams 1998).

Bag limits, where the number of individuals that can be taken in a given period is limited, have also been used for both vertebrate and invertebrate resources.

Limiting the level of fishing effort within a fishery to reduce exploitation pressure, while common in industrial fisheries, is often inappropriate or politically unattractive in artisanal fisheries. Indeed, in the subsistence/artisanal fisheries of tropical nations, the number of fishers involved (sometimes thousands) may make limiting access difficult, particularly when so many families and communities rely upon subsistence fishing. The approach is more feasible in areas like Australia, where the number of licences given to fishers can be limited and enforced (Cooper 1997).

Limits on the minimum and/or maximum size of caught individuals that can be retained have been used, with limits being set relative to key life history stages, such as the size at which a species becomes mature. There can be issues with enforcing such a measure, particularly in artisanal fisheries where fishers do not have specific landing sites. For vertebrate resources, the potential increased mortality of individuals returned after capture must also be considered. Size limits have been used in the Pacific region for invertebrate resources. Fiji operates a minimum legal size for export of dried sea cucumbers (sandfish (*Holothuria scabra*)), while Australia and other countries have set minimum legal sizes on whole holothurians. Minimum size limits are a logical approach for this species, since smaller individuals are not sexually mature, and fetch a lower value upon export. Indeed, control of minimum sizes can be easiest at the export stage, where commercial pressures can lead to more sustainable exploitation. Size limits have also been implemented for giant clam species, frequently combined with a ban on commercial harvesting and export, to increase the likelihood that individuals can reproduce at least once before harvesting.

Gear restrictions can be effective. These measures can include banning the use of destructive fishing practices such as explosives and poisons. Limits on the mesh size of traps or hook sizes can allow individuals of particular sizes to escape a trap, or increase the size of individuals taking hooks. A ban on the use of scuba gear or hookahs that can otherwise increase fishing pressure through increased time on the seabed can also be implemented. However, all these approaches can only work if enforcement is sufficient (Mahon and Hunte 2001).

Perhaps the best-known suggestion for management of reef fisheries is the use of marine protected areas (MPAs), designated areas that are closed to fishing. Closed areas have been shown to increase fish numbers, biomass and diversity within them (Roberts 1995). To sustain fishing near the closed area, however, an MPA may need to be very large or exhibit considerable migration of fish from within the MPA ('spillover'). This is needed to compensate for fishing that occurs often just on the border of the area, so located in order to maximize any resulting spillover benefits. MPAs are much less effective for highly mobile and in particular migratory species, which include many of the fish that are most valuable and vulnerable to fisheries.

While considered effective for maintaining biodiversity and protecting habitats, the effectiveness of MPAs for the protection of fishery resources is not always clear (e.g. Hilborn et al. 2004) and, although evidence for the success of this approach is increasing (Roberts, 2007), it cannot be viewed as the de facto answer for reef fisheries management. Threats to the coral reef ecosystem and hence reef fisheries, such as climate change, land-sourced run-off pollution and extreme weather damage, will not be reduced by the establishment of an MPA, for example, but may increase the resilience of the system to natural disturbance (Mellin et al. 2016).

Permanently closed areas may act as a 'source' of larvae to bolster the resources in surrounding exploited areas of reef. In this case, design of closed areas must be carefully considered (e.g. Schill et al. 2015). They must be sited upstream from suitable settlement habitats in prevailing currents. In the case of giant clams, spawning is triggered by the spawning of upstream neighbours, and hence spawning success may depend on the overall density of individuals in an area. This is the well-documented 'Allee effect', which is the term given to the positive correlation between the size of a population and the ability of that population to continue to reproduce (or of growth of the individuals in that population in some cases); basically, below a certain size of population, breeding between them becomes difficult and then impossible as it becomes increasingly unlikely for diminishing numbers of eggs or adults to find sperm or mates, thus decreasing the population further.

For relatively sedentary species, such as some invertebrates, a closed area can lead to significant increases in size and number. However, since movement out of zones is relatively limited, closed areas will have marginal direct benefits for excluded fisheries. Large MPAs are therefore generally not easily accepted for social or economic reasons. Given the potential objections by fishers to permanently closed areas, where appropriate, closure to 'industrial' fisheries, while allowing artisanal fisheries to operate within MPAs, can reduce fishing impacts. Seasonal or rotational closures can also be considered, an approach similar to land-based farming techniques (see 'Sustainable reef octopus fisheries'). These approaches have been used for trochus and sea cucumber in the Pacific. Periods of harvest in particular areas are followed by 'fallow' periods to ensure the resource was able to recover, leading to a rotation of exploitation between specified areas (Figure 7.8). The long-lived nature of many exploited species means that the duration of closure may need to be quite long to allow stocks to recover before fishing resumes. In turn, the ability to manage fisheries through much shorter seasonal closures may be limited due to difficulties in enforcing a frequent closure (e.g. every month), while further difficulties are experienced where there is modest technical capacity for planning and monitoring; weak enforcement; complex access rights to fishing grounds; and high numbers of fishers (Purcell et al. 2016). However, where acceptable and effective MPAs can be developed, they can be very successful in substantially

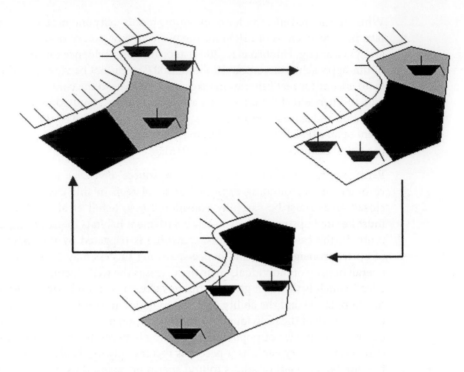

Figure 7.8   Diagrammatical representation of rotational closures. Dark-shaded areas are closed during one period. Light-shaded areas are potentially open to reduced fishing. White areas are fully open to fishing.

increasing both reef fish and invertebrate landings; indeed, the local communities become the best 'ambassadors' for the method.

More direct approaches have attempted to increase the resource available to fishers both by increasing the reef substrate and by directly replenishing exploited populations. Given that biodiversity and abundance of fish are related to the area and topographic complexity of reefs, a logical approach is to artificially increase both. Artificial reefs have been extensively employed in areas around the world, but their effectiveness in mitigating the problems of overfishing have been difficult to identify, due to the limited number of follow-up studies, and to the fact that even the largest artificial reefs are almost trivially small in comparison with natural reefs. Some have, however, been shown to increase lobster and octopus resource survival. To do this, the reef blocks have to be deployed in a particular way to provide these animals with suitable 'burrow' habitats.

Restocking of fish species, involving the release of hatchery-produced or small wild-caught fish, has focused on high-value grouper and snapper, as well as sedentary species such as giant clams and bêche-de-mer (see Section 7.1.2). However, the effectiveness of this restocking in restoring wild stocks is generally unclear, and should not be seen as replacing sound fisheries management practices. There are concerns over genetic issues, with the potential reduction in genetic diversity available to hatcheries as wild brood stocks decline, with resultant loss of productivity and viability of fry and fish, as well as the potential introduction of foreign genes into the wild population and the potential for a non-interbreeding population to outcompete local stocks. Released individuals are also likely to suffer high mortality, and the levels of release needed to have a positive impact on wild populations are likely to be massive. Restocking can only be used to supplement, rather than replace, the sustainable management of reef fisheries resources.

7.6 Controlling coral reef fisheries

There has been a range of methods used to manage coral reef fisheries. Politics, social systems and tribal hierarchies have historically been effective in successfully controlling the level of exploitation of reef resources. Current approaches include top–down control by government, the style most commonly used in developed countries for industrial fisheries. The alternative is through bottom–up control by local communities.

Management through top–down approaches can be effective where the governmental system is strong, and wealthy enough, to enforce the management approaches used (Figure 7.9). For example, Australia has successfully introduced licences on reef fisheries to limit exploitation levels. Management through a top–down approach may also be successful when the fishery is more 'industrial' in nature, or controls in the processing chain, such as that on exports, can be applied. In the Pacific Islands, trochus and bêche-de-mer are export commodities that pass through a limited number of centralized buying locations. For trochus, the potential number of button factories, or at least the number of machines in operation, could be limited. This uses the fact that it is easier to enforce controls on the fewer exporters when compared to the large numbers of fishers.

More commonly in artisanal fisheries, local control has proved effective in managing the large numbers of individuals exploiting resources. In parts of the Pacific and South East Asia, traditional customary marine tenure (CMT) systems have allowed communities, or community leaders, to exert controls over local marine resources. In Fiji, for example, a system of CMT called 'qoli qoli' has been in place for many years, often as long as living community elders can remember (Cinner et al. 2007). In this situation, a system of ownership of specific sites or areas exists for specific individuals, families or communities, who can make independent management decisions based on traditional beliefs and conservation concerns (Golden et al. 2014). Access to the resource can be controlled, while rights to fishing can be allocated (e.g. Ruddle 1996).

Figure 7.9   An example of control where strong and wealthy governments are present: a fishery patrol vessel in the Indian Ocean.
Source: Photo by Graham Pilling.

These systems can maintain strong controls through penalties (including confiscation of gear, boat or catch), compensation, public shaming or even capital punishment. These common-property arrangements limit marine resource use in many ways, through the closure of fishing grounds, gear restrictions, limited entry and the protection of spawning aggregations (Colding and Folke 2001). In this way, they use tools similar to those used 'modern' marine fisheries management, as detailed in 'Approaches to managing coral reef fisheries', but the management approach is more appropriate for the local situation.

However, traditional approaches are facing increasing pressure due to the modern market-driven economy (Mathews et al. 1998). There is therefore a need to improve the information available to the management system in an attempt to ensure sustainability. As a result, co-management approaches have been put forwards (e.g. Cinner et al. 2012). Under this system, advice and information from regional or governmental scientists is incorporated within local management practices. These approaches have proven generally successful in the Pacific. For example, in Vanuatu, the Port Vila fisheries department performed trochus surveys around the island, providing villagers with information and advice on why minimum size limits were desirable, where trochus refuges were best sited within fishing areas and the length of closures required to rebuild stocks. Decisions were then taken by villagers, allowing them to balance biological, social and environmental considerations and resulting in the successful management of trochus resources (Pomeroy and Berkes 1997; Johannes 1998). Increasing fisheries governance through greater user participation and the implementation of property rights may achieve more sustainable resource utilization at the community level (Cinner et al. 2016).

7.7 Future for reef fisheries

Millions of people and thousands of communities depend on coral reef fisheries for their livelihoods and as sources of protein (Sadovy 2005; Teh et al. 2013). However, pressure on reef resources continues to increase.

Newton et al. (2007) noted that anticipated human population growth will increase the pressure on coral reef resources. Estimates of population growth in island states suggested that, by 2050, the Earth would require an extra 196,000 km^2 of coral reef with the same productivity as those reefs studied in order to support the increased demand from those islands alone. Increased global human population growth will only add to the demands on reef resources from the global market.

Obviously, increased coral reef area, particularly of that magnitude, cannot be created. Nor does it seem likely that aquaculture will be able to completely fill the deficit. It is effective management of fisheries on existing coral reefs that is primarily needed to restore and increase productivity and prevent resource over-exploitation.

Impacts on reef resources will also be indirect, as fisheries landings will also be affected by the impacts of future climate change and other man-made impacts. Climate will directly affect biological and ecological dynamics. Increased water temperatures will affect the metabolic processes of fish, since they cannot regulate their body temperatures. Species and fish-size compositions around reefs and distribution and abundance of species may change as temperatures within areas move outside the thermal optima for particular species, but this change may improve the viability of other species. Increased frequency of coral bleaching events driven by increased water temperatures, extreme weather events and ocean acidification due to climate change (Hughes et al. 2003) and exacerbated by increasing human pressures such as pollution and fishing will also indirectly affect fish populations. Coral mortality from warming events may reduce species richness due to the reduction of habitat niches and live coral cover. If the structural complexity of reefs is maintained, the fish assemblage may be less affected than expected (e.g. Riegl 2002). However, as coral reef architecture collapses, the diversity and abundance of corallivores and planktivores may reduce (Graham et al. 2007). Indeed, coral reef death and the breakdown of reef architecture may accelerate with future increases in the frequency and severity of extreme events such as hurricanes. Fish species that form the focus of coral reef fisheries may not be directly affected by coral die-off, but may be affected by reduced recruitment of young, and the knock-on effect of the impact of coral bleaching on smaller fish through the food chain. These effects may take decades to become apparent (Graham et al. 2007).

The impact of these changes on coastal communities relying on fisheries for their livelihoods is difficult to estimate. However, recent research suggests that climate change combined with over-exploitation will reduce yields from coral reef fisheries by up to 20% by 2050, while climate change may make coastal aquaculture less efficient (e.g. Bell et al. 2013). In contrast, potential increases in macroalgal cover resulting from coral die-off may result in an increase in the abundance of some invertebrates, which may offer an alternative focus for fisheries, although the level and potential benefit from such fisheries is difficult to predict. Coastal communities will also be directly affected by climate change, as the physical effects on environment are likely to result in an increased risk of natural disasters and health problems in the coastal zone.

The Pacific Island countries and territories (PICT) provide an example. The contribution of marine resources to the diet of Pacific Islanders is more than three to four times the average global per capita fish consumption and is currently sourced primarily from inshore (e.g. reef) subsistence and artisanal fisheries, particularly in rural areas. However, increasing human population sizes will place further pressure on nearshore marine renewable resources, leading to increased exploitation pressure and reduced resource health (Bell et al. 2009). This, in combination with predicted negative impacts of future climate change on coastal fisheries, will create a widening gap between the protein requirements of growing PICT populations and decreasing inshore protein supply (Bell et al. 2013). As a result, regional policy interest has focused on the utility of offshore fish resources for food security as a cheap source of protein in order to try to fill that gap (Pilling et al. 2015).

The future for coral reef fisheries is therefore highly uncertain. However, all is not doom and gloom. Examples presented in this chapter show that reef fish can be sustainably managed in the face of serious pressures, although the level of those pressures appears to be increasing. The wide range of impacts experienced means that the solution to future challenges may lie outside the control of conventional fisheries management. A proactive approach to adapt to future pressures, taking into account the particular characteristics of the reef concerned and maintaining reef complexity while managing the consequences when complexity is lost, is needed (Rogers et al. 2014). For example, noting the link between grazing parrotfish abundance and Caribbean coral reef ecosystem health, harvest and fishery size-limit management approaches may be able to maintain coral resilience to climate change while allowing exploitation (Nash et al. 2015; Bozec et al. 2016). Maintaining fish population sizes should also preserve genetic variation, thereby enhancing the potential for genetic adaptation to climate change.

Generally, however, these levels may imply reductions in the harvest and, hence, food security that can be obtained from coral reefs. Given the levels of dependency on reef fisheries, and the multiple threats to them, reductions in the catch levels needed to move towards sustainability require the identification and support of alternative livelihoods for many of the people currently dependent upon them (Adger et al. 2005). McClanahan et al. (2008) suggest that, in areas less affected by coral bleaching, adaptive approaches should focus on integrated conservation and development, combined with livelihood diversification. In more critical locations, development strategies should not make local communities or industries more dependent on reef-based resources that are at risk. If it can be ensured that the future consequences of coral bleaching, population pressure and other stressors on the reef are considered within future coral reef fisheries management and that a holistic approach is taken, there is hope that the productivity of coral reef systems can be maintained.

8   Coral reefs in the modern world

Modern reefs began to appear in the Triassic period, developed through the Jurassic and flourished in the Cretaceous. In more modern times, they survived the transition through the ice ages, when much of their present morphology was laid down as they grew at greatly varying elevations as sea level rose and fell extensively. The last deglaciation led to the most recent geological epoch, the Holocene. Today, it has been seriously proposed that a new epoch be named, called the Anthropocene, to reflect the fact that biota is being changed by man-made influences in a way as great as occurred in any of the previous massive transitions. Indeed, the volume edited by Birkeland (2015) even uses the term in its title.

Many factors are responsible, but most have a trend that is increasing in severity and rapidity. There is a trend in some quarters to assign blame to the 'new' global issues arising from climate change and CO_2, but local impacts are commonly overriding, in the short term at least. For example, near Jakarta, 80% of variation in benthic community changes can be attributed to local impacts (Baum et al. 2015). Amongst the impacts affecting reefs are direct forms of pollution, including nutrient enrichment from sewage and land run-off, pollution from toxic substances, and sedimentation which comes from shoreline alteration, construction and development. Overfishing is widespread and invariably induces changes to the trophic system (Chapter 7), which in turn leads to algal domination on reefs and several other manifestations of decline.

All such factors differently affect different species, species groups or 'functions' of a reef. For many functions of a reef, there may be considerable redundancy of species while. for other functions, such as top carnivory amongst fish or of grazers, there may be little species redundancy—little 'resilience' in the system, or little 'insurance' for when things go wrong, as Mora et al. (2016) put it.

While impacts may all be caused ultimately by human activity, many are not within the local control of a single small area or even country. Examples include warming of sea surface water. At the same time the surface ocean is

The Biology of Coral Reefs. Second Edition. Charles Sheppard, Simon Davy, Graham Pilling, and Nicholas Graham, Oxford University Press (2018). © Charles Sheppard, Simon Davy, Graham Pilling, and Nicholas Graham (2018). DOI 10.1093/oso/9780198787341.001.0001

becoming more acidic or, more precisely, less alkaline. Numerically, these causes may not appear great (about 1–2 °C temperature rise, and a pH reduction of about 0.1) but this is more than sufficient to cause extensive stress and mortality to corals (Hoegh-Guldberg et al. 2007).

Chapters 8–10 describe today's major stressors, emphasizing where their impacts are synergistic, and describe examples of consequences. One factor is commonly ignored: that of human overpopulation. This 'elephant in the room' (Sale 2008) is now overwhelming many reef systems yet is still widely ignored for cultural, religious and political reasons. Incorporating this factor as a major driver of reef deterioration is still actively discouraged in many international fora, even though it is, of course, the main reason lying behind much habitat deterioration. However, it is difficult to solve an equation if one important term in it is ignored, and this must be resolved, not only for reefs but most of Earth's habitats.

Although synergies between stressors are important, as is the magnitude of each, a commonly overlooked point is that many impacts are pulsed, not continuous, and intervals between stressful events are shortening (Done 1999). Disease events are more frequent, spikes in temperature are more frequent and, on some coasts, intervals between episodes of severe sedimentation caused by coastal construction projects are shorter, all leading to reduced ability for recovery, both at the physiological level of organisms and at the level of community structure. It is increasingly frequent episodes of warming, for example, coupled with decreasing intervals between sedimentation events (and probably greater ambient levels of both) which led recently to the total demise of possibly the largest single reef in the Persian Gulf, one which stretched between Bahrain and Qatar (Sheppard 2008) (see Chapter 10, Figure 10.4). Increasing mean levels and increases in the number and frequency of extreme events all combine to generate the current alarming rate of reef destruction and loss of the resources they otherwise would provide.

8.1 Damaging impacts on reefs

8.1.1 Nutrient enrichment

Sewage and nutrients contained in terrestrial run-off are some of the most deleterious forms of pollution on coral reefs. Ecological responses by corals, soft corals and algae clearly follow gradients in water quality (Fabricius et al. 2005). Coral reefs are well adapted to thriving in very low nutrient concentrations and suffer when nutrients are raised in several ways. The results are, invariably, a decline of coral cover and a corresponding increase of fleshy algae. The ability of corals to compete successfully for space against much faster growing fleshy algae is to a great extent dependent on there being a very low concentration of dissolved nutrients in the water (Adey et al. 2000). Increases in nutrient concentrations remove the competitive advantage of calcifiers, increasing that of fleshy algae. As long as nutrients remain low and sufficient grazing takes place, coral continues to dominate, despite having growth and space occupancy rates several orders of magnitude lower than the algae. The period from the 1970s to the 1990s, especially, saw many examples of the collapse of reef ecosystems from direct and localized impacts, when a lack of sewage controls, for example, coupled with rapidly rising human populations on coral reef coasts saw a marked shift from coral to algal domination on reefs.

In addition, nutrients can inhibit coral calcification more directly. The phosphate ion has an inhibitory effect on the formation or deposition of the aragonite (calcium carbonate) crystals with which the coral constructs its skeleton (Muller-Parker and D'Elia 1997), and levels of phosphate above about 1 µM reduce calcification.

Other indirect effects of nutrient enrichment may also include stimulation of outbreaks of the coral predator, the crown-of-thorns starfish (*Acanthaster planci*; see Chapter 6, 'The crown-of-thorns starfish'). It has been shown (Brodie et al. 2005) that fourfold increases of nutrient discharges to the Great Barrier Reef during the rainy season correlate with phytoplankton blooms, which, in turn, permit larger zooplankton populations to develop at the same times and places as their *Acanthaster* larvae predators develop. Thus, more zooplankton means more food for the crown-of-thorns starfish larvae. The causal chains of events are becoming clearer and appear to lead directly from nutrient discharges to increased coral predation.

8.1.2 Industrial and physical impacts: Landfill, dredging, sedimentation

Landfill and dredging are two of the most harmful coastal activities with respect to coral reefs. This is partly because of direct burial of shallow reefs, but also because sedimentation plumes arising from the activity commonly spread for several kilometres, blanketing and shading reefs far from the original sites of dredging or construction. The scale of this activity can be enormous: the Saudi Arabian Gulf coast is now nearly two-thirds artificial, for example, and it may be no coincidence that coral reefs in this gulf are mostly dead (Sheppard 2016); warming events have been fatal, but susceptibility to disturbance is high (Bento et al. 2016; Buchanan et al. 2016). The driver for the activity is that the price of shallow reef areas usually rises if it has been converted into dry land for construction purposes.

Landfill sometimes is termed reclamation (by engineers doing it, at least). But this is a misnomer. The land is being 'claimed', not 'reclaimed', with the difference being that, in coral reef areas, the boundary of land and sea usually exists in dynamic equilibrium. Landfill forces an imbalance to both the boundary and depth profile of the coast; after landfill, currents and waves immediately work to 'reclaim' a stable configuration. This can lead to substantial erosion of the newly created land, requiring substantial shoreline armouring to protect against erosion, or to marked changes further along the coast.

Clearly, when shoreline habitats such as fringing reefs are buried, complete destruction of that reef occurs. Examples are numerous, the commonest driver being extension of the shoreline to gain additional land for construction. This may occur where land is otherwise scarce, such as beside the capital city of the Seychelles (Figure 8.1), and many island states such as Singapore, where nearshore reefs have become filled and 'fused' with the mainland.

Deliberate removal of patch reefs and fringing reefs for projects such as harbour construction or anchorages is also common. Most atoll lagoons naturally support numerous coral pinnacles (or 'bommies') which reach the surface but which have been removed for navigation reasons. But this may substantially reduce the bottom cover of stony corals in the lagoon, reducing not only diversity and organic productivity but also the production of sand that would replenish adjacent shorelines.

Figure 8.1   Landfill in the Seychelles. Seaward of the landfill are breaking waves at the edge of the reef flat.

Because of the high cost of landfill engineering, landfill for recreation purposes alone (Figure 8.2) is a more exotic use. Development for housing or industrial use is more common, and the most ambitious to date are the huge Palm Island and World Island complexes of Dubai (Figure 8.3). Impacts of the latter on all marine life in the area, not only coral reefs, are of potential concern and, although they are becoming well understood actions based on them lag severely (Vaughan and Burt 2016), something that is common to many parts of the world.

Impacts from landfill commonly extend over a greater area than the immediate development footprint. The material used commonly comes from excavating material from adjacent and deeper marine sites (commonly misnamed a 'borrow pit' in engineers' parlance). This destroys the excavated site, as well as the site where the dredged material is deposited. The material deposited as landfill is then layered or sculpted into the required shape

Figure 8.2   Fill on a reef flat, Jeddah, Red Sea coast of Saudi Arabia. When it was built, water exchange in the ponded areas was very poor, and the fringing reef itself was very heavily damaged.

Source: Photo taken from Jeddah telephone directory of the 1980s.

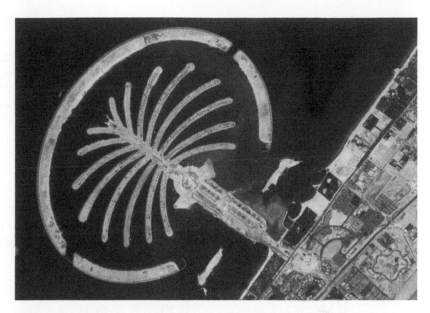

Figure 8.3   Landfill on and near coral reefs off Dubai, Persian Gulf. The development is known as Palm Islands. To the south is another Palm Island complex and, to the north, a large landfill area known as World Islands.
Source: Image from Google Earth.

on and around the reef (sometimes termed 'soil improvement'!). Both the excavation and the fill processes cause suspended sediment plumes, which are notoriously damaging to coral reefs. These plumes are the most difficult aspect to control effectively, and they may flow downcurrent for many kilometres. A large contribution to the problem comes from the method of excavation. Although responsible methods do exist, a common method is to use a suction cutter dredge, which cuts the substrate and then vacuums the material into tankers. Heavy particles settle in the tanks, while water and the finest particles are allowed to simply overflow back to the sea, until the larger particles have filled the tanks. Alternatively, if the sites of excavation and fill are adjacent to each other, the suctioned material is pumped directly to the fill site, again allowing massive overflow of excess water and very fine material until sufficient fill has been obtained. This spillover material commonly resembles liquid mud, which is extremely damaging to almost all adjacent marine life.

Effects of sediment are not easy to quantify in an ecologically meaningful way, although some physiological impacts are clear such as its adverse impacts on the reproductive ability of corals (Jones et al. 2015). Measurement of the optical quality of water is straightforward, but measurement of sediment settling rate is difficult and fraught with error, partly because of problems of later resuspension and further dispersal caused by water movement. Particle grain size is particularly important, with finer particles being most damaging, being transported furthest and remaining in suspension for longest. Because excavation includes layers below the surface, a much greater proportion of fine particles are suspended than would be the case after a natural storm. The tolerance of a smothered species is highly variable, with some colony shapes being more likely to trap falling sediment than others. Soft corals may also tend to shed sediment passively, while removal of sediment by stony corals is more active, using copious quantities of energy-requiring mucus.

Water around reefs is generally very clear, containing considerably less than about 5 mg of sediment per litre. This may cause settling onto the reef of perhaps $1-10$ mg cm^{-2} d^{-1} (Brown 1997b), with the lower end of this range being normal in unaffected reefs, and values above 10 mg cm^{-2} d^{-1} causing coral mortality (Rogers 1990). In terms of suspended sediment, $5-20$ mg L^{-1} may be regarded as causing stress, with over 20 mg L^{-1} being beyond the level of tolerance of many species. In terms of precipitating sediment, moderate sediment stress is caused by levels of $10-50$ mg cm^{-2} d^{-1}, while settlement greater than this may be classed as severe.

Harmful effects from sedimentation arise for two physiological reasons. First, an increased turbidity decreases light penetration, reducing the coral's photosynthesis and energy supply. At the same time that the coral is starved of energy, it needs a greater amount of it to actively slough off sediment which settles onto its surface. Organism physiology and resistance to sedimentation varies and, for all groups, too much sediment for too long results in organism stress and mortality.

All species can survive brief exposures of sedimentation, and all will experience increased turbidity following every storm. However, coastal engineering activities last far longer than storms, and typically release much smaller grain sizes, for longer. However, the size of any one development may not correlate well with its impact or with the changes it causes to current flows around or across a reef. Even a construction with a very modest footprint can cause substantial downstream effects, none more so than breakwaters and harbour wave barriers, which occupy little ground by themselves but whose design may greatly affect exposure, longshore drift and even mean seawater temperature in those places where water becomes partially impounded. Shores previously regarded as being distant from the development may suddenly receive much more sediment than they have been adapted to become starved of sand or experience increased wave energy and commence eroding. Numerous examples exist of undermined roads and buildings which were previously protected by a fringing reef but which lie downstream of even small but unwisely planned developments.

8.1.3 Chemical and oil pollution

Effects of metals, organic compounds and oil have been extensively studied. Metal analyses are amongst the easiest to carry out, and metals have well-known toxic effects. Several toxic metals accumulate up food chains, thus amplifying effects by the time humans consume the species. Many reef-dwelling species are used as monitors for toxic metals (Hutchings and Haynes 2000), and it is commonplace to find 'hotspots' of metal pollution in the vicinity of industrial areas. Different metals may impact on different stages of life cycles, and many are damaging at extremely low levels, affecting tissues such as nerves, in the case of mercury, or reproductive tissues, in the case of copper, for example.

Organic pollutants may be harder to analyse and many of the most bioactive ones are subject to rapid metabolism and so may be detected post hoc by their metabolites, or by their effects. Herbicides are potentially amongst the most toxic compounds in reefs, as they affect the zooxanthellae in corals as well as seagrasses and macroalgae. Several biocides appear in coral reef environments, from agricultural run-off and also more recently as a component of antifouling paints, following the ban on the extremely effective tributyltin. A 'booster biocide' irgarol has been a very effective ingredient of the newest range of antifouling paints, but it is also very toxic to photosynthesis of coral zooxanthellae, inhibiting the Photosystem II mechanism at water concentrations as low as 50 ng L^{-1}, or 50 parts per trillion. Antifouling products are designed to be highly toxic, so toxic effects on marine biota away from the ship's hull or other structure should be unsurprising.

Substantial work on the GBR has shown that numerous pesticides are discharged via rivers into Queensland's inshore waters, and these pollutants tend to accumulate in a range of places, from sediments to marine mammals. The concentrations of some have been demonstrated to cause deterioration in the condition of reefs, to the extent that they could compromise the World Heritage values of nearshore parts of the GBR (Hutchings and Haynes 2005).

Substantial work has been done on effects of oil pollution, reviewed thoroughly in NOAA (2001), along with many case studies and ways of restoring reefs after spills. Much of the earliest work was done in the Gulf of Aqaba, both on the oil itself as well as on the compounded effects of using oil dispersants and emulsifiers during subsequent clean-up processes. Chronically polluted areas exhibited higher mortality rates in some dominant corals as well as a reduced reproductive potential and reduced planulae settlement rates (Loya and Rinkevich 1980). Numerous studies have shown that dispersant chemicals multiply the harmful effects of oil on corals and other reef biota. A well-studied oil spill off Panama showed that harmful effects could extend to about 8 m depth, and coral cover was essentially halved after a few months (Guzman et al. 1991) with effects decreasing with increasing depth.

However, where oil simply floats over reefs, its impacts may be very limited. After the deliberate, huge oil spill in the Persian Gulf in 1991, oil passed
over most of the small reefs in that area, causing only very limited effects,
in contrast to the enormous and very damaging impacts on adjacent shorelines, where beached oil has persisted for many years. Large sheets of oil were
found to have sunk onto the seabed, creating anoxic conditions beneath.
It is assumed that the hot climate caused lighter fractions of the oil (which
includes many of the soluble, toxic components) to evaporate quickly, possibly before the slicks passed over reefs. Subsequently, an increase in density
of the remaining oil, made heavier no doubt with atmospheric dust and sand,
caused slicks to sink far offshore onto soft substrate. The number of slicks
known to have passed over coral reefs without causing noticeable impacts is
considerable. Damage may therefore depend partly on the origin and hence
chemical content of the soluble toxic components of the oil, while the state
of the tide also determines the amount of physical blanketing of very shallow
fauna such as that on the reef flat.

Because of these differing variables of duration and intensity of exposure,
different clean-up methods (including the use of different dispersants) and
oil origin, results from case studies have commonly been apparently contra-
dictory. Newer dispersants and greatly improved methods of dealing with
clean-up have resulted in substantial improvement in mortality figures. On
many shorelines, simply leaving the oil to the attentions of bacterial degradation has also been adopted, as it causes less damage than physical clean-up,
but varying circumstances dictate different approaches.

8.2 Other physical impacts

8.2.1 Structural stress from construction

Shortage of space for human expansion has been a problem on many coral
islands. On Malé, the capital of the Maldives, a shortage of space has led to
construction over almost the whole of the city's perimeter reef (Figure 8.4).
The result is that there is now no natural breakwater afforded by a reef flat
and, in the 1980s, this led to massive damage from a surge created by a distant
storm, leading to the need for a massive artificial breakwater to replace the
reef flat.

But of possibly greater concern today in such places are the effects of the
construction itself and later consequences from discharges. The discharge
of untreated sewage from the city over the reefs has essentially killed off
the reefs, which now are covered with algae and almost certainly no longer
accreting. The reef has thus moved from an accreting to a destructive phase
(Risk and Sluka 2000), and was probably killed almost completely in terms

Figure 8.4   Malé, capital of the Maldives. Construction has taken place to the edge of the surrounding reef flat and, to the south (left side in the photograph), the shore is now protected by an artificial breakwater.

of net growth in the early 1990s. Construction on the approximately 1×2 km island has been intense, using heavy equipment, pile drivers and blasting. Sewage discharges have increased, and the freshwater lens has all but disappeared (Risk and Sluka 2000). Many of the buildings themselves are built from coral which was excavated from reefs. Some relatively high-rise buildings were also constructed on the limestone foundations. As a consequence, cracks have been revealed which fissure the reefs. Some may be natural, which accreting processes would possibly have filled in had the reef been healthy, but there is evidence of additional cracking caused by the developments. Seismic work showed that the Holocene layer of reef is about 20 m thick, and that below it are the typical stacked layers of successive, earlier growth which form the foundation of the island. The entire island is part of the atoll rim, and Risk and Sluka (2000) have warned repeatedly that, if these deep cracks develop or reach a critical stage, possibly triggered by a distant earthquake or by more blasting or pile-driving, then parts of the rim could give way, drowning thousands of people.

8.2.2 Boat anchoring on reefs

Boat anchoring is clearly directly damaging to corals and other benthic fauna by direct crushing (Figure 8.5). The damage comes both from the anchor, especially when it drags, but more from anchor chains which pulverize a

Figure 8.5   Top: An anchor from a yacht in a popular Caribbean cruising ground tearing up sea fans and the reef-building coral *Montastraea* (British Virgin islands). Bottom: Damage caused by a fisherman's anchor in staghorn coral (Malaysia).

circular patch as the boats swing. Lagoonal corals include the most fragile forms, and lagoons are, of course, locations of choice for anchoring because their water conditions are calm. Anchoring a small, personal boat can cause 'halos' of cleared reef 10 m in diameter; larger ships may destroy anchoring arcs with an area of >3,000 m² from a single anchor drop in 1 day, requiring recovery periods of more than 50 years (Smith 1988).

This poses threats to the continued existence of reefs in some areas. Persistent anchoring along stretches of coast has resulted in numerous anchoring circles merging to leave continuous stretches of pulverized reef. Continued anchoring grinds those fragments down into ever finer sediment, which then causes the sediment effects noted already, both downcurrent and along the reef shoreward of the anchoring area, which is too shallow for boats yet supports rich corals. This occurs in several Caribbean islands where yachting is a major tourist industry but where regulation has been insufficient. The solution is in fact simple, and lies in provision of mooring buoys. When these are attached to the substrate, which itself involves negligible damage, no further damage need occur from anchoring. In several countries, recognition of the substantial damage caused has led to installation of sufficient moorings, as well as management measures that include penalties for anchoring.

8.2.3 Nuclear testing

The largest explosions to be conducted on coral reefs are, of course, fission and fusion bombs. Bikini, Eniwetok, Johnston, Fangataufa and Mororua, where the United States and France have conducted testing, are the best known. Studies have mostly focused on effects of radiation and human health, with fewer studies on impacts and recovery of the reefs. Many of the nuclear tests were surface or subsurface, causing cratering, surface water heating to 55,000 °C, shock waves of 30 m high moving at 80 m s⁻¹, and shock columns reaching to the lagoon floor 70 m deep. Coral fragments are reported to have landed on ships monitoring the blasts, and entire islets were vaporised, with substantial consequences and changes to the nature and grain size of enormous quantities of sediments. Cracking of reef rim structures has also been reported.

Richards et al. (2008) studied the effects of the Bikini Atoll testing programme on the corals of that atoll. While the immediate destruction of reef life was not recorded, the site previously had been the location of substantial investigations on the taxonomy of corals before the testing commenced. Of the 183 species that had been previously recorded, about 42 had been lost by the time of a survey conducted in 2005. While some of the losses may be due to random effects, some were of 'obligate lagoonal species' which disappeared and which subsequently have not recolonized from other areas.

8.3 Coral diseases

Many forms of 'ill health' of coral reef organisms have been recorded, but the pathology is complex and there are several still unknown causes and agents. In common with Weil (2004), the term 'disease' is reserved for afflictions by a known pathogen, while 'syndrome' is used for effects whose causes are still unidentified, or where the affliction is not caused by a pathogen but by, for example, warming. Thus, several syndromes may become diseases by this definition, as research identifies causative pathogens.

Weil (2004) documents and illustrates many Caribbean reef diseases, noting that the first known was by a fungus affecting reef sponges in 1938, killing over 70% of colonies in affected areas. Thereafter, reports increased throughout the world linking diseases to deteriorating water quality. It has been suggested that the greater number of reports was simply a function of increased research, but it is difficult to imagine that the reef researchers of the 1970s or earlier all simply overlooked such important and usually conspicuous events. In Guam, for example, sewage-derived nitrogen accounted for more than 48% of the variation in coral diseases (Reading et al. 2013); likewise, in Hawaii, sewage nitrogen increased diseases and decreased coral cover (Yoshioka et al. 2016).

Most main pathogenic groups have been associated with one or more diseases, including bacteria, slime moulds, protozoans, trematodes, ciliates and cyanobacteria. The presence and role of viruses is also being researched but, as with most of the other pathogenic groups, it is often unclear where these are the primary pathogens or some kind of secondary infection of already diseased tissues. Some pathogens have clear associations with, for example, human sewage, as in the case of the bacterium *Serratia marcescens*, to which is attributed 'white pox disease'. Several diseases appear to be caused by combinations of several pathogens and are known by their appearance, for example white band disease (WBD) (types I or II), white pox disease, red band disease, yellow blotch disease, purple band and several others (Figure 8.6). Several known pathogens appear to have been introduced by pollution, but others are caused by organisms known to be endemic but which are triggered into pathological condition by rising temperature or other stresses.

The pathogen assemblages impart the colour for many of these diseases but, in diseases where 'white' is part of the disease name, this colour is usually simply that of the underlying limestone skeleton, which shows through when original tissue is lost. Thus, in WBD, the white band of newly exposed skeleton moves along the coral skeleton; healthy coral tissue lies ahead of it, while behind it is a skeleton which is darkening again due to colonization by filamentous algae and other opportunistic organisms.

Reflecting partly the research done, but possibly also greater stresses from pollution, more than three-quarters of known reef diseases have been

Figure 8.6   Diseased corals. Top: A recently killed elkhorn coral (*Acropora palmata*), (British Virgin Islands). Bottom: A *Diploria* brain coral with black band disease. The band is visible slanting diagonally across the coral, the top half is dead and the band is creeping downwards, destroying the living polyps as it goes (Bermuda) (See Plate 14).

reported from the Caribbean (Rosenberg and Loya 2004). To date, associated pathogens have been identified for only about a third of the syndromes or diseases. Therefore, several of the wide range of names may indeed relate to the same pathogen occurring in different areas. In the Indo-Pacific, reports of diseases range from the Red Sea to the Pacific, including several syndromes not reported from Caribbean reefs (Willis et al. 2004).

The first major disease known to have widespread and profound consequences was WBD, sometimes conflated with white plague, which is similar but not identical. WBD first appeared in the 1970s. This affected the elkhorn coral (*Acropora palmata*; Figure 8.6, top), reducing the cover by this coral by 80%–90%. The pathogen has remained elusive despite many years of investigation, although it may be a form of *Vibrio*. WBD generally spreads up colony branches from basal areas to branch tips, while another type sees lesions occurring more variably.

Elkhorn coral dominated many reefs between the low water level and about 4–8 m depth, forming almost impenetrable forests with rich three-dimensional structure and orientating its massive branches into the oncoming wave direction. It thus provided substantial shoreline protection. Following the demise of elkhorn coral, its coverage in shallow water declined to a small fraction of former values throughout most of the Caribbean. Recovery of mature colonies did take place in many locations, but white pox disease then became evident from 1996 onwards. This is highly contagious, and coral tissue loss can spread at rates of over 2 cm d^{-1}, with patches eventually merging to cause total colony mortality (Sutherland and Ritchie 2004). The disease is caused by the gram-negative *Serratia marcescens*, which is a coliform bacterium associated with sewage from humans and other animals. Thus, one disease greatly reduced the largest branching corals of the Caribbean, while another now keeps it at it very low levels. Together, these diseases were the most important factors to have reduced this once dominant and most important primary shallow reef builder (Aronson and Precht 2001).

This disease (or suite of diseases) also eliminated much of the more finely branching *Acropora cervicornis*, turning this once common species into a rare coral in many parts of the Caribbean also. This species is relatively fragile, and depends to a great extent on dispersal through asexual fragmentation when its thin branches are broken by waves. Because it has low levels of sexual dispersal, its recovery has been equally poor and its populations remain low.

Black band disease (BBD) was the first coral disease to be studied, being reported in the early 1970s (Figure 8.6, bottom). This has been found in all tropical oceans and, in cooler water areas, it is more prevalent in warmer seasons. Its progression may be very slow, progressing over massive corals at <1 cm y^{-1}. The pathogen has been elusive because the black mass contains a wide range of microorganisms. Many of these may simply co-occur adventitiously, but a consortium of microorganisms may be involved, rather than a single pathogen (Richardson 2004). One element which consistently is found is *Oscillatoria*, a gliding, filamentous cyanobacterium.

Consequences of these and several other coral diseases are varied. Some do not appear to be of major importance at present. Some have caused major changes to reef structure, particularly in the case of the shallow water elkhorn coral, and more lately in the Caribbean in the important, deeper primary reef builders of the *Montastraea* group. In the Indo-Pacific, massive effects from diseases are less common, though. Coupled with diseases of other major reef organisms, and in some cases triggered or accelerated by warming, these diseases are changing the structure of many reefs in the Caribbean, turning large areas of once rich coral cover into areas with little cover and low diversity and productivity.

Microorganisms can affect reefs in ways beyond causing direct disease, as is entertainingly summarized by Rohwer et al. (2010). Such effects include, for example, simple anoxia at the coral tissue when it is blanketed by microbial-rich mucus covering the surface of a coral in areas where there is nutrient enrichment.

8.4 Diseases of other reef organisms

Other invertebrates and algae have also been affected by diseases. A wide range of diseases have now been identified (Peters 1997), some of which are proving to have far reaching consequences.

8.4.1 Red algae

Some members of the calcareous red algae are particularly important when it comes to reef construction, forming as they do the algal ridge and spur system which takes the brunt of oceanic waves. These algae suffer from a number of diseases. In the early 1990s, a bacterial pathogen was identified on Pacific reefs, recognized as bright orange, spreading patches. After infection, only the bleached white skeleton was left (Peters 1997). Another documented disease is coralline white band syndrome. The reasons behind the increase in diseases in the red algae are unknown but, given the importance of this group, it may become increasingly important as a change in acidification reduces the ability of these algae to deposit limestone.

8.4.2 Caribbean *Diadema*

One of the earliest reports of massive reef disease epidemics was of the grazing urchin *Diadema antillarum*, again in the Caribbean. This major grazing herbivore first succumbed off Panama in 1983, leading to suggestions that the waterborne pathogen entered the Caribbean via the Panama Canal or via ballast water of ships. Waves of mortality were then observed to radiate from that point, following roughly the pattern of water currents, until, after 1 or 2 years, adult populations of the sea urchin were reduced by 85%–100% (although juveniles were rarely affected). Recovery today is patchy but increasing.

These urchins were one of the most important reef grazers on Caribbean reefs. Some researchers suggested the mass mortality of these grazers was the cause of the subsequent increase in algae which came to dominate many reefs, although it is likely that several concurrent factors such as increasing nutrient and sewage run-off, and continued fishing of other herbivores, were locally equally important; indeed, some of the coral decline preceded the urchin mortality.

8.4.3 Caribbean gorgonians

Aspergillosis is a serious disease of Caribbean sea fans and is caused by the pathogen *Aspergillus*; where present, this disease kills large proportions of sea fans, including the conspicuous and iconic reef-dwelling *Gorgonia ventalina*. The fungal pathogen is widespread and salt tolerant and, in the last 20 years or so, it has become a serious cause of sea fan mortality (Figure 8.7).

Figure 8.7   The very abundant Caribbean sea fan *Gorgonia ventalina*. The bottom part is living, but the dark part of the fan is dead, decaying and covered with a red fungus (See Plate 15).

It appears as patches of purple or reddish brown on the sea fan blades, which later die, sometimes appearing black, but then abrading away. More recently, this disease has spread to other Caribbean gorgonians too, such as the spectacular *Pseudoptera*, amongst others.

The fungus is a natural inhabitant of soils, and it has been suggested that the disease comes from a massively increased introduction of the fungus into the marine environment. Local terrestrial run-off via rivers is one possible cause (Smith et al. 1996), and it has been isolated from the mouth of the Orinoco, which disperses water throughout the Caribbean (Smith and Weil 2004). Work has shown that injection from Saharan dust storms reaching the Caribbean is increasingly likely (Shinn et al. 2000; Garrison et al. 2003). The dust contains *Aspergillus*, along with numerous other known pathogens, and has been swept up by winds over northern Africa in quantities of hundreds of millions of tonnes over the last 25 years, and deposited over a wide area from the Amazon Basin to North America. Some of the near-synchronous mortalities over reefs have correlated with years of maximum dust transport. The effects of this dust also extend well beyond reef mortalities to a range of direct human health problems.

8.5 Climate change

8.5.1 Temperature rise

One of the most profound changes that have affected coral reefs in the last 20–30 years has been the widespread bleaching of corals, triggered by warming water (see 'Corals as archives of past climate'). Bleaching had been noticed on occasion since the 1870s, but the main causes were thought to be local, such as sewage discharges, sedimentation, terrestrial run-off, overfishing and industrial pollution. Increasing UV irradiation was a possible newer addition to the list. In the 1980s and 1990s, the focus was still on purported drivers of community change such as those described in Sections 8.1–8.4, as these, after all, caused the most marked observable effects in most places. During the 1990s, more than one influential and authoritative review scarcely noted warming as being an important issue at all. The remoteness of some coral reef areas, such as mid-oceanic atolls, from some of man's most extreme depredations meant that such areas were thought to be relatively immune from damage. This changed in the 1990s, when more severe temperature spikes, superimposed on top of an underlying rise in mean temperature, caused widespread damage throughout much of the coral reef world.

Corals as archives of past climate

Instrumental climate records from the tropical oceans are typically short (150 years or less) and not adequate to resolve the full range of natural climate variability on seasonal to centennial timescales. Scleractinian corals from modern reefs of the tropical and subtropical oceans provide an excellent archive of past climate variability, and may help to fill the gaps in the instrumental database.

Massive hermatypic corals are particularly well suited for climate reconstructions, as their uniform growth and longevity allows the development of long, continuous chronologies that extend over several centuries. Corals that have been used successfully for environmental reconstructions include massive *Porites* species and *Diploastrea heliopora* from the Indo-Pacific, as well as species of the massive genera *Diploria, Montastraea*, Orbicella and, to a lesser extent, *Siderastrea* from the tropical Atlantic.

Scleractinian corals form exoskeletons that consist of pure aragonite ($CaCO_3$), a common carbonate mineral. The living tissue, which is the site of calcification, occupies only a small fraction of the coral exoskeleton (typically, the outermost 1–5 mm). Coral growth includes linear extension and thickening of the skeleton. This results in a characteristic pattern of alternating high- and low-density bands, which are revealed by X-radiography (Figure, left). Each high- and low-density band pair represents 1 year of coral growth. Counting these density band pairs provides a precise chronology. Patterns of annual density bands may reflect changes in a number of environmental parameters (e.g. temperature, nutrient availability and light conditions). Typical growth rates vary between 0.5 and 1.5 cm y^{-1} (*Porites* spp., *Diploria* spp., *Montastraea* spp.) to 2–5 mm y^{-1} (*Diploastrea heliopora, Siderastrea* spp.).

Massive coral skeletons also carry a suite of so-called geochemical proxies in their aragonite skeletons, and these proxies provide excellent records of the environmental conditions in which the coral grew. Currently, the two most widely used coral proxies are the Sr/Ca ratio and the stable oxygen isotopes ratio ($\delta^{18}O$) in coral aragonite. The Sr/Ca ratio is an excellent tool for deriving high-resolution proxy records of past sea surface temperatures. Application of the Sr:Ca thermometer relies on the assumption that coral Sr/Ca varies predictably with temperature and that seawater Sr/Ca is invariant on millennial timescales, due to the long residence time of Sr and Ca in the ocean. In contrast, the oxygen isotope ratio of coral aragonite varies in response to temperature and changes in the $\delta^{18}O$ of seawater. The latter depends on the freshwater balance (evaporation and precipitation) and co-varies with salinity. Thus, at sites where seawater $\delta^{18}O$ variations are large, the combination of coral Sr/Ca and $\delta^{18}O$ measurements allows the reconstruction of sea surface temperature and changes in the freshwater balance and salinity. Other element-to-calcium ratios (e.g. Ba/Ca, U/Ca, Mg/Ca, Pb/Ca) in coral skeletons have been used to reconstruct environmental changes, with various degrees of success. Coral Ba/Ca ratios show great promise as indicators of terrestrial run-off. Pb/Ca ratios have been shown to record industrial pollution. However, other geochemical proxies such as U/Ca and Mg/Ca appear to be strongly influenced by biological effects and thus are of limited use as recorders of environmental changes.

Continued

Corals as archives of past climate (*Continued*)

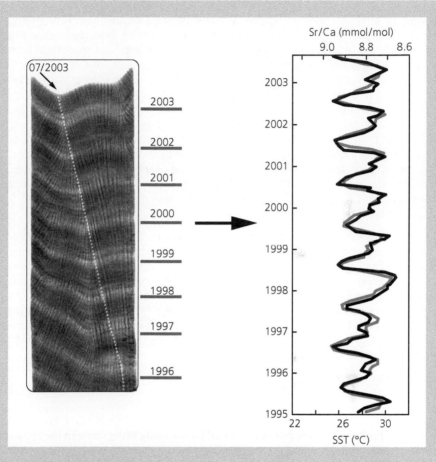

Figure. Massive corals show clear annual density bands, and can be easily sampled at monthly resolution. Left: An X-ray image of a *Porites* coral, showing typical annual density bands. The sampling transect for geochemical analysis is indicated by the white dots. Right: Sr/Ca ratios measured in coral aragonite provide monthly time series of ambient water temperature (black line). The quality of these proxy time series is comparable to that of instrumental records of sea surface temperature (SST; grey line).

The rapid growth of massive hermatypic corals allows the development of weekly to monthly resolved geochemical records, with a quality comparable to instrumental records of climate (Figure, right). This is of great importance, as the dominant modes of climate variability operate on a seasonal timescale, and other paleoclimatic archives (e.g. tree rings, lake sediments, stalagmites) typically only provide annual to decadal resolution.

Continued

Corals as archives of past climate (*Continued*)

To date, an extensive network of coral $\delta^{18}O$ and/or Sr/Ca records has been developed from the tropical Indo-Pacific and, to a somewhat lesser extent, from the tropical Atlantic Ocean. These records are typically 100–400 years long and have monthly to annual resolution. Most published coral records can be accessed via the NOAA Paleoclimatology Homepage (<http://www.ncdc.noaa.gov/paleo/paleo.html>). The coral proxy time series have been used to constrain the past behaviour of the dominant modes of the global climate system, including the El Niño–Southern Oscillation, the Asian monsoon, the Pacific Decadal Oscillation, tropical Atlantic variability and changes in hurricane intensity. Currently, a set of multicore reconstructions are being developed from the available coral time series, using statistical methods originally developed to improve the quality of historical climate data or for dendrochronological studies. In the near future, by using various techniques for cross-dating, it may be possible to produce multicore coral reconstructions that provide excellent records of tropical mean temperatures, extending over the past 500 years or even the past millennium.

Well-preserved fossil corals can be dated using radiometric methods (uranium series and/ or radiocarbon) and provide time windows of seasonal to decadal climate variability on geological timescales. However, the application of these methods is usually restricted to the past 130,000 years, because the aragonitic coral skeleton undergoes rapid diagenetic alteration. In rare cases, however, fossil corals have provided monthly resolved records of temperature variability, with these records going back as far back as 10 million years. Thus, fossil corals can provide important information on the dynamics of major climatic modes, such as the El Niño–Southern Oscillation, in times of different climatic boundary conditions, such as glacial–interglacial periods.

Dr Miriam Pfeiffer, University of Cologne, Germany.

Early inklings of problems caused by warming water came in the 1980s from a few sites, such as the Eastern Pacific. A serious event in the Eastern Pacific in 1982–3, causing high mortality, stimulated considerable search for causes (Glynn 1993, 1996), and the potential synergistic enhancement that warming would bring to 'traditional' pollution effects was recognized (Wilkinson 1996). Later, warming in the Persian Gulf in 1996 caused widespread mortality, an event which generated more interest, but in 1998 there was a warming event more severe than any previously recorded. In the Indian Ocean especially, corals bleached and then died over vast areas. The Caribbean was also affected (if not as noticeably as the Indian Ocean, because so many of its shallowest corals had already succumbed to disease anyway). In contrast, some Pacific islands were affected only a little or even not at all. In the worst affected sites of the Indian Ocean, mortality of corals was up to 90% over vast areas, with 100% being recorded on many shallow reefs. In 2005, it was the turn of the Caribbean, where widespread mortality affected almost all reefs. It was clear that the dominant ecological driver of reef condition had markedly changed. Then, in 2015–16, a further warming spike lasting several months caused approximately the same scale of mortality as the 1998 episode, so that, at the time of writing, a large proportion of the world's coral reefs have been bleached and then killed.

8.5.2 The progression of a bleaching event

The physiology of bleaching is described in detail in Chapter 4 but, briefly, it arises because of the expulsion of the coral's symbiotic zooxanthellae and/or the loss of their photosynthetic pigments. While most symbiotic associations 'are generally remarkably tolerant of variation in abiotic conditions . . . symbioses with *Symbiodinium* species are exceptional in that they commonly live in habitats at 1–2 °C below the temperature which triggers collapse of the symbiosis' (Douglas 2003). Cells of the alga *Symbiodinium* are deeply pigmented and, because the remaining coral tissue is mostly transparent, the white limestone shows through.

Whether a bleached coral recovers or dies depends on both the extent of the temperature rise above the usual value, and the duration for which the temperature is elevated. Exact values vary, but a measure of about 10 degree heating weeks (e.g. 1 °C rise above expected for 10 weeks) has become a standard, useful measure for estimating whether severe consequences will occur. Temperatures during the 'shoulders' of the peak are important also (McClanahan et al. 2007), as is previous warming history, when some sort of conditioning of corals to increasing temperature may occur.

Other environmental variables are important to bleaching too. The amount of light received by the coral is important. If, for example, strong winds fail in one year, the sea surface may become calm and flat, and, because a flat sea surface reflects much less light than a wave-covered surface, double the amount of photosynthetically active radiation penetrates through the surface of the water to reach corals. The two factors amplify each other, and bleaching occurs more readily. This may be observed during bleaching events, where many coral colonies are killed on their upper surface but not on shaded portions (see Chapter 4, Figure 4.8).

If the coral dies, its tissues slough off, leaving the limestone skeleton bare. This will rapidly become colonized by a wide range of organisms, two groups of which are especially important. First, algae quickly colonize the surface. Because these superficial algae contain chlorophyll, the coral skeletons darken again, and regain some of their former colour. From a distance they may be indistinguishable from living coral, so that attempts to track these events over remote areas by satellite have been hindered.

The second important group to colonize the dead surfaces are bioeroders. Without a live surface of carnivorous coral polyps to prevent settlement of such species, coral skeletons become eroded. Surface features disappear

Figure 8.8   Table corals (*Acropora cytherea*), which were mostly killed in a warming event of 2005. The dusty grey colour indicates dead surfaces, while darker surfaces indicate sections which are still living. These disintegrate after a year or two (Egmont Atoll, Indian Ocean).

and finely branched species disintegrate in a few weeks. Others may remain identifiable for several years. Table corals disintegrate from their edges, leaving a central stump which may persist for 5 years or more (Figure 8.8). An important consequence in ecological terms is that the reef increasingly loses its important three-dimensional structure so that, unless new corals appear, the surface of a heavily affected reef gradually takes on the appearance of a rather flat and lifeless limestone plain, lacking the three-dimensional structure which was so important in providing habitat for a high biodiversity.

8.5.3 Sea surface temperature curves and forecasts

Sea surface temperature data are collected at a range of different scales. In a wide range of sites, instruments record sea surface temperature at high resolution. Today, sea surface temperature data are also recorded by satellites which have a global coverage but also have a coarser spatial and temporal resolution. Some of the most useful work involving temperature blends a combination of both approaches.

In general, mean sea surface temperature started to rise in about the 1970s, but this was irregular, and several sites showed 3–5-year cycles which are related to various climatic cycles such as the El Niño oscillation and the Indian Ocean Dipole (Figure 8.9). Most of the warmest years in the last 137 years occurred very recently so that, even for those who persist in doubting that forecasts of continued warming are 'real', the historical record shows an alarming trend (Figure 8.10).

When historical records are combined with forecast data from the climate models (Sheppard 2003b), the future predicted temperature rise is significant (Figure 8.11), and trends suggest that, after 25–60 years, most reef areas will suffer mortality too frequently to be able to recover. This accelerating pattern of rise is seen throughout the oceans, although it is not completely universal, judging from present data. Donner et al. (2005) develop these analyses and remark that 'bleaching could become an annual or biannual event for the vast majority of the world's coral reefs in the next 30–50 years without an increase in thermal tolerance of corals of 0.2–1.0 °C per decade'. These predictions appear to be being borne out with the most recent work (van Hooidonk et al. 2016).

Following a warming event, the soft corals also die. Many of these alcyonarians contain symbiotic algae too, although they are not as dependent on the products of symbiosis as are stony corals (Fabricius and Alderslade 2001). However, many reefs support as many soft corals as stony corals in terms of substrate cover and, when these die, mostly they leave no skeletal traces.

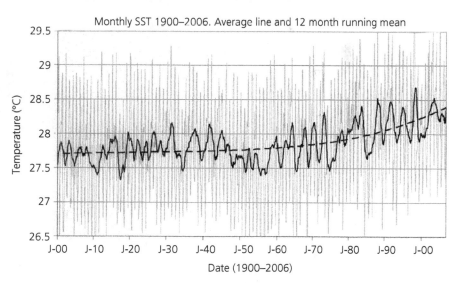

Figure 8.9   Sea surface temperature in the Chagos Archipelago in the central Indian Ocean, 1900–2006. The faint line represents monthly values, showing annual cycle. The dashed dark line is the line of best fit (fourth-order polynomial). The solid dark line represents the 12-month running mean, which shows a 3–5-year cycle which is linked to climate cycles; SST, sea surface temperature.

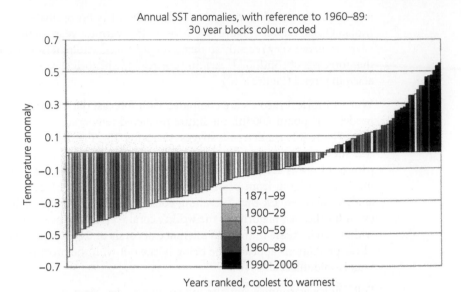

Figure 8.10 Trend of average sea surface temperatures 1871–2006 in the Chagos Archipelago, a central Indian Ocean site. Each year's temperature is averaged, and years are ordered (left to right) from coolest to warmest. Each 30-year block has a different shade of grey (lightest = oldest; darkest = most recent), but note that the final block has fewer columns (1990–2006 only). The y-axis is not absolute temperature but its difference from the 1960–1990 average reference temperature; SST, sea surface temperature.

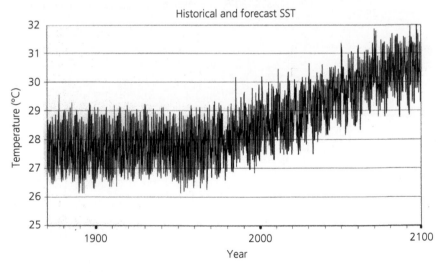

Figure 8.11 Blend of historical and forecast sea surface temperatures 1871–2100 in the Chagos Archipelago, a central Indian Ocean site; SST, sea surface temperature.

Source: Data is from Hadley Data centre (HadISST1 for historical data and the HadCM3 climate model for forecast data, blended as in Sheppard, C.R.C. (2003). Predicted recurrences of mass coral mortality in the Indian Ocean. *Nature* 425:294–7).

8.6 Acidification

A global change which may already be affecting coral reefs and which is likely to become increasingly important comes from acidification of seawater. Several clear patterns are now known at the physiological or organism level, but ecological consequences are still relatively speculative. They are likely to be substantial and, in human terms, permanent (Veron 2008). The effect arises, once again, from rising CO_2 concentrations in the atmosphere.

Aspects of the chemistry and physiology of CO_2 and the balance of this with bicarbonate and carbonate in seawater are described in Chapters 3 and 4. Briefly, before the industrial revolution, the relative proportions in seawater were 88% bicarbonate to 11% carbonate, in tropical waters at least, the remaining 1% being carbonic acid (Kleypas et al. 2006). With present levels of atmospheric CO_2, the proportion of bicarbonate has risen, so the balance has changed to 90% and 9%, respectively; in other words, there has been a rise in bicarbonate and a corresponding fall in carbonate. The doubling of CO_2 forecast for later in the century will change this balance further to 93% and 7%, respectively. This is the opposite of what is required for coral calcification, because the ability of corals to secrete limestone depends more or less directly on the carbonate ion concentration in seawater.

In the last two decades, about half of all produced CO_2 has been taken up by the oceans. Therefore, carbonic acid rises faster than the carbon can be sequestered or removed. Although atmospheric CO_2 levels have been higher in past millennia than they are today, and coral reefs flourished then (ISRS 2007), the present, rapid rate of increase exceeds the ability of the oceans to absorb it, since such feedback or buffering mechanisms such as the weathering of rocks cannot respond quickly enough. It is the unprecedented rate of increase in CO_2 levels in the atmosphere as much as the absolute amount that underlies the dangerous change in the carbonate saturation ratio. This change in pH changes the carbonate saturation state of the water. Figure 8.12 shows that for aragonite, which is the form relevant to coral calcification (the pattern for calcite is slightly different). Polar regions are affected first, with equatorial regions following. The saturation state is already becoming increasingly unfavourable for corals.

8.6.1 Slowing of reef calcification

Coral skeleton growth therefore will slow, and this has been demonstrated experimentally. Most studies which have examined responses of corals to these changes have shown that calcification will decrease by about 30% (with a fairly wide range in the estimates) by the time 30–50 years have passed. Ecologically, it is difficult to guess what the results might be to the reef system as a whole but, bearing in mind that this will happen along with seawater warming, it is feared that coral growth overall will markedly decline. Acidification and warming combined are likely to lead to catastrophic decline throughout the reef ecosystem.

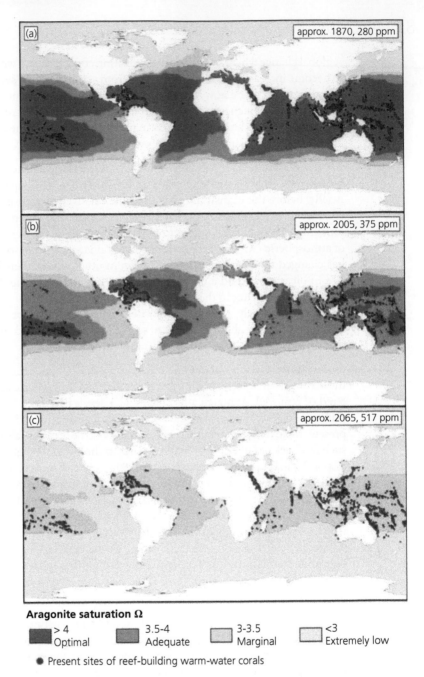

Aragonite saturation Ω

■ > 4
Optimal   ■ 3.5-4
Adequate   ☐ 3-3.5
Marginal   ☐ <3
Extremely low

● Present sites of reef-building warm-water corals

Figure 8.12   Acidification is expected to decrease water pH. Much of the ocean will become less alkaline and will slow and inhibit calcification. This figure shows the aragonite saturation state for (a) pre-industrial (c.1870), (b) present (c.2005) and (c) future (c.2065) oceans.

Source: From Schubert, R., Schellnhuber, H.-J., Buchmann, N., Epiney, A., Grießhammer, R., Kulessa, M., et al. (2007). *The Future Oceans: Warming up, Rising High, Turning Sour: Special Report*. German Advisory Council on Global Change. Available at <http://www.wbgu.de/fileadmin/user_upload/wbgu.de/templates/dateien/veroeffentlichungen/sondergutachten/sn2006/wbgu_sn2006_en.pdf> and <http://www.wbgu.de>.

In a few decades, reefs, which developed in water which was supersaturated with dissolved aragonite, will be bathed in water which is much less so. But the saturation state in the water column as a whole is not even, and there is a layer called the saturation horizon, below which calcium carbonate tends to dissolve and where calcification is impossible. This saturation depth is migrating upwards (albeit in a very irregular way) and, in the Southern Ocean, occasionally has reached near the surface. Deep forms of corals therefore, including those few, such as *Lophelia*, which form deep water reefs, may be affected before shallow and reef-constructing forms. It is not only corals which will be affected in this way. Crucially perhaps, experiments in Hawaii have shown substantial reductions in the ability of calcareous red algae to deposit limestone; these are the essential calcifiers in the most wave-exposed reef zones of many reefs (Jokiel et al. 2008; also see Chapter 9). Further, several major groups which presently sequester large quantities of limestone are planktonic (see Chapter 5), and it is also noted that various species (notably, small pelagic molluscs and coccolithophores) are already showing signs of pitting on their shells, suggestive of limestone solution rather than deposition.

Reef cementation processes are equally important to reef construction, and these will be affected too. Cementation on reefs bathed in the naturally lower alkalinity levels on the Pacific side of Panama has been related to the lower amounts of reef cementation there, compared to the higher alkalinity on the Caribbean side. This is correlated with weaker reef development on the Pacific side (Manzello et al. 2008).

While rising temperature will have its impacts first in many areas, in the longer term, acidification is likely to be of most pressing or immediate importance, since its effects are likely to last many millennia.

8.7 Sea level rise

Sea level has fluctuated for several reasons over recent millennia by up to 150 m (see Chapter 1, Figure 1.8). One important difference today is that the rate of rise is rapid, and also that modern humanity's use of the coastline is unprecedented.

In the twentieth century, sea level rose by an average of 17 cm because of warming, and estimates are clear that the rate of sea level rise will accelerate (IPCC 2007). Present rates of rise are 3 mm y^{-1}, and this is due as much to thermal expansion of water as it is to increased volumes of water originating from ice melt. However, the rise (or sometimes fall) seen at any one coast varies considerably around the world due to local land movements, and can be considerably more than this. Atolls, for example, grow on gradually submerging land and by a matching upwards growth of the coral reef which caps it. In such areas of atolls, the relative rise may be around 1 cm y^{-1} at present, and increasing. As noted by Hubbard et al. (2014), meteorological instability comes with rising temperatures, leading to increases in shoreline erosion, and numerous islands are currently experiencing difficulties in this respect, with some already being evacuated.

Although reefs grow vertically to the low tide level, and did so during the rapid and substantial Holocene transgression (the sea level rise of about 150 m up to about 8,000 years ago), their ability to continue to do so under present conditions of rising acidity and temperature is not at all certain. Vertical reef growth has been variously estimated as being up to 10 mm y^{-1} (Chapter 2). Some reefs are relatively solid throughout great thicknesses while others contain a substantial portion of loose coral fragments and other limestone rubble, perhaps sealed beneath a more solid cap of just a few metres deep but otherwise scarcely cemented at all. Overall, a vertical growth of 10 mm y^{-1} depends on the reef being in good condition and, when it is, growth and erosion in a healthy reef are closely balanced in favour of growth, which is of course why so many reach to approximately the mean low tide level.

Reefs may be able to keep up with sea level rise for several years to come, but it is possible that more and more reefs will fail to keep up with the surface as the sea level rises, although details are not well understood. In any case, effects may be masked by other changes which are taking place, one of which is horizontal erosion. Thus, while healthy reefs may continue to keep pace with a rise in sea level, as has been the case before, increasingly stressed and damaged reefs may become increasingly incapable of doing so. This is a process already observed in several parts of the Seychelles (Sheppard et al. 2005; also see Chapter 9).

When corals are killed for whatever reason, net erosion of the reef commences, which results in a reduction of their breakwater effect. The importance of the breakwater effect on reefs is marked, and can be seen when reefs have been removed previously by other reasons, for example by coral 'mining' on reefs around islands to obtain building material for houses. In several countries, this has lowered the surface of the reef by a further 0.5 m or more, and consequences of such changes are described in Chapter 9.

8.8 Cyclones, hurricanes, typhoons

Strong tropical cyclones (hurricanes in the Caribbean, typhoons in the Western Pacific and cyclones in Australia and the Indian Ocean) have marked effects on coral reefs. They cause immediate damage and physical

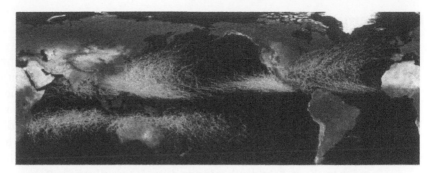

Figure 8.13   Global cyclone tracks, 1985–2005. Note absence of cyclones near the equator. Also note Cyclone Catarina in the South Atlantic; this storm, which occurred in March 2004, was the first of its kind known to strike Brazil.
Source: From Wikipedia, courtesy NASA.

restructuring when they pass over a reef, but coral reefs thrive and have evolved in cyclone-prone areas, so severe storm winds and waves are not at all prohibitive to life on reefs in the long term. Hurricanes are fuelled by the latent heat of water, needing warm >26 °C water to fuel them. To develop their initial 'spin', they also need to be located at a latitude where the relative speed of the earth's rotation changes sufficiently between their northern and southern sides; thus, they do not develop or strike closer to the equator than about 10° north or south (Figure 8.13), although equatorial reefs may feel the edges of many hurricanes and may experience storms of other kinds. In addition, there are decadal variations in cyclonic storm activity, and these are still not well understood (Emanuel 2005a).

Warming sea surface temperature and increased water vapour tend to increase the amount of energy available to drive hurricanes (Trenberth 2005). Important measures of this are the accumulated cyclone energy index and the power dissipation index. Although a warming world has been predicted to lead to increased hurricane activity, there is no clear pattern using simple measures. Overall frequency of hurricanes shows no trend, although there are trends of increasing destructiveness of tropical cyclones in recent decades (Emanuel 2005b), due to both longer storm durations (up to 60% longer over the past 50 years) and greater storm intensities or wind speeds. These measures correlate with tropical sea surface temperature. The overall result is that Category 4 or 5 storms have become more frequent, with the frequency having doubled over the past 50 years (Webster et al. 2005). The largest increases in frequency have been in the North Pacific, the Indian Ocean and the South West Pacific, with the least increase being seen in the North Atlantic area. Possibly significantly, until 2004 there had never been a recorded hurricane in the South Atlantic; but, that year, one struck the coast of Brazil.

8.8.1 Damage from storm energy

Wave energy and surge are the primary destructive forces behind a hurricane but, apart from the intensity of the storm, effects are modified by the angle at which the storm strikes the reef, by the storm duration (storms may move quickly or may stall) and by the shape and profile of the reef itself. While much breakage of corals and damage comes from water movement, a considerable amount comes from broken objects striking others. In addition, there may be a substantial increase in the amount of sediment which buries areas of reef or causes scouring. Blocks of reef may detach completely in water less than about 20 m depth, causing damage to well over half the corals present but, whereas shallow blocks may be tossed high on the shore, on steep slopes these may tumble downwards, stripping the reefs as they go and causing equal amounts of damage to double this depth or more.

Not surprisingly, shallow branching corals suffer the most. However, the main mode of reproduction of some species (such as the Caribbean's *Acropora cervicornis*) is by asexual dispersal so, if the fragments survive being vigorously tumbled, a colony's fragments may become widely dispersed, following which many will reattach and recommence growth. Massive corals, in contrast, may be simply scoured in situ.

Recovery from hurricane effects is variable. To return to an ecologically comparable condition might only take a very few years, although the structure of the reef may be changed in that the age distribution of corals will be different, with a younger median age.

Coral reef cover clearly recovers, given time; Caribbean reefs have been struck by such storms for millennia, and coral cover on most reefs up to at least the middle of the 1900s was high. Using nearly 300 Caribbean sites, Gardner et al. (2005) determined that hurricanes reduced coral cover across that region by about 17% on average in the year following a hurricane, with greater losses following storms of greater intensity. Further, there was no return to former values within at least 8 years. While it is known that coral cover declined from other causes also, reduction of coral cover in hurricane affected sites was significantly greater than sites not impacted by a hurricane. Overall, it was concluded that hurricanes have some effect on coral cover, but the general decline from diseases and warming, for example, was also marked, and it is probably the latter two factors which are preventing recovery from post-hurricane impacts today.

Because hurricanes stir up the surface layers of water and cause mixing with deeper layers, they leave behind them water which is significantly cooled. The degree to which this occurs is not well studied, but unpublished reports claim immediate reductions in surface water temperature of as much as 8 °C, which is clearly sufficient to reduce the surface water temperature below both that needed for hurricane maintenance and that which leads to coral bleaching. The problem in areas such as the Bay of Bengal and Gulf of Mexico is that those water bodies are too shallow to support a cool deep layer, so that hurricanes may massively intensify when they cross them. Where cooling does occur because of both the proximity of several hurricanes in a season and the existence of a deep cooler layer which becomes mixed with the surface, the extent and recovery of bleached corals can be significantly greater than on reefs where no such cooling took place (Manzello et al. 2007).

8.8.2 Tsunamis

Coral reefs were perhaps surprisingly little affected by the major tsunami of December 2005 (see Stoddart 2007 for collection of papers). The pattern of water movement was quite unlike that of a storm. The build-up of massive waves depended on a gently shoaling shelf leading up to the coast and, where this was the case, the water surge was, of course, enormous. But the damage to reefs was to a great extent caused by the large quantity of debris, from tree trunks to parts of buildings, which was dragged back from the land onto the reefs, destroying corals by impact and abrasion. Many corals were turned over by this and by the force of water. Further destruction came from sediment deposits, and from substantial suspended sediments which settled on corals. Mostly, however, this was shed or lost within a few days.

Atoll groups generally have a very steep seaward slope and so were not affected nearly as much as continental islands, where shoaling is gradual. While many patches of coral islands were washed over, with considerable removal of vegetation and deposition of reef rubble and corals onto land where this took place, damage was relatively minor in extent in most areas.

The greatest effects on reefs, however, came from the associated uplift of land in several areas, especially of Indonesia (Figure 8.14). These reefs were simply elevated many centimetres to a position above sea level which, of course, resulted in the complete mortality of the reef life in exposed areas.

8.9 Synergies, stasis and feedbacks

Many of the pressures on reefs act synergistically, and several show positive feedback loops (Ateweberhan et al. 2013). Coral bleaching, acidification and diseases are expected to interact synergistically, and negatively influence survival, growth, reproduction, larval development, settlement and post-settlement development of corals. Negative feedback loops exist, of course, whereby a normal or healthy condition tends, when disturbed, to return to its original condition. Taking a conceivable but imaginary example of a negative feedback loop, a one-time disease which reduces the number of herbivores might lead to an increase in algae, more food subsequently

Figure 8.14   Reef flat in Indonesia, uplifted above sea level by the earthquake that caused the Boxing Day tsunami.

Source: Photo by Robert Foster.

for herbivores and a return to normal herbivore status (Chapter 6 describes some of the complexities involved). In contrast, a positive feedback loop leads not to a restoration to a mean value but to increasing divergence away from it. This can come about because the stressor is too severe or acts for too long, or because its effects are magnified by additional stressors in synergistic ways. The results of such events lead to 'runaway' conditions. Many examples now exist, such as mortality of corals from disease and bleaching, subsequently increased bioerosion coupled with overfishing, leading to decreased herbivory, all of which serve to drive a reef towards an increasingly poor condition. The principles of hysteresis and alternative stable states (Chapter 9) suggest that it takes more than a simple lifting of one or two of the driving pressures to permit the ecosystem to return to its previously valuable condition.

It has been argued that there is no such thing as stasis in ecology and that growth, evolution and changing climate over the millennia have caused reef ecosystems to adapt and change continually. While undoubtedly true, it is the present rate of change together with the severity with which it is impacting reefs, as well as a rise in the number of different pressures acting on them more or less simultaneously, which gives rise to such concern over the future of reefs today. The enormous consequences of the 2015–16 warming in particular sound many warnings about the consequences of warming on coral reefs.

9 Consequences to reefs of changing environmental stress

9.1 Ecological consequences of environmental impacts

9.1.1 Alternative stable states, thresholds, phase shifts and hysteresis

Dominance by coral on a healthy reef is seen as a 'stable state', one which is self-perpetuating in the sense that negative feedback maintains corals in their dominant position. But this is stable only within certain environmental tolerances. When forcing factors, such as nutrient enrichment from sewage or land run-off, exceed a certain threshold or when grazers are removed (or both together), then algae are stimulated and fleshy algae increase to dominate the reef instead. A new ecological state is reached which is also stable, and it is extremely difficult for the system to revert to its initial coral-dominated state. The movement of an ecosystem which is forced from one state to another in this way, especially by several stressors acting synergistically, can be viewed as a ratchet rather than a freely moving wheel (Birkeland 2004).

This can be depicted (Figure 9.1) with a ball settled in a conceptual or energy valley. The coral state is the ball in the high valley and, with small displacements, it will return to the same valley. But many combinations of features can push the ball over a threshold (meaning over the intervening ridge in the diagram) into the adjacent 'algal valley', which is dominated by seaweeds unpalatable to grazers. There may be several ways the ball moves across into the lower valley. First, the 'push' might be greater, above the threshold or ridge, or the height of the ridge may itself be reduced, so that a smaller push can cause the ball to fall into the algal valley. The kinds of drivers that might be involved include increased nutrients which stimulate greater algal growth, or removal of herbivores. The schematic is simple and has a strong explanatory and illustrative power, especially if a concept of energy state is added

The Biology of Coral Reefs. Second Edition. Charles Sheppard, Simon Davy, Graham Pilling, and Nicholas Graham, Oxford University Press (2018). © Charles Sheppard, Simon Davy, Graham Pilling, and Nicholas Graham (2018). DOI 10.1093/oso/9780198787341.001.0001

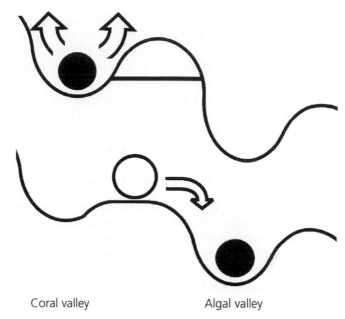

Coral valley   Algal valley

Figure 9.1   Conceptual diagram of different ecological 'states' of a coral reef. Top diagram: Two valleys exist: one coral dominated (left), and one algal dominated (right). The reef is coral dominated (position of the ball); disturbances tend to restore the ball to the 'coral valley'. A healthy reef tends not to be pushed hard enough to go over the ridge to the algal valley. Lower diagram: Very severe 'pushes' or lowering of the ridge between valleys (by overfishing, sewage inputs, etc.) push the ball into the algal valley (lower sketch). When in the lower valley, it is extremely difficult to return to the higher, coral valley.

to the y-axis, but the simplicity of the diagram belies the complexity of the multiple ecological conditions and drivers which act together.

The lower, algal valley (algal-dominated stable condition) might be more stable than the coral valley (healthy, coral-dominated reef), once that position is reached. In other words, the algal state might, as in the illustration, lie below the coral valley, so that it is much more difficult to return to coral domination than was required to move to algal domination. In reality, many details remain largely unanswered. Numbers which potentially might be added could relate to concentrations of dissolved nutrients or to a measure of grazing pressure (such as densities of urchins, or amount of fish biomass remaining on a reef), but we are a long way from being able to calculate such numbers. What we can do is note that such environmental factors very often do drive a reef from being coral dominated to algal dominated.

There are very few examples of a reef being seen to move from an algal-dominated state back to coral domination. In many instances, a return may not even be possible (Hughes et al. 2005). Unfortunately, a modest reduction in nutrient enrichment or easing of fishing pressure does not result in an automatic reversal to the coral-dominated state. That desirable event has, however, been noted in a bay in Jamaica (Idjadi et al. 2006); moreover, it occurred without changes in fishing pressure (Precht and Aronson 2006) but is, at best, extremely rare. This asymmetric phenomenon is described by another conceptual diagram: the hysteresis curve (Figure 9.2). The hysteresis effect means that simple reversal of the stressor does not simply allow the system to backtrack along the same slope of the graph down which it travelled. Instead, considerably greater removal of the stressor must take place before coral dominance is regained. The backtracking slope, in other words, is different to the forwards-tracking slope.

The suggestion that the hysteresis effect is real is strongly supported by a model (Mumby et al. 2007) which applied many different coral cover values and grazing intensities, by urchins and parrotfish, to Caribbean reefs (see also Chapter 6, 'Coral reef models: Hasty conservation action beats procrastination to enhance reef resilience'). The hysteresis effect was marked but, more importantly, levels of grazing were enumerated. Relating this to real reefs, it appears that the grazing needed to keep algae suppressed was near the upper level of what was actually achievable by parrotfish so that, on reefs where urchins had been removed (such as by diseases, as occurred in the Caribbean), any further removal of parrotfish would result in the system becoming heavily algal dominated. However, models are simplified reflections of a real reef, and many other factors are usually either not accounted for or are held at some constant value in order to test the effects of just one or two examined variables.

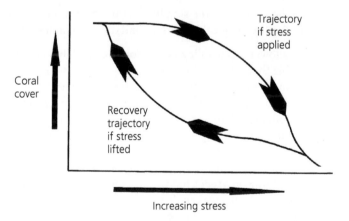

Figure 9.2   Hysteresis effect and application and release of a 'stress'. This shows the direction of movement of a condition because of addition of a stress. In this case, coral cover is the measure of the health of a reef. As stress is applied, coral cover 'holds' and then declines. If the stress were to be released, coral cover does not follow back up the same trajectory, but stays low for a long time first.

Phase shifts have also been known to drive systems to dominance by a different coral. Cores taken from two different reefs in the Caribbean region (Aronson et al. 2004) showed that, during the previous 2,000–3,000 years, dominance in one case was by branching finger corals of the genus *Porites*, and in another by *Acropora cervicornis*. In both cases, departures from this dominance over many centuries were rare and brief. Within the past 20–30 years, however, both sites were heavily impacted. In the case of the *Porites* reef, changes in water quality caused coral demise while, in the *Acropora* case, white band disease (WBD) caused mass mortality. Both sites changed to dominance by a leafy, fast-growing 'lettuce coral' (*Agaricia tenuifolia*), such that they now resemble each other despite their different starting points. This alternative state has been stable for some years now, and 'reflects the degradation of Caribbean reefs under increasingly stressful combinations of natural and human perturbation' (Aronson et al. 2004).

Algae appear rapidly on bare substrate where there is sufficient nutrient and sufficient space and when they are not removed by grazers. It is debated whether algal domination comes primarily from removal of herbivores or from opportunistic growth on the increased substrate which appears after corals are killed from other causes, such as disease. Aronson and Precht (2006) argue that global change and diseases are the main cause of mortality and that macroalgae growth simply follows on. Others (Jackson et al. 2001; Pandolfi et al. 2005) place greater emphasis on overfishing, which permits algae to grow faster and more extensively and to overgrow corals. Another factor, which is certainly required, is sufficient nutrient in the water to permit or stimulate faster algal growth. Nutrient enrichment is probably a common requirement for almost any scenario; in several reefs systems where nutrients remain extremely low and where there is no fishing, there may be no algal growth even after corals are heavily killed (Sheppard et al. 2002). Further, it is equally commonplace to observe abundant algae growing where there is a high level of nutrient enrichment, such as near cities, even where there are abundant grazing fish, which presumably do well from their rapidly growing algal food source. The elimination by disease of another herbivore, *Diadema*, is also blamed for triggering algal growth at the expense of corals, but a great deal of the recent decline of Caribbean corals (Gardner et al. 2003) took place before that event and, subsequently, much further loss has occurred from coral diseases and warming. Further, when new substrate was made available in one Caribbean site following a severe hurricane, algae colonized the newly available substrate (Rogers and Miller 2006) and, once established, slowed or prevented new coral colonization. In the latter case, removal by the hurricane of the branching structure created by corals in shallow water also removed the habitat for many grazing fish, thus removing a crucial control on algal cover.

In almost all these examples and others, it seems very likely that generalities can be applied, but with caution, because of exceptions and, while one or another factor might appear to be the obvious driver forcing a particular reef change, probably different factors are present to different degrees in different locations. It has been argued that nutrients cannot be responsible for increased algal growth when nutrient concentrations in the water are not raised or only slightly so, but this overlooks the fact that benthic algae take up nutrients extremely rapidly, and uptake by plankton, which are then removed from the system by currents, is even faster. Thus, nutrient concentration is a less sound estimate of algal fertilization than is nutrient input or turnover.

The stressor causing a shift in the ecosystem may not be the same as the stressor which subsequently maintains it in its changed condition. Increase in substrate by, for example a storm or a disease, may trigger the massive shift but, once algae have become established, their permanence and the subsequent prevention of coral recovery may be more attributable to simple physical exclusion coupled with lack of grazing. While there is often no unequivocal and universal single cause of coral decline, the exact reasons may be important if successful attempts are to be made to restore an algal-dominated condition to a coral-dominated one. For example, new regulations may address prohibition of fishing, or reduction of nutrients, but one might be useless without the other. The clear fact is that drivers act synergistically (Ateweberhan et al. 2013) and different drivers may be more important in different locations, depending upon their relative intensities and upon which processes (triggering or maintaining the change) they act.

9.2 Changes to the main architectural species

9.2.1 *Acropora* die-off in the Caribbean

One of the most striking changes in the Caribbean region has been the virtually total removal of the genus *Acropora*. For this genus, the Caribbean is species poor, with just three species (one being a hybrid of the other two), compared with over 100 species in the Indo-Pacific, but the two main species were once extremely important. *Acropora palmata*, the well-known elkhorn, forms the largest coral colonies of all *Acropora*, being dominant in shallow water to 3–4 m depth. Figure 9.3 shows part of a GIS atlas (see 'Remote sensing on coral reefs') depicting a 0.5 × 0.5 km section off the small island of Anguilla in the early 1990s, illustrating where a large elkhorn reef has been reduced to rubble from disease (Chapter 8). In Anguilla alone, 435 ha of elkhorn reef were essentially extirpated by disease in the 1980s, with live coral cover of those shallow areas rarely exceeding 2% (Sheppard et al. 1995).

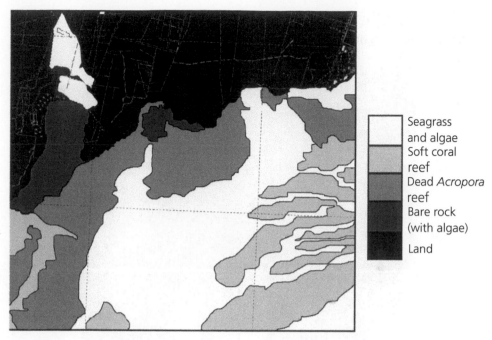

Seagrass
and algae

Soft coral
reef

Dead *Acropora*
reef

Bare rock
(with algae)

Land

Figure 9.3   Simplified section of a GIS map of Anguilla, Eastern Caribbean, showing the distribution
of dead elkhorn reef. The region shown is approximately 500 m² in size. The areas
shown in mid-grey represent reefs that were probably all elkhorn reefs before being
killed in the 1980s. The areas in darkest grey (i.e. consisting of bare rock with algae)
represent reefs that may also have been elkhorn reefs but have decayed through erosion
sufficiently so as not to be recognizable as such. The areas in pale grey represent reefs
dominated by soft corals (Gorgonacea) but with some limited quantities of the reef-
building *Montastraea* group. White indicates areas containing various mixtures of
seagrass and sandy substrate with some seagrass.

Adapted from Blair Myers, C., Sheppard, C.R.C., Mathesen, K. and Bythell, J.C. (1994). *Habitat Atlas
of Anguilla*. NRI.

Remote sensing on coral reefs

Because tropical waters are usually clear, satellite images have been widely used in
coral reefs studies, and many standard ecological techniques are used in conjunction
with them in order to 'ground-truth' results. There is a wide range of satellites with
which users can obtain images or data, and the use of satellites in observing tropical
waters has been thoroughly analysed (Mumby et al. 1997). Remote-sensed images
have global coverage and have become essential for observing areas of reef tract at a
broad range of scales. The range of resolution achieved by satellite imaging is wide,
ranging from several kilometres, for examining whole oceans, to 1 m, for examining
sections of reef.

Continued

Remote sensing on coral reefs (*Continued*)

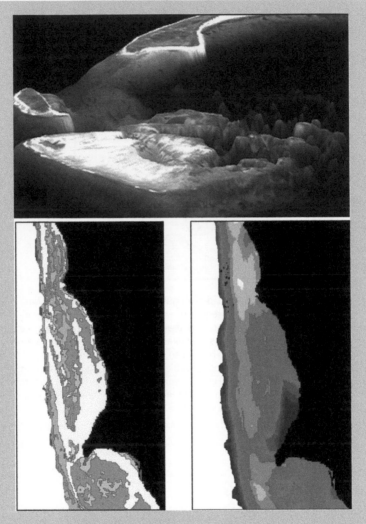

Figure 1. An example of a very fine resolution image. The remote-sensed image was from the IKONOS satellite (resolution 2 m), coupled with a bathymetric profile obtained by echo-sounding. The processed satellite image was then draped over the 'wireframe' bathymetric image to produce the final image, which can be rotated in three axes. The location is the mouth of the Diego Garcia atoll. Bottom: Cozumel Island, Mexico. Processed images resulting from a combination of aerial photography and Landsat, with extensive 'ground-truthing'. **Figure 2.** Left: A habitat map which reduces the complex scene to 15 habitat classes (for details of classes, see Rioja Nieto, R. and, Sheppard, C.R.C. (2008). **Figure 2.** Right: A beta diversity map which divides the area into ten zones of increasing diversity (darker to lighter categories are increasing habitat beta diversity, from lowest (darkest) to highest (lightest)). The land to the right of both images is set to black.

Source: Top: image by Dr Sam Purkis, Nova University, Florida; bottom: images by Dr Rodolfo Rioja Nieto.

Continued

Remote sensing on coral reefs (*Continued*)

One important complication regarding reef monitoring comes from the optical qualities and the depth of water overlaying reefs. Water absorbs red light first, with blue light penetrating deepest. Green light penetration lies somewhere between the two but varies according to the amount of organic matter and plankton in the water. Such characteristics must be adjusted for before the amount of coral, seagrass, algae or sand on the reef and seabed can be interpreted. Direct inspection of selected parts of the image provides the 'ground-truthing' needed to verify, in small but hopefully well-chosen and representative locations, what the various colours which the satellite has recorded over a much greater area represent.

Early work showed that the accuracy of interpreted images commonly was less than 50% and that aerial photographs were much better (Mumby et al. 1997), although, now, interpretation accuracy from satellites has greatly improved. There is still a trade-off between the number of different 'classes' a scientist attempts to distinguish and the accuracy of each of those classes. Commonly, the number of defined, discrete substrate types around and on coral reefs may be reduced to between 4 and 20, depending on need.

Where water depth is unknown, accuracy is reduced but, where depth is known, its optical effects can be corrected for. Thus, additional data inputs such as bathymetry and the optical quality of the overlying seawater itself can now greatly enhance the accurate mapping of both topography and habitats on a coral reef (Purkis 2005). Developments, not only in the types of sensors used but in the imaginative interpretation of them, are rapid. Results from processed remote sensing can be draped over a 'wireframe' topographical map of bathymetry obtained from echo-sounding, to produce a three-dimensional habitat map (Purkis and Riegl 2005). In the Figure 1 (top), a 2 m resolution image obtained by the satellite IKONOS image has been layered onto such a three-dimensional frame. This image has been used to examine fish habitats and distribution (Purkis et al. 2007) by relating topography to known fish habitat preferences.

Further, once the nature of an area is well understood, maps can be made not only of habitat types but also of concepts such as beta diversity (Rioja Nieto and Sheppard 2008) (see Figure 2, right).

These techniques enormously extend our knowledge of reef systems in ways which cannot practically be done in other ways, due to constraints of cost or access in remote areas. Monitoring by remote-sensed images is one of the main keys to successful management of whole reef systems and, coupled with field inspections and measurements at finer scales, forms one basis of managing the world's reef systems.

Prof. Charles Sheppard, University of Warwick, UK.

Ecologically, consequences have been marked (the physical consequences of the loss of breakwater effect are described in 'Role of reefs in wave energy reduction'). Loss of the shallow and strongly branched *Acropora palmata* causes loss of the physically robust, forest-like three-dimensional reef structure, which is replaced progressively by fields of rubble formed from bioeroded or smashed branches (the 'stump and boulder zone' (Blanchon et al. 1997)), then by fine rubble and sand and, finally, by just the bare, underlying limestone substrate of a much older reef. That substrate has proved to be hostile to significant further coral colonization, partly because of the 'liquid sandpaper' effect of mobile rubble and sand, and partly because, in several places, it becomes quickly colonized by fast-growing algae which keep corals excluded. But, even if algae do not develop, the shallow area's high-energy regime means that there are a limited number of other coral species which could inhabit that shallow zone anyway; the elkhorn was the only very successful species there. Thus, the area remains devoid of the iconic, forest-like structure which so characterized Caribbean reefs. Because of the solidity and thickness of the branches, the dead skeletons may persist for many years. Sometimes their removal is hastened by storms or hurricanes, but, where this does not occur, stumps may be seen as long as 20 years after mortality.

A rich invertebrate and fish fauna used to be found in this zone but has now all but vanished. Because some stands extended over 100 m to seaward where topography was suitable, and because some entire shallow bays used to be literally filled with elkhorn coral, the losses of many fish and invertebrates, and even of some nursery grounds, have been substantial. In the same way that, when a forest is cut down, the birds and monkeys are lost too, many fish and invertebrates disappear when their forests of *Acropora* are 'felled' by disease or warming. The massive losses also can change current flows in bays, thus causing erosion of adjacent seagrass and mangrove stands.

Deeper in the reef's structure, the finely branched *Acropora cervicornis* (see Chapter 2, Figure 2.6) likewise provided dense thickets at between about 5 and 25 m depth, depending on water clarity. This species is fragile, and uses fragmentation more than sexual reproduction as an effective means of dispersal. But studies on the ecological loss of such thickets are few; more commentary has focused on the loss of the reef-building or framework role of this species than on the ecological consequences of its demise.

9.2.2 *Montastraea* and *Orbicella* in the Caribbean

A major reef-building or framework group of corals in most Caribbean reefs is composed of the similar genera *Montastraea* and *Orbicella*, which comprise four hemispherical species which grow at between approximately 4 and 25 m depth. These form massively solid lumps that add enormously to the structure of reefs throughout the depths at which they are found. These corals have not suffered from disease to anywhere near the same extent as have *Acropora* species, although a range of diseases affect them in many places. However, these corals are susceptible to mortality from warming. In late 2005, particularly strong warming caused almost all *Montastraea* and *Orbicella* in the north-eastern Caribbean to bleach (Figure 9.4) and, in some countries such as the British Virgin Islands, this was followed by

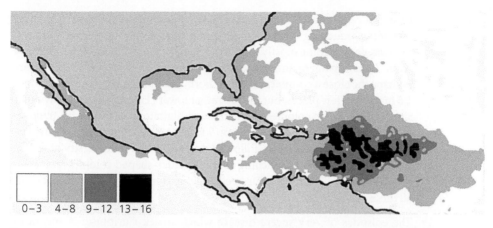

0–3 4–8 9–12 13–16

Figure 9.4   Map of degree heating weeks in the Caribbean region for 18 October 2005 and the previous 12 weeks. The pool of warmed water over the eastern Caribbean, over the Windward and the Leeward islands and extending into the Atlantic is clear.

Source: NOAA Coral Reef Watch Program NOAA National Environmental Satellite Data and Information Service. See <http://www.ospo.noaa.gov/Products/ocean/cb/dhw/index.html> for current maps of degree heating weeks.

mortality of approximately 30% of those colonies (but, fortunately, recovery of the remaining 70%). Such bleaching and mortality in this area was unprecedented, and it extended to the deepest extent of these reefs, over 20 m deep. Thermal stress of this level would, if it was a natural fluctuation, be a >1,000-year event without the warming trend (Donner et al. 2007) but, with current trends, this degree of warming is likely to become a biannual event in 20–30 years. It may be speculated that the warming that year would have caused mass mortality to shallow *Acropora*, but there was little *Acropora* surviving, because of earlier disease episodes.

From all causes and with all species, coral cover in the Caribbean has declined an average of 80% over previous values (from 50% to 10%; Gardner et al. 2003). Much of this was formerly due to loss of *Acropora*, but is now due to the loss of *Montastraea* and *Orbicella* as well.

9.3 Changes on Indo-Pacific reefs

9.3.1 Indo-Pacific shallow assemblages

In overall terms, a similar decline has been recorded for Indo-Pacific reefs, where the 'estimated yearly cover loss based on annually pooled survey data was approximately 1% over the last 20 years, and 2% between 1997 and 2003' (Bruno and Selig 2007), such that, today, coral cover is approximately half of its state a century ago. In the Indo-Pacific, there is no species with the sheer size of the branching Caribbean elkhorn, although the counterparts in shallow water again come mostly from the genera *Acropora* and *Isopora*. In many parts, *Isopora palifera* and branching forms such as *Acropora pharaonis* are perhaps the closest counterparts to elkhorn, developing stands that are about 1.5 m tall in a few places, but mostly rather lower (Figure 9.5). In many other sites, very crowded colonies of much lower, digitate species such as *Acropora humilis* and members of the genera *Stylophora* and *Pocillopora* are typical dominants of exposed areas, followed deeper by a larger diversity of forms, most conspicuously of branching corals, boulder-shaped colonies and table-shaped species of *Acropora*. On reef slopes and in lagoons, many species of all these can be found.

The episodes of very severe disease which struck Caribbean *Acropora* in the 1980s did not occur in the Indo-Pacific, although the same or similar diseases do exist there. Mass mortalities did come from waves of crown-of-thorns starfish, which appeared in plague proportions in several places from the Red Sea to Pacific islands (Moran 1986; also see Chapter 6, 'The crown-of-thorns starfish'), and of the coral-eating mollusc *Drupella*, whose numbers increased where diseases occurred or where other stressors such as fishing were heavy (Schuhmacher 1992; McClanahan 1994). Much attention was given to these in the 1980s and early 1990s.

In the Persian Gulf, where water around reefs is nowhere deeper than about 40 m and mostly no more than 10 m deep, warming destroyed substantial swathes of staghorn coral in both 1996 and again in 1998. The latter episode was the more severe of the two and affected most reef areas heavily (Chapter 8), and this was repeated in 2015. Mortality effects extended much deeper in several oceanic places, to as deep as 35 m or more on many

Figure 9.5   Shallow branching corals of the Indo-Pacific. These are smaller than their Caribbean counterparts but form equally robust communities at the seaward edges of many reefs. These also were nearly eliminated in some areas by recent warming events. Left: *Acropora palifera*. Right: *Acropora pharaonis*.

mid-ocean atolls. It is presumed that the extent of the depth where mortality was high is due to different depths of thermoclines and thicknesses of the warm-water layers, although little direct evidence is available. Within the killed zones, however, mortality of corals and soft corals was very high, and nearly total near the surface (Sheppard and Obura 2005).

Following the mortality, coral cover initially started to recover very slowly on the denuded areas. In many places, synergistic effects of other stressors have meant that recovery has been substantially impeded. In some places, such as the Seychelles (Sheppard et al. 2005), warming episodes recurred, killing most of the new recruits that had settled. In some sites where no other stressors exist, coral cover slowly became re-established—although the situation following the 2015 warming has again caused massive mortality. However, even where cover did recover in shallow zones, there is still a major change in the habitat, and this change will persist for many more years. This is because juvenile corals of the dominant species in shallow, turbulent water have a largely encrusting form during their early years, following which incipient branches then develop, so that it will clearly take several more years before any significant three-dimensional structure re-forms.

Less attention has been paid to other coral species, with a few notable exceptions. Many reports noted that the genus *Porites* was particularly immune from the effects of warming. The reasons are not yet known. Similarly, several faviid corals survived well. The effects of this differential mortality may lead to changes in the coral assemblage. Reefs of the Persian Gulf are affected by more severe environmental conditions than most areas are, so events there may presage those of wider areas. Persian Gulf reefs have a very simple zonation pattern, with a shallow 3–5 m of branching and tabular *Acropora* forms, followed deeper by an assemblage of *Porites* and faviids, especially *Favia, Playtgyra* and *Favites*. From the two warming episodes of 1996 and 1998, very little *Acropora* was left in shallow zones in Qatar and the United Arab Emirates, with the impact being so severe that extensive exploration of reefs formerly thickly carpeted with it revealed no live coral at all (Sheppard and Loughland 2002). These, ironically, included the reefs and corals which 40 years previously had been noted by Kinsman (1964) as being those with the highest known tolerance to high temperature, thus indicating the severity of the recent warming events. Along with the corals, no invertebrates whatever were found in some reef areas in the years immediately following the mortality. The formerly rich stands of branching coral were progressively reduced to piles of rubble. Yet, below a few metres depth, corals thrived and were conspicuously dominated by *Porites* and faviids. Over the next few years, a few *Acropora* reappeared, but most noticeable was the enormous increase in juvenile faviid corals (Figure 9.6). In most places, *Porites* remains the dominant form but, given the continued increase in faviid colony size, even without further recruitment, faviids are likely to assume the dominant position on these reefs below about 4 m depth.

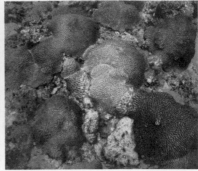

Figure 9.6   Juvenile faviid corals, which have enormously increased in the Persian Gulf following the mortalities of the 1990s, on the vast areas of substrate left devoid of corals. Left: *Favia* spp. with some *Platygyra*. Right: *Platygyra* spp.

What such changes mean to the large number of coral reef fish and invertebrates is not yet known. Assuming the situation continues, the structure of the reef may be assured, as faviids are composed of denser limestone than both *Porites* and *Acropora*. They are, however, generally smaller, forming small mounds rarely more than 50 cm in radius in the gulf, and so will not create the important three-dimensional structure once provided by the *Acropora*.

9.3.2 Changes to reef fish

Following the coral mortalities from the 1990s and onwards, little difference in fish populations was observed initially, and mostly reef fish showed no equivalent, immediate massive mortality. It is possible that increases in algal cover in some places helped to sustain good populations of grazers. However, in the following 3–5 years, reports appeared describing a decline in a wide range of fish groups (Wilson et al. 2006). Obligate corallivores such as several species of butterflyfish disappeared from badly impacted sites, following which much more substantial changes to fish populations were recorded. In the 1–3 years after the coral mortality, impacts were largely limited to species that depended on corals, either for food or for shelter (Figure 9.7); indeed, some obligate corallivores are thought to have become locally extinct in areas such as the Seychelles and Papua New Guinea, and presumably in many other places where research has not been undertaken. Fishes with specialist requirements were more likely to be affected than generalists (Munday 2004). At One Tree Island on the Great Barrier Reef, fish species which associate closely with live corals also showed reduced recruitment compared to rates in years prior to the mass mortality, and assemblage structure of fish recruits changed too (Booth and Beretta 2002). Pomacentrid densities dropped off at sites where there had been coral mortality, but not at sites where the coral remained alive.

Figure 9.7   The harlequin filefish (*Oxymonacanthus longirostris*) is an Indo-Pacific fish which feeds on *Acropora* coral polyps, and is also dependent on *Acropora* colonies for shelter. With *Acropora* coral being so threatened by climate change, this is a precarious specialization.
Source: Photo by Nick Graham.

With the greater passage of time, more species which were less obviously associated with corals were seen to become affected in several parts of the Indo-Pacific, and even specialized fish sometimes took several years to show population effects (Pratchett et al. 2004, 2006). When they did, however, the results were profound in some areas, such as the Seychelles (Graham et al. 2006). There, the reefs showed a major phase shift from rich corals to algal-dominated rubble and this change in habitat had a marked effect on coral dependent fish species, notably monacanthids, chaetodontids and pomacentrids, the possible local extinction of probably four species, and several more reduced to critically low levels. The loss of species richness was not associated with loss of corals per se, but with the loss of physical structure; this explains why the short-term impacts in fish assemblages may appear negligible until the structure begins to collapse. Changes in many groups react only when the complexity of the reef begins to decline.

The change in biomass of fish does not necessarily follow the same decline as that of diversity, at least in the first few years, and, in the Seychelles, biomass remained relatively stable, despite the decline in other parameters (Graham et al. 2006). This is attributed to the main herbivorous species being long-lived, large and old, that is, recruited before the coral mortality in 1998.

The different life histories and habitat requirements of the many different species of fish, and their different size ranges, make it unwise to generalize, but it appears as though a 'lag' of several years may pass before effects of the habitat change brought about by the coral mortality trickle across to the fish populations. Chapter 6 describes effects on fish in more detail.

9.3.3 Effects of bioeroding species

There is no doubt that bioerosion is both important and too often overlooked in reef deterioration. It may result in reduction of several kilograms of limestone each year, on every square metre of reef, a rate greater than rates of deposition. When sedimentation or pollution kills corals and exposes greater areas of limestone for attack, bioerosion will increase. There may be greater rasping of the surface by parrotfish grazing, or greater activity by boring species, and these may remain partly unseen until the coral skeleton collapses. This is exemplified by the sponge genus *Cliona*, whose presence is revealed on dead corals only by their inhalant and exhalent pores and by a thin but often brightly coloured surface film of tissue over massive corals (see Chapter 2, Figure 2.14). Consequently, the interior of affected coral colonies becomes hollowed out. This may be particularly important in accelerating the erosion of dead elkhorn skeletons (Figure 9.8), hastening removal of this shallow branching species once they have been killed by WBD or other means.

Figure 9.8   A Caribbean elkhorn completely smothered by the brown boring sponge (*Cliona tenuis*). The interior of the colony is being eroded away, leading to early collapse.

Sediment-laden water which is not contaminated by high levels of fertilizers may deter grazers while still allowing other species to bore into the limestone. With nutrient enrichment, this situation may change to encouragement of algal growth, and hence of greater grazing of the surface. Fishing pressure on grazers influences the amount of direct rasping of the rock by parrotfish, while fishing pressure on the predators which eat rasping urchins has the opposite effect. At present, relationships appear far from simple and, while mechanisms can be clear, an overall relationship between human impact and bioerosion is extremely complex.

9.3.4 Role of reefs in wave energy reduction

The breakwater effect of reefs is crucial and, when reefs deteriorate, their ability to attenuate wave energy diminishes. Unfortunately, wave erosion and bioerosion continue even when accretion is impaired, and this can have serious consequences for shoreline protection (Sheppard et al. 2005).

Examples of consequences from increased wave energy are numerous. In the Caribbean, the once dominant wave breaks of elkhorn reefs lay seaward of many shorelines composed of relatively soft rock, such as old limestone reef and, where this was the case, many examples of erosion are now seen. Roads have been undermined and hotels, beach restaurants and homes have become damaged or have needed expensive strengthening. In at least one case, an oil facility has been threatened. The loss of elkhorn reefs, in other words, has required expensive shoreline armouring to avoid considerable economic loss.

On coral islands, the whole shoreline is made from soft limestone rock, having been constructed over preceding centuries by the same species which now protect them. When they suffered substantial mortality, the breakwater effect at the edge of the reef became reduced. Four main elements to shoreline protection by reefs may be considered, not all of which apply in every location (Figure 9.9). First is the most seaward barrier, which comes from shallow reef slope corals, where waves initially break. Second, immediately shoreward of these in many (but not all) locations, is the algal ridge and a spur and groove system (Chapter 1), which takes the brunt of the waves. Third is the reef flat, which lies at or near mean low water level and which waves must cross to reach the shoreline. Finally, the fourth is the shoreline itself, most of whose sand is biogenic, as it comes from corals and other calcareous organisms. Sand is continually generated on a reef and commonly is pumped onto the beach at the same rate that waves remove it, so that stable beaches generally are in some kind of equilibrium. In this zone, shoreline plants may help to stabilize the beach by means of their extensive root systems.

With mortality of the shallow corals, many of these elements start to fail. The energy reduction from the seaward-most corals is reduced or removed.

Figure 9.9   Natural breakwater provided by a reef fringing an island. The main barriers to wave energy are provided by (1) the corals reaching the surface on the seaward slope, (2) the algal ridge, where present (lower right photo), (3) the width and depth of the reef flat (upper right photo) and (4) shoreline plants which bind and stabilize the poor soils and sand (lower left photo).

Corals on the reef flat die, thus dropping the level of the top of this natural breakwater (Sheppard et al. 2005). With the death of corals, there may initially be an influx of greater quantities of sand caused by the disintegration of the increased amount of dead coral skeletons, resulting in drifts of sand on shores, but this is later followed by a shortage of sand because of the loss of these important sand producers.

A study in the Seychelles estimated the increase in wave energy striking shorelines as their protecting reefs decayed. The decay followed coral mortality from warming water and, perhaps, by several other direct human disturbances as well, and the work resulted from concerns about shoreline changes manifested by erosion, by quantities of sand appearing on some sections of roads in some areas and by areas where sand was disappearing. The study examined the widths, coral cover and thickness of dead and disintegrating reefs which fronted 14 shorelines (Sheppard et al. 2005).

The reef flat corals were progressively disintegrating (Figure 9.10), eroding down to the surface of the underlying reef flat pavement. As they disappeared, wave energy at the shore was estimated to have increased in the decade up to

Figure 9.10   Reef flat in the Seychelles; here, the previously thriving corals are now dead. The distance between the upper surface and the water surface (right arrow) has a strong influence on the amount of wave energy transmitted across the reef flat to the shore. Unless substantial new recruitment and survival occurs, continued erosion will drop the reef surface by the additional distance of the left arrow (about 75 cm in this example at Mahé).

the study (2004) by one-third or more. Further, with continued decay of the killed reef flat corals and with no new coral settlement, the next decade was predicted to see a further, even greater attrition of coral structure and a correspondingly greater increase in transmitted wave energy, expressed as the percentage of offshore wave energy reaching the coast. The average figures were 7% a decade ago, rising to 12% in 2004 and, more worryingly, estimated to rise to ~18% by about a decade in the future, following the complete disintegration of the dead corals (Figure 9.11). In terms of change, the rise was about 35% over the previous decade, and a further 75% increase was expected in the next decade, assuming progression of the current trend. The rate of deterioration was predicted to accelerate markedly.

The main driving factors were the drop in reef surface caused by the coral mortality, with corresponding reduction in friction on the reef flat caused as rough corals gave way to smooth surfaces of sand or bare rock. Also important was the change in shape at the outer edge of the reef, as this interacted with factors controlling wave 'set-up' at the reef crest. Global sea level rise is a component of this model, but this was shown to be much less important than the 'pseudo-sea level rise' caused by lowering of the surface of the reef due to

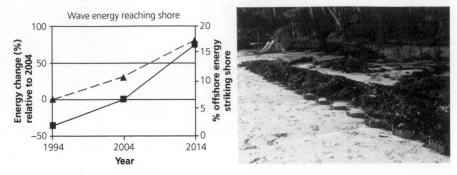

Figure 9.11   Changes in the amount of energy striking the shores in the Seychelles, following mortality of reef flat corals. Left graph: Two measures of energy change: the solid line and the left y-axis show the per cent change in energy reaching the shores, in relation to 2004, which is the reference point; the dashed line and the right y-axis indicate the percentage of offshore wave energy reaching the shores. All data are average values of measurements taken on 14 different reefs on three islands. Right photo: Shorefront 'armouring' using pile-driven interlocking steel containment slabs disguised with vegetation. This is designed to protect the shore; behind it, an eroded step is visible. Hotel rooms are seen 10 m inland.

loss of corals. Of key importance in wave energy reduction was the overall width of each reef flat, but this is not expected to change.

In the Maldives, Perry and Morgan (2017) found that the 2016 mortality caused a reduction in coral cover by 75%, causing the reef carbonate budget to drop from strongly net positive (nearly 6 kg $CaCO_3$ m^{-2} y^{-1}) to net negative (-3 kg $CaCO_3$ m^{-2} y^{-1}), thus causing a 10× reduction in net growth and forcing the reefs into a phase of net erosion. This will 'limit reef capacity to track IPCC projections of sea level rise, thus limiting the natural breakwater capacity of these reefs and threatening reef island stability' (Perry and Morgan 2017). In the Chagos Archipelago, in 2015, calcification rates that had recovered from 1998 but were studied before the 2016 warming showed average calcification values of nearly 4 $CaCO_3$ m^{-2} y^{-1} but with some areas showing considerably less. This was considered to be sufficient to conclude that 'maximum potential accretion rates around Chagos average 2.3 mm yr^{-1} across all sites, but are higher (mean: 4.2 mm yr^{-1}) within *Acropora*-dominated sites, compared to *Porites/Pocillopora* dominated sites (mean: 0.9 mm yr^{-1}).' In 2016, however, cover dropped substantially (R. Roche pers. comm.). The same authors also caution about the decline of corals threatening reef growth in the Caribbean (Perry et al. 2015).

Loss of reef calcification therefore leads to a pseudo-sea level rise which must be added to the 'actual' sea level rise. Such a rise may occur when any factor lowers the height of the protecting reef. One classic example is seen on Malé, where reef limestone was for a long time the only natural rock available for construction (Figure 9.12). Reef flats have been excavated for building

Figure 9.12   Homes built entirely of coral quarried from shallow reefs in Malé, the capital of the Maldives.

materials in many countries, which resulted in lowering of those reefs by perhaps 0.5 m, with similar although unquantified loss of protection.

Constructed by calcareous red algae from the genera *Porolithon* and *Lithothamnion*, reef edges lie in the strongest surf zones. As they are a key element of shoreline protection, their maintenance in healthy form is essential. Present knowledge is scarce, but it seems likely that the warming which is so important in the case of corals is not an important factor in the case of these algae, which can tolerate strong sun in periods of calm seas and low tides. However, ocean acidification is extremely important to their continued ability to deposit limestone, reducing limestone deposition substantially, and even reducing the mass of free-living crustose coralline algae by 250% (Jokiel et al. 2008). Further, some diseases are now reported to have affected these algae in several places. Research is badly needed on these remarkably strong but so far much neglected components of many coral reefs.

9.4 Sizes of the coral 'reservoir'

Because corals require good water exchange, any shoreline construction which restricts the flow of localized currents is likely to result in a gradual or possibly sudden selective mortality of the less tolerant species. With modest restriction of water flow, several species which require clear water and good water exchange will disappear, and more sediment tolerant species will appear in greater number. More often than not, the result is a decline in cover and diversity and, where the works cause increased sediment also, there is a progressive elimination of sensitive species without corresponding increase of resistant species, because the presence of mobile sediment is likely to inhibit coral settlement and growth.

Not all man-made constructions need have harmful effects. Important also to the long-term survival of corals is the scale of any development relative to the size of the reservoir of species nearby. Figure 9.13 shows a development in the Red Sea at Yanbu, north of Jeddah. The construction extended a large ship terminal to the edge of the fringing reef, into deep water, and is one of several such constructions near this new city. The terminal is solid fill for much of its extent, as opposed to being built on pilings which permit long-shore water and sediment flow. Solid fill jetties are cheaper to construct than piers on piles are, but restricted water flow also leads to lethally high water temperatures where the water becomes ponded. But, in this case, when built, there was almost no other development or coastal disturbance for 200 km either side of the location, and there was a large quantity of rich fringing reefs and offshore patch and barrier reefs too. Although the construction has clearly led to the accumulation of sediments to the south

Figure 9.13   Terminal at Yanbu, Saudi Arabia, crossing the Red Sea fringing reef. Shortly after the terminal was constructed (in the late 1970s and early 1980s), coral diversity and cover were still very high in this area, in part due to the huge quantity of healthy corals on the otherwise untouched Red Sea reef (See Plate 16).

(right of the terminal), the terminal was isolated and so did not have any noticeable impact on the coral cover and diversity of the region (Sheppard and Sheppard 1985); this was presumably partly because of the huge supply of coral larvae in the immediate vicinity.

9.5 Changing food chains and trophic balances

Maintenance of a stable food web is essential to the survival of diverse ecosystems. Coral reefs developed in nutrient-poor waters; indeed, the reef is the classic low-nutrient-but-high-diversity, high-productivity ecosystem. How it functions at the system scale has been one of the most discussed aspects of reef ecology. The key seems to lie in the close symbiosis that exists between cnidarians and their zooxanthellae. It has been suggested that, when you look at a healthy reef covered with coral, you are really looking at a field of captive unicellular algae. Bare patches of rock always have a film of free-living, small, filamentous forms (with a good diversity of other usually small forms) but an observer does not see vast swathes of macroalgae on a healthy reef, unless there are disturbances which have distorted the ecosystem.

Changes in the trophic balances caused by the common stressors can be broken down into a few basic principles. First, algae can be fertilized and stimulated into faster growth when nutrients become available to them. Second, algae can grow much faster than corals. Third, when free-living algae do well, they occupy substrate rapidly, preventing further coral growth; they may smother corals and, by occupying space, also prevent juvenile corals from settling. Fourth, phosphate inhibits coral calcification. Fifth, fishing removes higher levels of the food chain, with varying results: if herbivores are removed, there is less algal predation, for example, but, conversely, removal of carnivores may lift predation of herbivores (Chapter 6). Many of these impacts have complex impacts lower down the chain, and these are not always obvious.

There are several areas of the world where macroalgae and corals coexist in natural communities; these are usually at the higher-latitude extremes of coral ranges. In Western Australia, at higher latitudes, *Sargassum* and the kelp *Ecklonia* and numerous other macroalgae coexist with corals, as they do also in Oman, which is at a low latitude but is cooled by upwelling. But, in such places, corals are sparser, usually do not construct reefs and are referred to as coral assemblages rather than reefs. These locations and others such as in the Eastern Pacific, south-eastern Australia, South Africa or Florida are at the extreme limits of the natural range of reefs and most of their component coral species. They lie at the extreme ends of the ranges of cool-water macroalgae too and, in such areas, the algal and coral communities overlap. It is surely the case that research should be devoted to such areas in the search for clues about coral growth in nutrient-richer waters, but much less has been done in such regions than has been devoted to corals (or macroalgae) well within their normal ranges.

The ranges of corals today might be expanding towards higher latitudes as a consequence of warming. Some are now reported further north along Florida and south along eastern Africa and Australia, for example, and this purported migration has geological support (Precht and Aronson 2006). But the salvation of coral reefs generally does not lie in those limited movements, no matter how interesting they may be, and Muir et al. (2015) show that attenuation of light at higher latitudes (whatever the temperature) means that extensive expansion towards the poles could not happen because of this reason alone.

If algae simply displace corals, the total primary production on an area of reef may not change much, although the basis of that primary production changes beyond recognition, from the cnidarians to macroalgae and plankton. Increased plankton drives the fauna towards filter-feeding species, adding yet another layer of complexity to what is becoming a very changed 'reef'.

There comes a point when the limestone structure, which perhaps still supports many corals, is no longer a coral reef, and there may be several sensible definitions of when that point occurs. A geologist may favour the view that it occurs when limestone depletion becomes greater than limestone deposition. An ecologist may view the percentage cover of corals and determine that cover of less than a certain threshold marks a reef's demise. There is no clear value for the latter, however; note that the average coral cover for Indo-Pacific reefs is now only about 20%, while that in the Caribbean is now only 10% (Gardner et al. 2003; Bruno and Selig 2007). Many of these reefs still function as reefs, have a 'reef-like' trophic structure and certainly remain valuable for tourism. Yet, many are increasingly becoming eroding limestone platforms, dominated by algae and filter-feeders and with hardly any fish at all. It has been said that life on many coral reefs is reversing in an evolutionary sense, returning to a state more resembling that of the Pre-Cambrian than of today—that they may be on the 'slippery slope to slime' (Pandolfi et al. 2005).

10 The future, human population and management

Two groups of issues driving the damage to coral reef function are increasingly reported. The first comprises what can be called local drivers: sedimentation, sewage, overfishing, and the like. The second comprises those resulting from CO_2 rise: namely, warming water and acidification. The first may be immediate, can be short term, if moderated, and can often be influenced by human intervention on a single governmental scale. The second group, the CO_2-driven factors, seem at the moment to be intractable and, while they are slower-acting, they are much longer-lasting. It is likely that reefs will be the first major ecosystem in the modern era to become ecologically extinct. That the possibility is being raised at all, even with all due scientific caution, is a measure of how concerned scientists are about the future of coral reefs. One-third of the 845 main reef-building species are predicted to be at significantly increased risk of extinction (Carpenter et al. 2008). This is not only a prediction of the distant future; many reefs are already well along this path, with about one-third already being reported as irrecoverably damaged more than a decade ago (Wilkinson 2004) and updates of trends showing that this is increasing. Fish are being affected correspondingly (Wilson et al. 2006; Graham et al. 2007). This ecosystem is under severe stress.

Problems with isolated parts of the coral reef world have been reported for 130 years or so (Glynn 1993), but only in the past two decades have widespread and massive declines threatened the system as a whole. Based on present evidence and trends, few will survive more or less unchanged, most will survive as grossly distorted and stressed relics, and a good proportion will no longer be anything like a recognizable reef. Many factors combine to create problems. Here, one of the most intractable drivers, warming, is used to illustrate some of the issues. The issue of demands on resources is amplified, as are some possible solutions.

Preserving just 10% of reefs worldwide will require limiting warming to 1.5 °C (Frieler et al. 2013). The trend is not a smooth downward path,

The Biology of Coral Reefs. Second Edition. Charles Sheppard, Simon Davy, Graham Pilling, and Nicholas Graham, Oxford University Press (2018). © Charles Sheppard, Simon Davy, Graham Pilling, and Nicholas Graham (2018). DOI 10.1093/oso/9780198787341.001.0001

however. Many stressors occur as episodic pulses, so recovery periods between the pulses are an important issue too. The existence of thresholds and synergies means that more unforeseen interactions almost certainly will surprise us; even though warnings were sounded a decade ago about the importance of warming (e.g. Glynn 1993), many scientists were still taken by surprise at the magnitude of the changes in the late 1990s. In relation to reefs, Knowlton (2001) said: 'These quintessentially nonlinear relationships are common but nevertheless often surprising: when the thermostat is turned up one notch, people tend to expect one notch's worth of additional heat, not a house in flames'. Since the major warming event of 1998, several more impacts have had lesser but cumulative effects, notably the mortality in the eastern Caribbean of up to a third of major reef builders in late 2005. The 2016 warming appears to have been as severe in its consequences as was the 1998 event, resulting in, for example, a collapse of cover in observed Indian Ocean atolls (Perry and Morgan 2016), while the cover on the Great Barrier Reef (GBR) also declined substantially; at the time of writing, a major programme currently being worked on gives preliminary results of very severe effects; for example, Hughes (2016) recorded that over 90% of the reefs in the GBR bleached, leading to 50%–80% mortality on its northern reefs.

10.1 The timescale available

In the case of temperature, it is summer peak values rather than annual means that cause mortality. (Cold water episodes have occasionally killed large amounts of coral too, but these are not providing us with the same cause for concern.) Annual sea surface temperature ranges across the world are very variable. In the open ocean, seasonal temperature fluctuations might not exceed about 3 °C, from 26 to 29 °C, for example, while, in enclosed areas, winter cooling and summer heating may cause temperatures to range from 20 to 33 °C, a difference of 14 °C. Many shallow embayments see even greater extremes.

Using probability curves, it is possible to estimate the probabilities of lethal temperatures recurring at any future date. These dates were computed at several transects running north–south in the Indian Ocean (Sheppard 2003b) and for 36 sites in the Caribbean (Sheppard and Rioja-Nieto 2005). A recurrence more frequent than every 5 years was chosen as an 'extinction point' because 5 years is the approximate age at which many corals can reproduce; thus, if mortality on the 1998 scale occurs more frequently than this, it would lead to general extinction. In the Indian Ocean, reefs just south of the equator are forecast to reach this point in the 2020s or 2030s. The extinction date recedes into the future north and south of this tropical belt, but nowhere does it extend beyond about 2080. In the Caribbean, it is suggested that, on a coarse scale, mortality will progress as a wave over the next couple of decades from the east or south-east towards the west or north-west. Other estimates from the Indo-Pacific, focusing on Australia in some cases (Hoegh-Guldberg 1999, 2004), are similar. These results have been supported, with even more pessimistic results, by van Hooidonk et al. (2016) for different IPCC scenarios.

The prognosis is not good for corals in a time when seawater is warming, when intervals between dangerous spikes are becoming shorter and when multiple stresses, such as a gradual lowering of water pH and, in many places, continued pollution from industry, shoreline construction, nutrient enrichment and overfishing, compound the problem.

But partial respite may come from several causes. The deeper zones may remain unscathed in many areas where shallow reefs were badly impacted, and these contain reservoirs of breeding adults of many species, from which repopulation could occur (Sheppard et al. 2008), although this may not apply to several key reef builders which live only in very shallow water.

Second, water in very large bodies may not warm as much as the previously mentioned graphs predict. In the Pacific, water appeared to become thermally capped, for a time at least, such that the Western Pacific Warm Pool (Figure 10.1) did not rise as much as in other areas, with the likely result that coral reefs in such locations may either escape the usual thermal consequences to some degree, or may at least have longer to acclimatize (Kleypas et al. 2008). Temperatures in the Warm Pacific Pool area naturally average about 29 °C, which is close to the suggested 'thermostat' limit, but they have warmed by a smaller amount than cooler parts of the oceans have. The mechanism for the thermostat, according to this argument, may be increased evaporation, which itself requires latent heat of evaporation, taking thermal energy from the surface of the ocean. The warm water may also increase cloud cover, which reduces insolation and causes increased winds, which lead to more evaporation; both of these act as negative feedback for warming. Prolonged periods of warm water may give the coral–algal symbiont time to acclimate a little, but the degree to which this is possible is so far unknown. At present, this is far from a settled argument. An absence of reported bleaching does not indicate a lack of bleaching, evaporation cannot produce an effective regulating mechanism on its own and clouds probably are not regulators unless there are departures from the balance between cloud greenhouse and cloud albedo effects (Pierrehumbert 1995). This argument suggests that the main factor in tropical climate is the clear-sky water vapour greenhouse effect and that 'evaporation does not serve as a thermostat for the tropical climate . . . it merely acts as a buffer to keep the sea surface temperature from ever getting much larger than the low-level air temperature'. The debate continues.

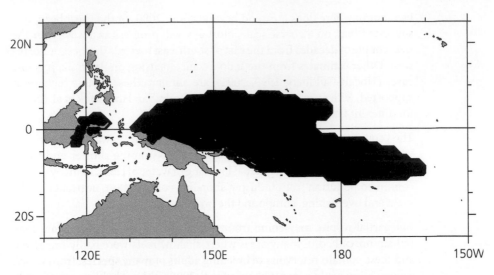

Figure 10.1   The Pacific Warm Pool; the area encompassed by white line is the area where average yearly temperatures in the period from 1950 to 2006 are above 28 °C.

Source: Figure courtesy of Ruben van Hooidonk.

10.1.1 Possible adaptation to stressors

One important question is whether corals can adapt or acclimate to these changes. As just noted in the context of the Pacific and the warm pool, it is likely that, given sufficient time, some acclimation to warm water can and has occurred. However, in the case of not only warming but of acidification and nutrient enrichment, there are no suggestions so far that reefs can acclimatize to anything like the amount needed. Regarding temperature rise, some acclimation clearly has happened. Several coral species in oceanic sites are killed when temperature reaches, say, 30 °C yet thrive in bays where temperatures are at least 2 °C warmer than that every year (Figure 10.2). Further, many species may survive warming spikes in a warm atoll lagoon, while the same coral species is killed on exposed, cooler seaward reefs. Water in lagoons warms more every year than open, oceanic water, so presumably corals used to lagoonal water have acclimatized to some degree. But this merely begs the question, what is the nature of acclimatization? One possibility lies in the nature of the zooxanthellae, not the coral. The strain known as Clade D is particularly resistant to warmer temperatures (see Chapter 4), and this clade may have become more common in waters which regularly warm (Baker et al. 2004; Rowan 2004). This subject is currently undergoing considerable research.

If the solution for warm adaptation lies in propagation of heat-resistant algal clades throughout the world's corals, then there is the question of time needed for this to happen. Present estimates suggest that there are just a few decades for corals to achieve this. Whether this is possible remains to be seen.

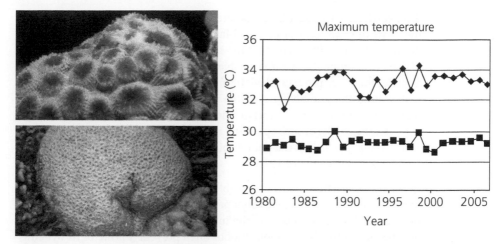

Figure 10.2   Coral adaptation. Right: Warmest summer month temperatures in the Arabian Gulf (top) and the central Indian Ocean (bottom). The gulf is about 4 °C warmer than the central Indian Ocean. Left: This species of *Favia* coral survives well (top) at 32–34 °C every year in the Arabian Gulf but was killed (bottom) by the 30 °C spike in the central Indian Ocean in 1998.

In biogeographic terms, some species are expanding their range towards the poles. Caribbean *Acropora* species are a good example (Precht and Aronson 2004). These grew further north along the Florida coast in the early Holocene, when temperatures were warmer than they are today, and then contracted south; but, now, with seas warming, they are extending north again. Similar effects are being reported northward along Japan, and south along the coasts of both East Africa and Eastern Australia. While this does not come close to compensating for losses in the warmer, tropical belt, it does bring possible advantages to some species.

If temperatures in the past have been warmer than today, then why is present warming causing the observed mortality? The answer appears to be today's rapid rate of change, an answer which applies also to the problems arising from the increase in acidification and of other local and synergistic stressors. Acclimation may be a fairly rapid process at some levels, but evolutionary processes of Metazoa generally take too long for current rates of change.

10.2 The elephant in the room

Many aid agencies and research programmes in the world's poorer, reef-dependent countries are now acutely aware of the importance of human population numbers. This awareness has arisen at the most senior or political levels (most people implementing projects on the ground understood this decades ago) because of the failure of so many reef-related projects, and the waste of huge sums of money. Many developing countries currently do not have the capacity to conduct the research needed to identify the causes of their environmental resource problems, let alone solve them, yet many aid agencies focus on 'poverty alleviation' and 'capacity building' and are reluctant to fund the underlying research needed for these activities. Criticism that such scientific work is a sort of 'scientific imperialism' is common and, in any case, the results of such spending rarely produce the kinds of immediate and striking results that the politics of the aid and financial systems require (Hancock 1993; Sheppard 1996).

In many international fora, sensible discussion of human population numbers does not take place (Sheppard 2003c). It may be recognized that pressure of human numbers is the main cause of reef (and other) habitat decline, but cultural or religious taboos often prevent this factor from being included effectively, to the extent, even, of excluding it as a topic in international fora. Yet, the increasingly crowded coastal zones of many tropical countries are clearly and closely linked with a decline in reef resources and condition. Many such places have human population doubling times of 15 years; on coasts around several cities where wars or inland poverty are rife, it is sometimes only 5 or 6 years, due to migration. If the 'human numbers' part of the equation is not addressed, then, naturally, the equation cannot be solved. Instead (at international fora at least), discussion may be deflected into technical fixes, networking, discussions, further meetings, and so on—indeed, anything except the underlying drivers which cause the problems. The phenomenon—also called the 'ostrich factor' (Hardin 1999)—means that human population is ignored to the point where technical fixes are commonly overwhelmed. It is argued (correctly) that affluent living styles are a problem, but population increase is also, and that tends to be faster in poorer countries. Although more than one donor agency now strives for sustainability, this is destined not to succeed if due recognition is not given to the increasing demand for resources. However, several countries now do (sometimes reluctantly) install units or even ministers for 'child spacing', in order to fit eligibility requirements for receiving aid. The problem is that the world is now interconnected to the degree that societies can no longer 'externalize' their problems. There is nowhere left to externalize to (Jameson 2008).

Wars can greatly increase pressures on reefs. Examples include the displacement of people in Sri Lanka southward to coastal regions; the flight to the coast in Mozambique as a result of its (now finished) 18-year civil war, which rendered many inland parts unsuitable for agriculture; and the surge of population on the Yemen coast when large numbers of expatriate workers were expelled from oil-rich countries when the Yemen government was insufficiently supportive of the Iraq war. In such cases and others, demands on the supposedly free reef services swiftly overwhelm what those reefs can supply. There remains the belief that the sea is a supply of free food, as, indeed, it would be, if managed properly. This entire issue, the elephant in the room (Sale 2008), has to be addressed adequately before recovery of reefs can substantially occur.

10.2.1 Shifting-baseline syndrome

The 'shifting-baseline syndrome' was coined in relation to fishing (Pauly 1995) to describe the starting or reference point of an ecosystem (in this case, fisheries) against which future change would be measured. It was pointed out (Sheppard 1995) that the same problem applied to benthic baseline coral reef surveys which purported to observe the 'natural' condition before some major construction changed it. The 'baseline' concept is still popular in environmental impact assessments (EIAs), so that the amount of damage or stress caused can be estimated. There is, however, a major problem with the method, which has led to much of the global deterioration seen. That is, any baseline ecological condition measured today has probably itself drifted far from its 'true' (perhaps pre-industrial) condition. Each generation of observers has a natural tendency to assume that the 'best' reef they see represents the 'baseline' condition, whether or not it has in fact drifted ecologically, trophically and in terms of its species composition, towards a changed and deteriorated condition over the preceding century or two. There is no easy means of determining true condition, so the existing 'best' condition substitutes for the need. The reference baseline, in other words, shifts inexorably closer to various thresholds.

From the 1990s, attempts were made to estimate the amount by which 'pristine' ecosystems had in fact changed. Jackson (2001) pointed out that 'the persistent myth of the oceans as a wilderness blinded ecologists to the massive loss of marine ecological diversity caused by overfishing and human inputs from the land over past centuries'. He points out that the word 'natural' usually means the way things were when the present generation of scientists first saw them. Thus, changes have occurred to reefs which only now are being recognized for what they really are: namely, in most parts of the world, massive distortions from their condition of a century ago. The importance of this is that the remaining resilience of the system may be a great deal less than an ecologist might hope—the elastic has been stretched so far already (without us really knowing it) that it has less scope to stretch further before snapping than we think it does.

In reef ecosystems, this is now recognized. We know that many reefs of today are poor shadows of those of decades or centuries ago, and ill-informed management of them often consists of no more than aspiring towards a condition which is itself heavily degraded. In many places, reef assemblages are merely bumping along the x-axis of any graph which depicts condition with time.

Coral reefs lend themselves to long-term examination because they are relatively easily preserved in fossil records, and some in the Caribbean have shown the marked degree to which they deteriorated between pre-Columbian times and the 1980s (Jackson 2001). Stressors in days after Western discovery were mainly driven by fishing: of grazers, which reduced algal dominance; of sea cucumbers, which consumed detritus; of turtles, which grazed seagrass beds; and of shellfish, which had multiple roles, including filtering seawater. It is now estimated that, from the seventeenth to the nineteenth centuries, the quantities taken were scarcely credible, so substantial impacts are not surprising. Then, increasingly, land run-off of both toxins and fertilizers from increasing agriculture and from disturbed sediments caused further changes. It appears that, for Caribbean reefs at least, the stage for their distorted state as seen today was probably set in the nineteenth century, when so many herbivorous fish were removed (Jackson 2001). But urchins are effective grazers too, and built-in ecological redundancy meant that no gross changes were recorded (although observations were extremely scarce in any case) until the urchins succumbed to disease in the 1980s. Then, with no grazing, with no ecological redundancy remaining and with increasing nutrients entering the sea to promote algal growth, large sections of the Caribbean reef system essentially collapsed. Later episodes of coral mortality, probably from introduced pathogens, increasing physiological stress and, most recently, from warming temperatures, simply added to reef demise in the Caribbean. In other reef areas, these different stressors have acted with different degrees of importance or effect, but the same groups of stressors seem to be common throughout.

Coastal resource use therefore (not only fishing) has been and remains a major driver of ecological change on reefs, and any management attempt to conserve reefs without attention to regulating the extraction of protein or other resources cannot possibly work. Nor can it work without effective controls on nutrient and pathogen discharges into the surrounding waters. One main mechanism for achieving this in the short term lies in protected and properly managed areas.

10.3 Protected areas and coral reefs

At a local or country level, one of the most hopeful solutions lies in the establishment of strictly protected marine areas, where management is targeted and where enforcement is effective. There are, of course, many marine parks or protected areas in existence, but very few are effective, and many permit fishing; since fishing is one of the main ecologically distorting pressures on a reef, such areas can hardly said to be 'protected' effectively. This partly explains why, for example, the Caribbean region has about 500 marine reserves of one kind or another and yet the region has lost about 80% of its coral cover. Strong arguments have been made that only total protection in substantial areas will have the required effect, that effect being not only protection of reefs within the area but also replenishment by 'overspill' of adjacent areas due to the passage of both adults and larvae from the protected nurseries into adjacent areas. There are now several examples of reef areas where fish catches have been shown to rise substantially in areas adjacent to protected areas, providing benefits to local people.

Much more protection needs to be implemented. Today, only about 4% of the world's oceans are protected and, of the world's coral reefs, only 6% are effectively managed, 21% are ineffectively managed and 73% lie outside any MPA (Burkeet al. 2011). But, in this sense, management can only respond or prevent local impacts, and none are immune from the global effects of CO_2 rise. Further, only one-tenth of that number is at low risk from other extraneous forms of impact and pollution from upcurrent areas (Mora et al. 2006). Viewed beside these figures, the amount which needs to be fully protected seems daunting. An early estimate that 10% of habitats need to be protected (Souter and Linden 2000) was followed by a recommendation in 2003 by the World Parks Congress, which estimated the requirement to be 20%–30% (Mora et al. 2006). Now, estimates today state that 20%–40% must be completely protected if there is to be good chance of allowing the reef ecosystem as a whole to be conserved (Roberts 2007). Every revision has been upwards.

The sizes and spacing of the protected areas also needs careful calculation. About half of the present MPAs are smaller than 1–2 km². This is inefficient. They should be approximately 10–20 km in diameter and spaced about the same distance apart to achieve maximum effect, because of factors connected with propagule dispersal and 'loss' of larvae. Because coral reefs are scattered geographically, an optimum network of properly protected areas might consist of MPAs, each of which having an area of 10 km² and spaced about 15 km apart. This would require another 2,500 MPAs fully protected from all exploitation. With their own index of effectiveness, Mora et al. (2006) calculated that the 2% of reefs that currently lie in MPAs are insufficient in this respect and that, even if all 'paper parks' became effective, the amount would still fall short of what is needed.

The benefits to all sectors of such protection are proven (Gell and Roberts 2003; Roberts 2007), although there is resistance to the concept from sectors with vested interests. Examples of an improvement in fish catch from spillover effects now exist, but it is not the scientific argument that convinces fishing communities but rather examples from other fishing villages where it has been applied. Extracted protein from crustaceans, shelled molluscs and octopus, as well as fin fish, has been shown to double or increase many fold when some areas are protected from fishing, and it takes only 2–5 years for this to happen (Gell and Roberts 2003; Roberts 2007). One reason is simply that protected areas provide a refuge for breeding adults; in addition, the size of a mature adult greatly affects egg output, such that a 10 kg adult grouper produces many times more eggs than do ten groupers of 1 kg each. These eggs and larvae 'spill over', providing greater returns, while maintaining the breeding stock.

Instead of exploiting the concept of increased egg production from larger adults to the full, it is the breeding adults that are often exploited preferentially (they are the largest adults)—and are extinguished, one after the other. In the case of grouper and snapper, for example, breeding aggregations are commonly targeted, eliminating the brood stock for all following years. The targeting of spawning grounds partly stems from ignorance, but also partly stems from the fact that the seas are 'commons', so that there is fear amongst fishermen that, if they do not get the fish, someone else will. Some successful protection has come from providing different sorts of ownership of reefs to particular communities on adjacent shores (see Chapter 7). As Roberts (2003) remarked,

> We can get more out of the sea by leaving some of it alone . . . We can obtain more seafood not by creating enormous aquaculture facilities and in the process destroying marine habitats to make way for them, but by simply allowing our oceans to produce it. In essence, reserves could help us turn back the clock by 200 years as they replace critical fish refuges that have been whittled away by fishing.

10.3.1 Connectivity and selection of areas

Choosing where protected areas should be is rarely simple. Many species of corals and fish have very wide ranges, while, in some groups, up to 53% of species have very restricted ranges (Roberts et al. 2002). In any location where several groups show a high endemicity, that area may be termed a 'biodiversity hotspot'. The concept is potentially confusing because, while some definitions or uses might suggest that the existence of many species should constitute a biodiversity hotspot, some very low-diversity areas exist which also contain a high percentage of species which occur nowhere else. The Brazilian region of the Atlantic, and the Red Sea, are two examples. Such areas, as well as species-rich areas, might be regarded as being priorities for protection.

Centres of endemism predominate in locations which are comparatively isolated by a combination of geographical distance and ocean current patterns, especially non-reversing currents. Roberts et al. (2002) identified the 18 richest multitaxon centres of endemism and found that they included 35% of the world's reefs, and between 58% and 69% of restricted range species. The richest ten centres of endemism covered just 15% of the world's reefs, yet included approximately half of all the restricted range species in the groups considered. It is clear where efforts need to be placed, even though they may present considerable practical challenges. Many of those centres are at considerable and continuing risk of damage and loss from burgeoning human populations and, even where some areas with low human populations exist and where substantial sets of endemic species are found, such as the cash-rich countries of the Middle East, the very poor control on damaging coastal construction and unbridled development means that reefs and associated biota are becoming rapidly diminished. High endemism alone, however, is not the only criterion; areas such as the Chagos Archipelago, the 'Chagos stricture' as Veron et al. (2015) have called it, have no or almost no coral endemics because they are a stepping stone on the main east–west highway of Indian Ocean coral dispersal and, because of this, remain critical to Indian Ocean diversity.

Issues of connectivity (species and genetic) are likewise important in any coherent design of protected area networks. Surface currents transport larvae. Connections occur from an upcurrent location to a downcurrent location, but not generally the other way around. Currents may reverse in an annual cycle, in which case, the connection will reverse, although a direction of flow must coincide with a period of reproduction by sessile organisms (like corals) if the connection is to be effective. Thus, even where currents do reverse annually, larval dispersal may still be predominantly or only one way. Connection thus affects the resilience of a site, or its ability to recover from an episode of damage. This, in turn, should affect the design of large-area or multinational MPAs (Roberts 1997). Yet, some elements of geography appear to conspire against such networks: Sheppard and Rioja-Nieto (2005) showed that ocean warming is likely to result in the Caribbean in a wave of reef extinction which travels from the south-eastern side of the sea to its north-western side. The gross currents in this region flow broadly from the south-east to the north-west too, before turning northward to form the Gulf Stream. While many local eddies occur, this means that the most vulnerable region—the eastern Caribbean, including the Windward and Leeward Island chains—lie broadly upcurrent in the Caribbean region. Two problems have resulted from this. First, some countries in these island chains have acknowledged problems with their reefs and hope that recruitment of species from other countries may one day restore them. But there are few or no upcurrent connections to supply those reefs. By the same token, destroyed reefs in the Windward and Leeward chains will no longer be rich larval sources for downcurrent reefs, which, in the case of the Caribbean, is all the rest of it. The Florida Keys fortuitously form the target for most of the broad-scale Caribbean currents before they deflect northward, which may be one reason why it retains a relatively high diversity despite is current abuse and relatively low coral cover of <5% in many parts (Dustan 2000). Despite this, two decades ago, Dustan warned: 'Data collected since 1974 presents a grim picture for the future of the reefs of the Florida Keys. Recovery from the present severe ecological degradation is probably not possible within a human lifetime'. This refers to a downcurrent area which receives larvae from most of the Caribbean and is in the world's richest country, with one of the world's best resourced coral reef states. Today, NGOs and government agencies throughout the poorest parts of the world are trying to change the habits of starving people with respect to how they treat their own reefs in their attempts to feed their burgeoning populations. It is an uphill struggle.

MPAs do not necessarily afford any protection from damage from climate change (Graham et al. 2008). Selection of locations was made often years ago, based on various criteria not usually including resilience to climate change effects. Future management efforts need to include identification and protection of regional refugia which are most resilient to climate change.

10.4 Environmental assessments

Any development judged likely to cause environmental disturbance or which fits certain criteria has to be preceded, in many countries, by an EIA. Variants of the term exist, such as 'environmental statement' and so on, commonly interchangeable but with generally similar and sometimes specific legal meaning. In its basic essentials, the process identifies the value of the environment, identifies which aspects are likely to be affected, estimates risks associated with proceeding with the development and should propose ways to mitigate any damage. Ideally, it will have the power to veto a development completely, or put in place mechanisms to temporarily halt work (a 'stop order') until a problem disappears. An example of the latter might be to halt dredging of the seabed next to a reef if mechanical screens fail to contain a sediment plume, until suspended sediment loads subside. Parallel with this in some cases might be a 'social impact assessment' to consider the economic and social values or depredations to local people following a proposed development.

An essential condition for an effective EIA is that it is performed before a project starts, that is, well before a time when it would become impossible to relocate a proposed site if the assessment showed it would be a disastrous choice. Also essential is that those conducting the EIA are not connected with either the contractor who stands to gain or the commissioning authority or government. Many EIAs fall down on several of these points (Sheppard 2003a).

Many countries have well-developed systems for implementing EIAs, with clearly laid-out protocols. Many do not, or, if they do, their EIAs are as much ignored as they are implemented. Where this happens, many truly absurd examples of EIAs can be found (Sheppard 2003a). Unfortunately, many coral reef areas are in countries which ignore them or claim overriding need so that the proposed development proceeds. The costly downside is always for the local people.

Even with the best designed and meticulously executed EIAs, it is common for a judgement to be made that, although a small amount of environmental impact will occur (it would be difficult to imagine a significant coastal development that did not involve any), the development's benefits far outweigh any damage. With one or a few such projects in a large, unscathed area, this may, indeed, not threaten the integrity of the site, as noted in Chapter 9 and Figure 9.13. But, cumulatively, many such adjacent projects will have an accumulative effect. This is the 'shifting baseline' again. Given that many impacts are synergistic and that many stressors have a threshold beyond which the resilience of a reef system is compromised, even though any one development may indeed have no marked impact by itself, many done simultaneously or consecutively can have an irreversible impact. Examples are commonly found on those coral reef-fringed shores where ribbon development along the coast extends progressively out from the original coastal fishing village, each stage requiring bigger and deeper facilities for servicing the needs of the growing community. In such cases, it is rare in many countries for there ever to have been an EIA of the whole plan stretching several decades in the future. More likely, each tiny increment has its own EIA, if, indeed, it has one at all. It is not unknown for a reef which has been dead for a few years and which has become principally a non-accreting algal bed to be used as the starting reference site for the next EIA in the series.

10.4.1 Offset schemes

One solution to incremental damage, which occurs with even a good EIA mechanism, may lie in offset schemes. This is the relatively new (in implementation, if not in principle) mechanism by which impacts deemed to be unavoidable can be counterbalanced by environmental improvement elsewhere. The intention is that the net environmental change in a country, region or stretch of coast is neutral, or even an overall improvement. Commonly, the offset work is designed to offset the residual damage which will occur after all reasonable or affordable mitigation is done (McKenney 2005).

The principles of equivalence are probably the most difficult to address. No two sites are identical, so it is difficult to judge how much 'repair' needs to be done to one damaged site to provide a suitable offset to damage which will occur because of a new development. Offsets may, in fact, be applied to quite different kinds of habitat. Other complications arise in connection with the question of the duration of the benefits gained, for example whether benefits from a restored site should be in perpetuity or only for the duration of the damage which has been forecast to occur in the construction site. In effect, the entire system might require two EIA procedures: one for the proposed development and another for the site scheduled to be restored from its degraded state. Most examples come from US wetlands and estuaries, although the principle is being slowly adopted by more countries.

One major drawback concerning coral reefs is that it has not been easy, nor even possible, to restore damaged reefs to any ecologically significant extent, due to the fact that impacted coral reefs show a hysteresis effect and fall into alternative ecological states from which it is extremely difficult to recover (Chapter 9). Given that impediment, only offsets directed to other habitats such as estuaries or seagrass bed may even be possible. This may do little for the global condition for reefs.

Offset schemes appear to be enormously valuable in several contexts, both to other kinds of habitat, such as wetlands or forests, or for other 'commodities', including the best known of all: carbon offsets, for which entire trading mechanisms have sprung into being. But the answer to reef degradation, at present, seems to be simply not to cause it in the first place, given the difficulties and uncertainty of restoration.

10.5 Costs, prices and values

The social costs of degraded reef systems, to those who depend on them, are obvious, shortage of food and reduction of food security being foremost. Numbers of deaths resulting from this are almost impossible to quantify but, undoubtedly, already number in the millions. The linkage of this huge number with environmental degradation may be obscured by many factors: cause of death is usually recorded (although it is not recorded at all in many reef-containing countries) as being from a particular disease, but many of those diseases only kill a patient who is already severely malnourished. For example, 54% of child deaths in 2000 and 2005 were related to malnutrition, although the final illness that killed each child could be a named disease instead (Figure 10.3). As populations increase, this proportion is likely to increase also.

At the national level, the issue of food security becomes important. 'In general the countries that succeeded in reducing hunger were characterized by more rapid economic growth and specifically more rapid growth in their agricultural sectors. They also exhibited slower population growth' (FAO 2003). The number of areas which have significant coral reef sectors but which are failing in this respect is increasing, and these areas include a significant proportion of the one billion (and rising) people who are malnourished at any one moment. Achievements in protecting a small protected area which can support a given number of people are set to naught when the number of people requiring food from it doubles in 10 years, as is commonly the case.

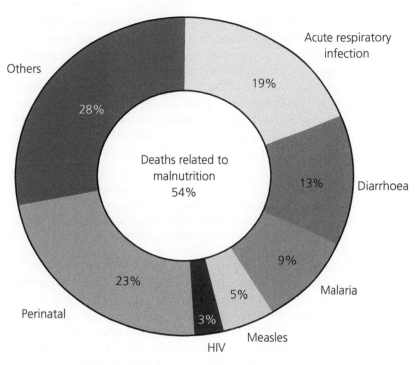

Figure 10.3 Major causes of death amongst children under 5 years worldwide, 2000.
Source: Case-specific mortality, EIP.

10.5.1 Economic costs

Unwise and short-sighted use of reefs is expensive, and costed examples are only recently becoming available. It is increasingly thought that assigning dollar costs to reef (and other) resources is necessary to convince authorities of the importance of the habitat's continued survival. Although 'ecology' and 'economy' have the same Greek root (*oikos*, meaning 'home'), the two disciplines have diverged to an alarming degree, even though they should work hand in glove. The cost of the use of destructive fishing methods in Indonesia was estimated 20 years ago at being up to £1.2 million km^{-2} (Cesar et al. 1997). The costs of the 1998 warming was as much as USD 8 billion over the following years (Cesar 1999; Wilkinson et al. 1999), made up of loss of food, coastal erosion and loss of tourist revenues. Indeed, examples are increasing of impacted tourism in several reef areas, and even of tourists asking for money back when the purported reefs they paid to see turn out to be dead.

Assigning financial costs to damage on reefs is notoriously difficult. The basis for most such costing comes from values identified by Costanza et al. (1997). Until this kind of work was done, it was not appreciated by many that the 'goods and services' provided by nature as a whole considerably exceeded the sum of all countries' gross national product. For coral reefs specifically, they are estimated (in 1997 prices) to supply over USD 6,000 ha^{-1} y^{-1} which, summed by their estimate of total area, was calculated to be USD 375 × 10^9 annually. Development of different mechanisms for costing continues, although many of the goods and services have still not been properly captured by available methods (Moberg and Folke 1999). Costanza et al. (2014) have revised upwards their suggested valuation of coral reefs to a value 50 times higher, to about a third of a million dollars per hectare per year 'due mainly to a decrease in coral reef area and the substantially larger unit value for coral reef'. They emphasize that

> valuation of ecoservices (in whatever units) is not the same as commodification or privatization. Many ecoservices are best considered public goods or common pool resources, so conventional markets are often not the best institutional frameworks to manage them. However, these services must be (and are being) valued, and we need new, common asset institutions to better take these values into account.

As losses of reefs progressively highlight additional costs which were not fully appreciated until they were gone, these values increase more rapidly than initially expected. They also overlook, of course, the human cost. However, by working on the principle that governments of the world use dollar accounting rather than ecological assessment methods, translation of ecological costs to economic prices may help create a better appreciation of what is being lost. It does face the criticism that it is more attuned to *price* than to *value*; just as the value of a Rembrandt is greater than the price of its canvas and paint, so the value of a functioning reef system is greater than the price that its components might fetch in an aquarium shop. Nevertheless, such costing is becoming increasingly effective in liberating many authorities from the tyranny of the short-term spreadsheet.

There is a substantial body of work now investigating the prices that can be assigned to ecosystems and, more generally, to the environment. Partly, this has come about because of rapidly rising costs to insurance companies of damage caused by severe environmental conditions. It also comes about from simple estimates of what tourists can supply. One example will suffice: the Maldives depends heavily on diving tourism on its coral reefs, and Anderson (1998) has calculated that, whereas one captured shark has a one-time value of about USD 35, leaving it alive to attract divers brings in about USD 3,300 per year for the life of the shark and, of course, allows it to breed more sharks. There should be no contest.

10.6 The 'slippery slope to slime'

'Today, the most degraded reefs are little more than rubble, seaweed and slime' (Pandolfi et al. 2005). The 'rise of slime' has more recently been quantified in the Caribbean (de Bakker et al. 2016), where, in parts of the

Netherlands Antilles, cyanobacterial mats have become the dominant benthic component of coral reefs down to 40 m depth. The biomass of predatory and herbivorous fish in coral reefs which are very remote or which are strictly protected from intensive fishing are one to two orders of magnitude greater than those in intensively used reefs; correspondingly, remote or protected reefs have less of what a healthy reef does not need (the 'slime' noted, mats of cyanophytes, diseases, smothering sediments, etc.). Some reefs are affected by neither overfishing nor harmful run-off, and these are the essential global controls that are needed to provide a reference baseline (hopefully only a slightly shifted one) for reef management in the future.

Environmental management has become a discipline in its own right and is increasingly important. D'Agata et al. (2016) showed that travel time between an area and a sizeable human population is the most significant factor controlling the biomass of a reef's top predators, a factor scarcely affected by 'management' of any kind at all. Reefs have low yields despite a high productivity because 'diverse and intense interactions consume the production' (Birkeland 2015), and most of the high production is consumed within the ecosystem itself, such that only about 1% of the gross primary productivity of the reef could be available to humans before the system collapses. Too often, management ignores this and is little more than tinkering blindly with a poorly understood system with, as often as not, management goals being overridden by adjacent pressures of short-term food security or limited economic gain. Most importantly, the creation of a 'management plan' is often considered to be an end in itself, with the plan itself, once made, either not being implemented or being implemented ineffectually. A management plan is scored in national statistics as a success; a protected area may become one of many notorious 'paper parks', being unprotected and where no further monitoring is done to gauge the area's subsequent recovery or deterioration. Some activities which are known to damage a coral reef and which should be prohibited may be exempted during negotiations between management and interest groups that precede a development; this commonly includes fishing, which, of course, is one of the main drivers of reef deterioration. In other words, the labels 'protected', 'park' or 'reserve', to name a few, may involve some controls, but may not include some of the important ones, for historical, traditional or economic reasons.

10.7 The future for reefs

What is the possibility that reefs and their species can adapt to the environmental stressors they are experiencing? Geographically, refugia clearly exist to some extent in upwelling areas, deep areas, locations adjacent to favourable oceanographic conditions, and so on. Riegl and Piller (2003) remark: 'Calculations show that if the top 10 m of the ocean became inhospitable to corals, still 50.4% of the coral area would remain intact in the Red Sea and

99% in South Africa.' The numbers, if the top 20 m becomes inhospitable, instead, are 17.5% for the Red Sea and 40% for South Africa. Regarding thermal tolerance, there is a little evidence that coral juveniles are non-specific regarding which clade of zooxanthellae eventually dominates them, and so can potentially increase their component of the relatively heat-tolerant version known as Clade D or perhaps of low-light tolerant versions.

But, while several mechanisms clearly exist, most curves of sea temperature rise, whether from instruments or satellite series, show that the current warming trend began in the 1970s and, in the last 30 years, there has been little adaptation on an ecological scale, suggesting that longer may be required for most of these mechanisms to have much effect and that they are simply insufficient. In any event, the 2016 warming and mortality showed that acclimation had been very limited and was insufficient. In the case of acidification, there is no evidence yet that corals can somehow counter their reduced ability to calcify, and the geological record suggests that they cannot (Veron 2007). There is likewise no evidence that reefs can thrive and develop in the presence of significant sewage pollution and land run-off.

Estimates of reef loss are susceptible to interpretation: what constitutes loss? Measurements of cover may be useful. On a regional scale, in the Caribbean, average coral cover has declined from about 50% to 10% in the last three decades (Gardner et al. 2003). Of Caribbean reefs, 11% are categorized as 'lost' and 16% 'severely damaged', and the authors of that study suggest that 'given current predictions of increased human activity in the Caribbean, the growing threat of climate change on coral mortality and reef framework building, and the potential synergy between these threats, the situation for Caribbean coral reefs does not look likely to improve in either the short or the long term.'

In the Indo-Pacific, too, where coral cover is declining, Bruno and Selig (2007) suggest that, while there are many examples of reefs in good condition, these are now anomalies which 'currently represent less than 2% of reefs in the Indo-Pacific'. Even on the comparatively very well-managed and researched GBR, coral cover is declining greatly (Hughes 2016).

Quite obviously, reefs cannot continue declining at this rate and survive in anything like a meaningful way. Although technical fixes might in some cases be possible, and much can be done to reduce impacts, a key factor in any reversal must be greater awareness by authorities, but this is easy to say and difficult to ensure, notwithstanding huge increases in available information. For example, one very large reef in the Persian Gulf, possibly the gulf's largest in area (Figure 10.4), has a remaining coral cover of well under 0.1%, with small surviving fragments of corals spaced at intervals of not less than 20–50 m apart. The shallow corals died in warming events in 1996 and 1998, but deeper corals in the gulf survived. Sedimentation, however, appears to have been greatly increased by very substantial engineering works in the last few years and, today, almost no corals throughout the depth of this large reef remain alive. Perhaps five species of faviid or siderastreid corals could be found alive. But the most alarming point about this reef is that officials in the country concerned, even those in charge of monitoring, were unaware of, or disinterested in, the demise of this huge and rich habitat (Sheppard 2008, 2016), a not-uncommon event in many countries which either cannot or do not take an interest in their natural environment. The first step must be to demonstrate the value of, and irreversible nature of, their loss. It is clear that this not a scientific issue any longer but rather it is a political problem and failure. In the Persian Gulf, for example,

> it is a mistake, perhaps, to think that 'management' of coral reefs, or indeed any of the other marine habitats, is even possible in the first place! In one sense it is simply conceit or hubris to consider that we can manage such a complicated habitat, because we cannot successfully manage even its component species or industries (such as fishing) satisfactorily. The management that we can do, in contrast, is management of people's interactions with the habitat. This may not be such an appealing concept, but probably is much more realistically aligned with what managers could do, and it is certainly what they should be focussing on. (Sheppard 2016)

Figure 10.4　Demise of a reef. The large reef, Fasht Al-Adhm, lies between Bahrain and Qatar in the Persian Gulf. It was killed over the last few years by temperature plus sedimentation.

Extinction comes in two forms, which should not be confused. In ecological terms, a coral reef may be dead, but it might be the case that all species once known to live there could be found somewhere if a search was extensive enough. If that was so, species extinction might not have occurred, leaving a possibility of recovery. Functional extinction occurs well before extinction of most of the component species (Sheppard 2007). It is functional extinction which provides by far the greatest loss to humans in all cases except unusual ones, such as where a species is discovered to contain a biochemical of great value. An adjunct of the latter might be termed economic extinction, in which harvesting the diminished numbers is no longer economically sensible.

The next question is, if a handful of hardy survivors can be found, can the reef they live on one day recover, given some decent management? This question is crucial for inhabitants of many parts of the world, as well as for determining how and where large sums of money should be spent in aid and restoration projects such as the creation of artificial reefs (see 'Restoration of reefs'). This answer is not easy either. Edwards and Gomez (2007) point out that much of any significant recovery has to come from within the system and, if conditions remain unfavourable, the trajectory will continue downwards, whatever restoration work is attempted. Species extinction remains possible after functional extinction, however. The Allee effect explains how more than just a very few scattered colonies are needed for reproduction to occur—gametes of each sex simply will not meet in the water column below a certain population density in the case of corals—so, given a decline below a certain but usually unknown threshold, a 'regional extinction' certainly becomes possible.

Restoration of reefs

Artificial reefs and coral transplantation

Artificial reefs have a long history (Schuhmacher 2002) and a wide range of applications, from attracting fish to enhancing reef appearance in order to attract recreational divers. The range of materials used to construct artificial reefs is large and, while some of the materials are useful, others are useless or even damaging. The use of appropriate reef material results in improved biogenic construction and increased three-dimensional structure, with increased diversity of fish and invertebrates. Experience over many years, but with too little sound experimental work, has led to several successful solutions, but also to many examples where the purported construction of an artificial reef was no more than cheap or disguised garbage disposal. Many artificial reefs never had any chance of working but were used for public relations exercises by organizations wishing to show 'green' credentials; some of these 'public relations' reefs have wasted money and done more harm than good in ecological terms, largely when taking corals for transplanting to the target area denuded source areas, with the harvested corals promptly dying in the process, or

Continued

Restoration of reefs (*Continued*)

when, for example, as a cheap alternative to proper disposal, thousands of used tyres or vehicles, or tonnes of surplus concrete, were dumped offshore as 'artificial reefs'.

Some striking successes have been achieved by small-scale examples, such as artificial reefs placed outside the windows of underwater restaurants, but these have little conservation value, although presumably they provide good educational value and raise environmental awareness.

A key criterion for the success of an artificial reef is that it must be located in a suitable site. If the cause of the original coral mortality persists, newly introduced colonies are unlikely to survive any better than the original coral. If the cause of the mortality was episodic, such as a sediment plume from construction, then conditions may become suitable again. But, especially where reefs have been killed, the seabed may be covered with substantial amounts of sediment, much of it originating from erosion of the previous coral. This 'liquid sandpaper' is a major factor preventing new larval recruitment. Sometimes the original reef becomes blanketed by thick sediment so that, even where water quality has reverted to a satisfactory state, there is a new problem of finding suitable substrate. Artificial reefs, therefore, must elevate the substrate above the abrasive or smothering layer. This technique has had some success in areas which never supported reefs in the past, such as sandy areas adjacent to other reefs.

Realism and the areas involved

Examples of poor planning are not difficult to find, but knowledge of their failure can help inform the process for others (Precht 2006). A not-uncommon kind of project might involve the creation of a few hectares of artificial reef, but this has a high cost (higher, in some cases, than the cost of providing wardens to monitor and protect a natural reef hundreds of times larger). Some projects have been disastrous, with dumped tyres breaking loose and rolling about, causing damage. In more than one or two areas, many structures that had been dumped earlier to create artificial reefs had to be removed, as little life settled on them; the costs of removing these structures were high and the associated movement caused scouring over a wide area.

Key issues with artificial reefs

Artificial reefs may assist in recovery of an ecosystem that has been degraded or destroyed but are always a poor second best to preservation of the original habitats (Edwards and Gomez 2007). Both preservation and 'nurseries' are now probably required to reverse the trend of coral reef decline (Rinkevich 2008). It is important to determine which species are most likely to survive, especially if environmental factors have changed, such as turbidity. Fast-growing, branching species are preferred because of the faster success seen with them.

Sediment stabilization may be all that is needed in clear-water situations. This may be achieved with steel mesh, or interlocking concrete blocks (Figure, top). These must be properly anchored, or one storm will undo several years of work and coral growth. 'Reef

Continued

Restoration of reefs (*Continued*)

Figure 1. Two experimental attempts to stabilize substrates and encourage the return of coral growth; both examples are in the Maldives. Top left: Interlocking concrete blocks and transplanted corals. Top right: Steel mesh, which is proving less successful than the concrete blocks. **Figure 2.** 'Reef balls', produced by the Reef Ball Foundation, provide relatively a high surface area elevated above the substrate, attracting the settlement of numerous species in the right conditions. Many such structures placed together can demonstrably increase biota in an area which may previously have been reduced to shifting rubble.

Source: Photo by T. Barber.

balls' produced by the Reef Ball Foundation (<http://www.reefball.org>) have been one of the more successful innovations (Figure, bottom). Based on concrete, but with chemical and pH balancing and, in some cases, with abundant limestone chips, they provide a

Continued

Restoration of reefs (*Continued*)

large area of rough, elevated substrate. Such structures are hollow, providing essentially a three-dimensional space. If inappropriate material is used, such as smooth concrete blocks, there may be little attachment by marine fauna.

Correct placing of blocks is essential in order to optimize 'island' effects yet permit species to move from one to another, while minimizing the number of placed structures and, hence, cost for a given area. Some trials have found that placement in groups, with each group spaced from other groups at the range of typical visibility, permits good recruitment.

The use of 'electric reefs' is another approach, only partly tested but with some of the reefs showing promise (Schuhmacher 2002). The principle is to construct a steel structure or mesh which then becomes a cathode. Electrolysis in seawater's dilute solution of carbonate and bicarbonate results in the deposition of limestone on the cathode. Energy requirements and voltages are low—about 300 watts for 100 m^2 of structure—and the resulting rock is very hard. In addition, it has been found that corals grow well on the 'biorock'. More experimentation needs to be done, as not all attempts have been successful, but the method may become a useful tool in a world of declining coral.

Prof. Charles Sheppard, University of Warwick, UK.

Reef ecology, function and structure are currently moving beyond recoverable thresholds in a large and increasing proportion of reefs, leaving areas which, although still labelled 'reef' on charts, have greatly reduced biological and economic value. It is difficult today to find reefs that can serve as any kind of control site or which can serve as a reference to what a reef 'should' look like. Yet, to be able to have such reefs is crucial so that those making an attempt at recovery or restoration know what to aim for. Multiple avenues need to be pursued to ensure the survival of this ecosystem and, even though vast sums are needed, the sums are very much less than those which will otherwise be lost.

The future of reefs is not optimistic, given human inertia, current trends and present systems of accounting for short-term price rather than long-term value. There are positive notes, however.

First, we know that sites which are free of most direct human disturbances, because of either their remoteness or good management, can recover from climatic change disturbances much more easily than sites affected by both (Downing et al. 2005; Sheppard et al. 2008). There are some encouraging indications in many countries that there is a growing realization of the need to ensure that this ecosystem survives, and this might encourage better regional and local management of those stressors which are locally controllable.

Second, long ago Johannes (1998) outlined several Pacific island methods of community-based coral reef management; some of these had been important in the past when populations were low but fell into decline or disuse when global markets become too tempting; however, they are now seeing a renaissance in some areas. It is clear that such methods have an important place but, as noted by Birkeland (2004), it may be unrealistic to hope that these can be extended to all the world's coral reefs. Instead, as also stressed by Birkeland (2004), human population pressure and associated ecosystem demands are key elements which need to be addressed, together with a pro-active rather than a reactive view of reef management.

Buddemeier et al. (2004) remarked that 'research into adaptation and recovery mechanisms and enhanced monitoring of coral reef environments will permit us to learn from and influence the course of events rather than simply observe the decline'. While clearly true, the fact is that there are about a thousand agencies, NGOs and other bodies that are concerned with reefs and associated aspects and, despite numerous local successes, given the continued decline of reefs, these organizations can hardly be said to have succeeded so far, despite the expenditure of billions of dollars. This means that the existing system does not work sufficiently, something which few of these organizations are prepared to recognize. Precautionary approaches are needed. The starting point must be the application of clear and determined efforts to protect at least 30% of the world's reefs from all forms of exploitation.

Figure 10.5   A healthy mid-reef slope (5–15 m depth) in the northern Red Sea, very near a shipping terminal in Yanbu, Saudi Arabia. Careful construction, and not too much of it, will enable coral reefs to survive for many years.

Accompanying this must be reduction of stressors in all possible ways, starting with the sedimentation and sewage which impact those protected areas. Without these local controls, reefs will continue their decline. And, of course, reduction of atmospheric CO_2 is, as noted earlier, crucial too, although it is probably hardest to achieve. Almost all commentators have made the clear point that it is not too late for coral reefs and that the decline can be arrested, so that reefs can be allowed to recover to their once vibrant state (Figure 10.5). With such controls in place, we could, at the very least, buy two or three decades in which to try and solve climate change impacts, whose control is essential in the medium term.

Bibliography

Acker, K.J. and Risk, M.J. (1985). Substrate destruction and sediment production by the boring sponge *Cliona caribbaea* on Grand Cayman Island. *Journal of Sedimentary Research* **55**:705–11.

Adam, M.S., Anderson,R.C. and Shakeel, H. (1997). Commercial exploitation of reef resources: Examples of sustainable and non-sustainable utilization from the Maldives. *Proceedings of the 8th International Coral Reef Symposium* **2**:2015–20.

Adams, T.J.H. (1998). The interface between traditional and modern methods of fishery management in the Pacific Islands. *Ocean and Coastal Management* **40**:127–42.

Adams, T.J.H. and Dalzell, P. (1995). Management of Pacific Island inshore fisheries. In: 'T. Summerfield (ed.), *Proceedings of the Third Australasian Fisheries Managers Conference*. Fisheries Management Paper 88. Fisheries Department of Western Australia, pp. 225–37.

Adey, W.H. (1975). The algal ridges and coral reefs of St Croix, their structure and Holocene development. *Atoll Research Bulletin* **187**:67.

Adey, W.H., McConnaughey, T.A., Small, A.M. and Spoon, D.M. (2000). Coral reefs: Endangered, biodiverse, genetic resources. In: C.R.C. Sheppard (ed.), *Seas at the Millennium*, Vol. 3. Elsevier, pp. 33–42.

Adger, W.N., Hughes, T.P., Folke, C., Carpenter, S.R. and Rockström, J. (2005). Social-ecological resilience to coastal disasters. *Science* **309**:1036–9.

Allemand, D., Tambutté, É., Girard, J.P. and Jaubert, J. (1998). Organic matrix synthesis in the scleractinian coral *Stylophora pistillata*: Role in biomineralization and potential target of the organotin tributyltin. *Journal of Experimental Biology* **201**:2001–9.

Allemand, D., Tambutté, É., Zoccola, D. and Tambutté, S. (2011). Coral calcification, cell to reefs. In: Z. Dubinsky and N. Stambler (eds), *Coral Reefs: An Ecosystem in Transition*. Springer, pp. 119–50.

Alongi, D.M., Trott, L.A. and Pfitzner, J. (2007). Deposition, mineralization, and storage of carbon and nitrogen in sediments of the far northern and northern Great Barrier Reef shelf. *Continental Shelf Research* **27**:2595–622.

Amann, R.I., Ludwig, W. and Schleifer, K.H. (1995). Phylogenetic identification and in situ detection of individual microbial cells without cultivation. *Microbiological Reviews* **59**:143–69.

Anderson, R.C. (1998). Economics of shark watching in the Maldives. Available on the CD-ROM distributed with R.S.V. Pullin, R. Froese and C.M.V. Casal (eds), (1999), ACP–EU Fisheries Research Initiative. Proceedings of the Conference on

Sustainable Use of Aquatic Biodiversity: Data, Tools and Cooperation. Lisbon, Portugal, 3–5 September 1998, Fisheries Research Report Number 6.

Andrews, J.C. and Pickard, G.L. (1990). The physical oceanography of coral-reef systems. In: Z. Dubinsky (ed.), *Coral Reefs*. Elsevier, pp. 11–48.

Annis, E.R. and Cook, C.B. (2002). Alkaline phosphatase activity in symbiotic dino-flagellates (zooxanthellae) as a biological indicator of environmental phosphate exposure. *Marine Ecology Progress Series* **245**:11–20.

Aronson, R.B., Macintyre, I.G., Warnick, C.M. and O'Neill, W.O. (2004). Phase shifts, alternative states and the unprecedented convergence of two reef systems. *Ecology* **85**:1876–91.

Aronson, R.B. and Precht, W.F. (2001). White-band disease and the changing face of Caribbean reefs. *Hydrobiologia* **460**:25–38.

Aronson, R.B. and Precht, W.F. (2006). Conservation, precaution and Caribbean reefs. *Coral Reefs* **25**:441–50.

Atema, J., Gerlach, G. and Paris, C.B. (2015). Sensor biology and navigation behaviours of reef fish larvae. In: C. Mora (ed.), *Ecology of Fishes on Coral Reefs*. Cambridge University Press, pp. 3–15.

Ateweberhan, M., Feary, D.A., Keshavmurthy, S., Chen, A., Schleyer, M.H. and Sheppard, C.R.C. (2013). Climate change impacts on coral reefs: Synergies with local effects, possibilities for acclimation, and management implications *Marine Pollution Bulletin* **74**:526–39.

Ayre, D.J. and Hughes, T.P. (2000). Genotypic diversity and gene flow in brooding and spawning corals along the Great Barrier Reef, Australia. *Evolution* **54**:1590–605.

Babcock, R.C., Bull, G.D., Harrison, P.L., Heyward, A.J., Oliver, J.K., Wallace, C.C., et al. (1986). Synchronous spawnings of 105 scleractinian coral species on the Great Barrier Reef. *Marine Biology* **90**:379–94.

Babcock, R.C. and Davies, P. (1991). Effects of sedimentation on settlement of *Acropora millepora*. *Coral Reefs* **9**:205–8.

Babcock, R.C., Milton, D. and Pratchett, M.S. (2016). Relationships between the size and reproductive output in the crown-of-thorns starfish. *Marine Biology* **163**:234.

Baghdasarian, G. and Muscatine, L. (2000). Preferential expulsion of dividing algal cells as a mechanism for regulating algal–cnidarian symbiosis. *Biological Bulletin* **199**:278–86.

Baillon, S., Hamel, J.-F., Wareham, V.E. and Mercier, A. (2012). Deep cold-water corals as nurseries for fish larvae. *Frontiers in Ecology and the Environment* **10**:351–6.

Baird, A.H. and Morse, A.N.C. (2003). Induction of metamorphosis in larvae of the brooding corals *Acropora palifera* and *Stylophora pistillata*. *Marine and Freshwater Research* **55**:469–72.

Baker, A.C. (2001). Ecosystems: Reef corals bleach to survive change. *Nature* **411**:765–6.

Baker, A.C. (2003). Flexibility and specificity in coral-algal symbiosis: Diversity, ecology, and biogeography of *Symbiodinium*. *Annual Review of Ecology, Evolution, and Systematics* **34**:661–89.

Baker, A.C., Starger, C.J., McClanahan, T.R. and Glynn, P.W. (2004). Corals' adaptive response to climate change. *Nature* **430**:741.

Banaszak, A.T., LaJeunesse, T.C. and Trench, R.K. (2000). The synthesis of mycosporine-like amino acids (MAAs) by cultured, symbiotic dinoflagellates. *Journal Experimental Marine Biology Ecology* **249**:219–33.

Barnes, D.J., Chalker, B.E. and Kinsey, D.W. (1986). Reef metabolism. *Oceanus* **29**:20–6.

Barnes, D.J. and Devereux, M.J. (1984). Productivity and calcification on coral reefs: A survey using pH and oxygen electrode techniques. *Journal of Experimental Marine Biology and Ecology* **79**:213–31.

Barott, K.L., Venn, A.A., Perez, S.O., Tambutté, S. and Tresguerres, M. (2015). Coral host cells acidify symbiotic algal microenvironment to promote photosynthesis. *Proceedings of the National Academy of Sciences of the United States of America* **112**:607–12.

Baum, G., Januar, H.I., Ferse, S.C.A. and Kunzmann, A. (2015). Local and regional impacts of pollution on coral reefs along the Thousand Islands north of the megacity Jakarta, Indonesia. *PLOS ONE* **10**:e0138271.

Baums, I.B., Devlin-Durante, M.K. and LaJeunesse, T.C. (2014). New insights into the dynamics between reef corals and their associated dinoflagellate endosymbionts from population genetic studies. *Molecular Ecology* **17**:4203–15.

Beanish, J. and Jones, B. (2002). Dynamic carbonate sedimentation in a shallow coastal lagoon: Case study of south sound, Grand Cayman, British West Indies. *Journal of Coastal Research* **18**:254–66.

Bell, J.D., Ganachaud, A., Gehrke, P., Griffiths, S.P., Hobday, A.J., Hoegh-Guldberg, O., et al. (2013). Mixed responses of tropical Pacific fisheries and aquaculture to climate change. *Nature Climate Change* **3**:591–9.

Bell, J.D., Kronen, M., Vunisea, A., Nash, W.J., Keeble, G., Demmke, A., et al. (2009). Planning the use of fish for food security in the Pacific. *Marine Policy* **33**:64–76.

Bell, J.J. (2008). The functional roles of marine sponges. *Estuarine Coastal and Shelf Science* **79**:341–53.

Bellwood, D.R., Goatley, C.H.R., Cowman, P.H. and Bellwood, O. (2015). The evolution of fishes on coral reefs: Fossils, phylogenies, and function. In: P.F. Sale (ed.), *Ecology of Fishes on Coral Reefs*. Cambridge University Press, pp. 55–63.

Bellwood, D.R., Hoey, A.S., Bellwood, O. and Goatley, C.H.R. (2014). Evolution of long-toothed fishes and the changing nature of fish-benthos interactions on coral reefs. *Nature Communications* **5**:3144.

Bellwood, D.R., Hoey, A.S. and Hughes, T.P. (2012). Human activity selectively impacts the ecosystem roles of parrotfishes on coral reefs. *Proceedings of the Royal Society of London. Series B, Biological Sciences* **279**:1621–9.

Bellwood, D.R., Hughes, T.P., Folke, C. and Nyström, M. (2004). Confronting the coral reef crisis. *Nature* **429**:827–33.

Bellwood, D.R. and Wainwright, P.W. (2002). The history and biogeography of fishes on coral reefs. In: P.F. Sale (ed.), *Coral Reef Fishes: Dynamics and Diversity in a Complex Ecosystem*. Academic Press, pp. 5–32.

Benavides, M., Houlbrèque, F., Camps, M., Lorrain, A., Grosso, O. and Bonnet, S. (2016). Diazotrophs: A non-negligible source of nitrogen for the tropical coral *Stylophora pistillata*. *Journal of Experimental Biology* **219**:2608–12.

Bentis, C.J., Kaufman, L. and Golubic, S. (2000). Endolithic fungi in reef-building corals (Order: Scleractinia) are common, cosmopolitan, and potentially pathogenic. *Biological Bulletin* **198**:254–60.

Bento, R., Hoey, A.S., Bauman, A.G., Feary, D.A. and Burt, J.A. (2016). The implications of recurrent disturbances within the world's hottest coral reef. *Marine Pollution Bulletin* **105**:466–72.

Benzie,J.A.H. (1999). Major genetic differences between crown-of-thorns starfish (*Acanthaster planci*) populations in the Indian and Pacific oceans. *Evolution* **53**:1782–95.

Bergquist, P.R. and Tizard, C.A. (1967). Australian intertidal sponges from the Darwin area. *Micronesica* **3**:175–202.

Berkelmans, R. and van Oppen, M.J.H. (2006). The role of zooxanthellae in the thermal tolerance of corals: A 'nugget of hope' for coral reefs in an era of climate change. *Proceedings of the Royal Society of London. Series B, Biological Sciences* **273**:2305–12.

Berkelmans, R. and Willis, B.L. (1999). Seasonal and local spatial patterns in the upper thermal limits of corals on the inshore Central Great Barrier Reef. *Coral Reefs* **18**:219–28.

Bernal, M.A., Floeter, S.R., Gaither, M.R., Longo, G.O., Morais, R., Ferreira, C.E.L., et al. (2016). High prevalence of dermal parasites among coral reef fishes of Curaçao. *Marine Biodiversity* **46**:67–74.

Berner, T. and Izhaki, I. (1994). Effect of exogenous nitrogen levels on ultrastructure of zooxanthellae from the hermatypic coral Pocillopora damicornis. *Pacific Science* **48**:254–62.

Bertucci, A., Moya, A., Tambutté, S., Allemand, D., Supuran, C.T. and Zoccola, D. (2013). Carbonic anhydrases in anthozoan corals: A review. *Bioorganic and Medicinal Chemistry* **21**:1437–50.

Birkeland, C. (1982). Terrestrial runoff as a cause of outbreaks of *Acanthaster planci* (Echinodermata: Asteroidea). *Marine Biology* **69**:175–85.

Birkeland, C. (2004). Ratcheting down the coral reefs. *BioScience* **54**:1021–7.

Birkeland, C. (ed.) (2015). *Coral Reefs in the Anthropocene*. Springer.

Blackall, L.L., Wilson, B. and van Oppen, M.J.H. (2015). Coral: The world's most diverse symbiotic ecosystem. *Molecular Ecology* **24**:5330–47.

Blair Myers, C., Sheppard, C.R.C., Mathesen, K. and Bythell, J.C. (1994). *Habitat Atlas of Anguilla*. NRI.

Blanchon, P., Jones, B. and Kalbfleisch, W. (1997). Anatomy of a fringing reef around Grand Cayman: Storm rubble, not coral framework. *Journal of Sedimentology Research* **67**:1–16.

Blank, R.J. and Huss, V.A.R. (1989). DNA divergency and speciation in *Symbiodinium* (Dinophyceae). *Plant Systematics and Evolution* **163**:153–63.

Blank, R.J. and Trench, R.K. (1985). Speciation and symbiotic dinoflagellates. *Science* **229**:656–8.

Blum, S.D. (1989). Biogeography of the Chaetodontidae: An analysis of allopatry among closely related species. *Environmental Biology of Fishes* **25**:9–31.

Booth, D.J. and Beretta, G.S. (2002). Changes in a fish assemblage after a coral bleaching event. *Marine Ecology Progress Series* **245**:208–12.

Bourne, D.G. and Munn, C.B. (2005). Diversity of bacteria associated with the coral Pocillopora damicornis from the Great Barrier Reef. *Environmental Microbiology* **7**:1162–74.

Bozec, Y.M. and Mumby, P.J. (2015). Synergistic impacts of global warming on coral reef resilience. *Philosophical Transactions of the Royal Society B*, **370**:20130267.

Bozec, Y.M., O'Farrell, S., Bruggemann, J.H., Luckhurst, B.L. and Mumby, P.J. (2016). Trade-offs between fisheries harvest and the resilience of coral reefs. *Proceedings of the National Academy of Sciences of the United States of America* **113**:4536–41.

Bradley, D., Conklin, E., Papastamatiou, Y.P., McCauley, D.J., Pollock, K., Pollock, A., et al. (2017). Resetting predator baselines in coral reef ecosystems. *Scientific Reports* 7:43131.

Briggs, J.C. (1999). Coincident biogeographic patterns: Indo-West Pacific Ocean. *Evolution* 53:326–35.

Brodie, J.E., De'ath, G., Devlin, M., Furnas, M.J. and Wright, M. (2007). Spatial and temporal patterns of near-surface chlorophyll *a* in the Great Barrier Reef lagoon. *Marine Freshwater Research*. 58:342–53.

Brodie, J.E., Fabricius, K.E., De'ath, G. and Okaji, K. (2005). Are increased nutrient inputs responsible for more outbreaks of Crown of Thorns starfish? An appraisal of the evidence. *Marine Pollution Bulletin* 51:266–78.

Brodie, J.E. and Mitchell, A. (1992). Nutrient composition of the January 1991 Fitzroy River Plume. In: G.T. Byron (ed.), *Workshop on the Effects of the 1991 Floods*. Queensland National Parks and Wildlife Service and Great Barrier Reef Marine Park Authority Publication Workshop Series 17. Queensland National Parks and Wildlife Service and Great Barrier Reef Marine Park Authority, pp. 56–74.

Brown, B.E. (1997a). Coral bleaching: Causes and consequences. *Coral Reefs* 16:S129–38.

Brown, B.E. (1997b). Disturbances to reefs in recent times. In: C. Birkeland (ed.), *Life and Death on a Coral Reef*. Chapman & Hall, pp. 354–79.

Brown, B.E., Ambarsari, I., Warner, M.E., Fitt, W.K., Dunne, R.P., Gibb, S.W., et al. (1999). Diurnal changes in photochemical efficiency and xanthophyll concentrations in shallow water reef corals: Evidence for photoinhibition and photoprotection. *Coral Reefs* 18:99–105.

Bruno, J.F. and Selig, E.R. (2007). Regional decline of coral cover in the Indo-Pacific: Timing, extent, and subregional comparisons. *PLOS ONE* 2:e711.

Buchanan, J.R., Krupp, F., Burt, J.A., Feary, D.A., Ralph, G.M. and Carpenter, K.E. (2016). Living on the edge: Vulnerability of coral-dependent fishes in the Gulf. *Marine Pollution Bulletin* 105:480–8.

Buck, A.C., Gardiner, N.M. and Boström-Einarsson, L. (2016). Citric acid injections: An accessible and efficient method for controlling outbreaks of the crown-of-thorns starfish *Acanthaster* cf. *solaris*. *Diversity* 8:28.

Buddemeier, R.W. (1997). Symbiosis: Making light work of adaptation. *Nature* 388:229–30.

Buddemeier, R.W. and Fautin, D.G. (1993). Coral bleaching as an adaptive mechanism: A testable hypothesis. *Bioscience* 43:320–6.

Buddemeier, R.W., Kleypas, J.A. and Aronson, R.B. (2004). *Coral Reefs and Global Climate Change: Potential Contributions of Climate Change to Stresses on Coral Reef Ecosystems*. Pew Center for Global Climate Change.

Buddemeier, R.W., Maragos, J.E. and Knutson, D.K. (1974). Radiographic studies of reef coral exoskeletons: Rates and patterns of coral growth. *Journal of Experimental Marine Biology Ecology* 14:179–200.

Buddemeier, R.W. and Smith, S.M. (1999). Coral adaptation and acclimatization: A most ingenious paradox. *American Zoologist* 39:1–9.

Burke, L., Reytar, K., Spalding, M. and Perry, A. (2011). *Reefs at Risk Revisited*. World Resources Institute.

Burriesci, M.S., Raab, T.K. and Pringle, J.R. (2012). Evidence that glucose is the major transferred metabolite in dinoflagellate–cnidarian symbiosis. *Journal of Experimental Biology* 215:3467–77.

Bythell, J.C. (1988). A total nitrogen and carbon budget for the elkhorn coral *Acropora palmata* (Lamarck). *Proceedings of the 6th International Coral Reef Symposium* **2**:535–40.

Caballes, C.F., Pratchett, M.S., Kerr, A.M. and Rivera-Posada, J.A. (2016). The role of maternal nutrition on oocyte size and quality, with respect to early larval development in the coral-eating starfish, *Acanthaster planci*. *PLOS ONE* **11**:e0158007.

Cairns, S.D. (2007). Deep-water corals: An overview with special reference to diversity and distribution of deep-water scleractinian corals. *Bulletin of Marine Science* **81**:311–22.

Campbell, S.J. and Pardede, S.T. (2006). Reef fish structure and cascading effects in response to artisanal fishing pressure. *Fisheries Research* **79**:75–83.

Caporaso, J.G., Kuczynski, J., Stombaugh, J., Bittinger, K., Bushman, F.D., Costello, E.K., et al. (2010). QIIME allows analysis of high-throughput community sequencing data. *Nature Methods* **7**:335–6.

Cardini, U., Bednarz, V.N., Naumann, M.S., van Hoytema, N., Rix, L., Foster, R.A., et al. (2015). Functional significance of dinitrogen fixation in sustaining coral productivity under oligotrophic conditions. *Proceedings of the Royal Society of London. Series B, Biological Sciences* **282**:20152257.

Carlos, A.A., Baillie, B.K., Kawachi, M. and Maruyama, T. (1999). Phylogenetic position of *Symbiodinium* (Dinophyceae) isolates from tridacnids (Bivalvia), cardiids (Bivalvia), a sponge (Porifera), a soft coral (Anthozoa), and a free-living strain. *Journal of Phycology* **35**:1054–62.

Carpenter, K.E., Abrar, M., Aeby, G., Aronson, R.B., Banks, S., Bruckner, A., et al. (2008). One-third of reef building corals face elevated extinction risk from climate change and local impacts. *Science* **321**:560–3.

Cates, N. and McLaughlin, J.J.A. (1979). Nutrient availability for zooxanthellae derived from physiological activities of *Condylactis* spp. *Journal of Experimental Marine Biology Ecology* **37**:31–41.

Cathalot, C., van Oevelen, D., Cox, T.J., Kutti, T., Lavaleye, M., Duineveld, G., et al. (2015). Cold-water coral reefs and adjacent sponge grounds: Hotspots of benthic respiration and organic carbon cycling in the deep sea. *Frontiers in Marine Science* **2**:37.

Ceccarelli, D.M. (2007). Modification of benthic communities by territorial damselfish: A multi-species comparison. *Coral Reefs* **26**:853–66.

Ceh, J., Kilburn, M.R., Cliff, J.B., Raina, J.B., van Keulen, M. and Bourne, D.G. (2013). Nutrient cycling in early coral life stages: *Pocillopora damicornis* larvae provide their algal symbiont (*Symbiodinium*) with nitrogen acquired from bacterial associates. *Ecology and Evolution* **3**:2393–400.

Celliers, L. and Schleyer, M.H. (2002). Coral bleaching on high-latitude marginal reefs at Sodwana Bay, South Africa. *Marine Pollution Bulletin* **44**:180–7.

Celliers, L. and Schleyer, M.H. (2008). Coral community structure and risk assessment of high-latitude reefs at Sodwana Bay, South Africa. *Biodiversity and Conservation* **17**:3097–117.

Cesar, H.S.J. (1999). Socio-economic aspects of the 1998 coral bleaching event in the Indian Ocean. In: O. Linden and N. Sporong (eds), *Coral Reef Degradation in the Indian Ocean: Status Report and Project Presentations*. CORDIO, SAREC Marine Science Programme, pp. 82–5.

Cesar, H.S.J, Lundin, C.G., Bettencourt, S. and Dixon, J. (1997). Indonesian coral reefs: An economic analysis of a precious but threatened resource. *Ambio* **26**:345–50.

Chappell, J. (1983). Evidence for smoothly falling sea level relative to north Queensland, Australia, during the past 6,000 yr. *Nature* **302**:406–8.

Charpy, L. (2005). Importance of photosynthetic picoplankton in coral reef ecosystems. *Vie et Milieu* **55**:217–23.

Charpy-Roubaud, C., Charpy, L. and Larkum, A.W.D. (2001). Atmospheric dinitrogen fixation by benthic communities of Tikehau Lagoon (Tuamotu Archipelago, French Polynesia) and its contribution to benthic primary production. *Marine Biology* **139**:991–7.

Chen, M.C., Cheng, Y.M., Hong, M.C. and Fang, L.S. (2004). Molecular cloning of Rab5 (ApRab5) in *Aiptasia pulchella* and its retention in phagosomes harboring live zooxanthellae. *Biochemical and Biophysical Research Communications* **324**:1024–33.

Chesher, R.H. (1969). Destruction of Pacific corals by the sea star *Acanthaster planci*. *Science* **165**:280–3.

Cheshire, A.C., Wilkinson, C.R., Seddon, S. and Westphalen, G. (1997). Bathymetric and seasonal changes in photosynthesis and respiration of the phototrophic sponge *Phyllospongia lamellosa* in comparison with respiration by the heterotrophic sponge *Ianthella basta* on Davies Reef, Great Barrier Reef. *Marine Freshwater Research* **48**:589–599.

Choat, J.H., Robbins, W.D. and Clements, K.D. (2004). The trophic status of herbivorous fish on coral reefs. II. Food processing modes and trophodynamics. *Marine Biology* **145**:445–54.

Choat, J.H. and Robertson, D.R. (2002). Age-based studies. In: P.F. Sale (ed.), *Coral Reef Fishes: Dynamics and Diversity in a Complex Ecosystem*. Academic Press, pp. 57–80.

Cinner, J.E., Huchery, C., MacNeil, M.A., Graham, N.A.J., McClanahan, T.R., Maina, J., et al. (2016). Bright spots among the world's coral reefs. *Nature* **535**:416–19.

Cinner, J.E., McClanahan, T.R, MacNeil, M.A., Graham, N.A.J., Daw, T.M., Mukminin, A., et al. (2012). Comanagement of coral reef social-ecological systems. *Proceedings of the National Academy of Sciences of the United States of America* **109**:5219–22.

Cinner, J.E., Sutton, S.G. and Bond, T.G. (2007). Socioeconomic thresholds that affect use of customary fisheries management tools. *Conservation Biology* **21**:1603–11.

Cloud, P.E. (1952). Preliminary report on the geology and marine environment of Onotoa Atoll, Gilbert Island. *Atoll Research Bulletin* **12**:1–73.

Coffroth, M.A. and Santos, S.R. (2005). Genetic diversity of symbiotic dinoflagellates in the genus *Symbiodinium*. *Protist* **156**:19–34.

Colding, J. and Folke, C. (2001). Social taboos 'invisible' systems of local resource management and biological conservation. *Ecological Applications* **11**:584–600.

Cole, A.J., Pratchett, M.S. and Jones, G.P. (2008). Diversity and functional importance of coral-feeding fishes on tropical coral reefs. *Fish and Fisheries* **9**:286–307.

Colella, M.A., Ruzicka, R.R., Kidney, J.A., Morrison, J.M. and Brinkhuis, V.B. (2012). Cold-water event of January 2010 results in catastrophic benthic mortality on patch reefs in the Florida Keys. *Coral Reefs* **31**:621–32.

Coles, S.L. (1992). Experimental comparison of salinity tolerances of reef corals from the Arabian Gulf and Hawaii. Evidence for hyperhaline adaptation. *Proceedings of the 7th International Coral Reef Symposium* **1**:227–34.

Coles, S.L. (2003). Coral species diversity and environmental factors in the Arabian Gulf and the Gulf of Oman: A comparison to the Indo-Pacific region. *Atoll Research Bulletin* **507**:1–19.

Coles, S.L. and Brown, B.E. (2003). Coral bleaching: Capacity for acclimatization and adaptation. *Advances in Marine Biology* **46**:183–223.

Coles, S.L. and Fadlallah, Y.H. (1991). Reef coral survival and mortality at low temperatures in the Arabian Gulf: New species-specific lower temperature limits. *Coral Reefs* **9**:231–7.

Conand, C. (1997). Are Holothurian fisheries for export sustainable? *Proceedings of the 8th International Coral Reef Symposium* **2**:2021–6.

Constantz, B. and Weiner, S. (1988). Acidic macromolecules associated with the mineral phase of scleractinian coral skeletons. *Journal of Experimental Zoology* **248**:253–8.

Cook, C.B. and D'Elia, C.F. (1987). Are natural populations of zooxanthellae ever nutrient-limited? *Symbiosis* **4**:199–211.

Cook, C.B., D'Elia, C.F. and Muller-Parker, G. (1988). Host feeding and nutrient sufficiency for zooxanthellae in the sea anemone *Aiptasia pallida*. *Marine Biology* **98**:253–62.

Cook, C.B. and Davy, S.K. (2001). Are free amino acids responsible for the 'host factor' effects on symbiotic zooxanthellae in extracts of host tissue? *Hydrobiologia* **461**:71–8.

Cook, C.B., Muller-Parker, G. and D'Elia, C.F. (1992). Ammonium enhancement of dark carbon fixation and nitrogen limitation in symbiotic zooxanthellae: Effects of feeding and starvation of the sea anemone *Aiptasia pallida*. *Limnology and Oceanography* **37**:131–9.

Cook, C.B., Muller-Parker, G. and Orlandini, C.D. (1994). Ammonium enhancement of dark carbon fixation and nitrogen limitation in zooxanthellae symbiotic with the reef corals *Madracis mirabilis* and *Montastrea annularis*. *Marine Biology* **118**:157–65.

Cooper, L. (1997). Western Australia bêche-de-mer management. In: S.B. Damschke (ed.), *Proceedings of the Sea Cucumber (Bêche-de-Mer) Fishery Management Workshop, Brisbane 8–9 December 1997*. Queensland Fisheries Management Authority, pp. 11–13.

Cornish, A.S. and DiDonato, E.M. (2003). Resurvey of a reef flat in American Samoa after 85 years reveals devastation to a soft coral (Alcyonacea) community. *Marine Pollution Bulletin* **48**:768–77.

Correa, A.M.S., Ainsworth, T.D., Rosales, S.M., Thurber, A.R., Butler, C.R. and Vega Thurber, R.L. (2016). Viral outbreak in corals associated with an in situ bleaching event: Atypical herpes-like viruses and a new megavirus infecting *Symbiodinium*. *Frontiers in Microbiology* **7**:127.

Corredor, J.E., Wilkinson, C.R., Vicente, V.P., Morell, J.M. and Otero, E. (1988). Nitrate release by Caribbean reef sponges. *Limnology and Oceanography*. **33**:114–20.

Costa, O.S., Leão, Z.M.A.N., Nimmo, M. and Attrill, M.J. (2000). Nutrification impacts on coral reefs from northern Bahia, Brazil. *Hydrobiologia* **440**:307–15.

Costanza, R., d'Arge, R., De Groot, R., Farber, S., Grasso, M., Hannon, B., et al. (1997). The value of the world's ecosystem services and natural capital. *Nature* **387**:253–60.

Costanza, R., de Groot, R., Sutton, P.D., van der Ploeg, S., Anderson, S.J., Kubiszewski, I., et al. (2014). Changes in the global value of ecosystem services. *Global Environmental Change* **26**:152–8.

Cowan, Z., Pratchett, M.S., Messmer, V. and Ling, S. (2017). Known predators of crown-of-thorns starfish (*Acanthaster* spp.) and their role in mitigating, if not preventing, population outbreaks. *Diversity* **9**:7.

Cox, E.F., Ribes, M. and Kinzie, R.A., III. (2006). Temporal and spatial scaling of planktonic responses to nutrient inputs into a subtropical embayment. *Marine Ecology Progress Series* **324**:19–35.

Craig, P., Green, A. and Tuilagi, F. (2008). Subsistence harvest of coral reef resources in the outer islands of American Samoa: Modern, historic and prehistoric catches. *Fisheries Research* **89**:230–40.

Crossland, C.J. (1984). Seasonal variations in the rates of calcification and productivity in the coral *Acropora formosa* on a high-latitude reef. *Marine Ecology Progress Series* **15**:135–40.

Crossland, C.J., Barnes, D.J. and Borowitzka, M.A. (1980). Diurnal lipid and mucus production in the staghorn coral *Acropora acuminata*. *Marine Biology* **60**:81–90.

Cunning R, Yost, D.M., Guarinello, M.L., Putnam, H.M. and Gates, R.D. (2015). Variability of *Symbiodinium* communities in waters, sediments, and corals of thermally distinct reef pools in American Samoa. *PLOS ONE* **10**:e0145099.

D'agata, S., Mouillot, D., Wantiez, L., Friedlander, A.M., Kulbicki, M., Vigliola, L. (2016). Marine reserves lag behind wilderness in the conservation of key functional roles. *Nature Communications* **7**:12000.

D'Elia, C.F. (1977). The uptake and release of dissolved phosphorus by reef corals. *Limnology and Oceanography* **22**:301–15.

D'Elia, C.F. and Wiebe, W.J. (1990). Biogeochemical cycles in coral-reef ecosystems. In: Z. Dubinsky (ed.), *Coral Reefs*. Elsevier, pp. 49–74.

Daly, R.A. (1910). Pleistocene glaciation and the coral reef problem. *American Journal of Science* **30**:297–308.

Darwin, C. (1842). *On the Structure and Distribution of Coral Reefs*. Ward, Lock and Bowden Ltd.

Davey, M., Holmes, G. and Johnstone, R. (2008). High rates of nitrogen fixation (acetylene reduction) on coral skeletons following bleaching mortality. *Coral Reefs* **27**:227–36.

Davies, A.J., Duineveld, G., Lavaleye, M., Bergman, M., van Haren, H. and Roberts, J.M. (2009). Downwelling and deep-water bottom currents as food supply mechanisms to the cold-water coral *Lophelia pertusa* (Scleractinia) at the Mingulay Reef Complex. *Limnology and Oceanography* **54**:620–9.

Davies, P.S. (1984). The role of zooxanthellae in the nutritional energy requirements of *Pocillopora eydouxi*. *Coral Reefs* **2**:181–6.

Davies, P.S. (1991). Effect of daylight variations on the energy budgets of shallow-water corals. *Marine Biology* **108**:137–44.

Davies, P.S. (1992). Endosymbiosis in marine cnidarians. In: D.M. John, S.J. Hawkins and J.H. Price (eds), *Plant–Animal Interactions in the Marine Benthos*. Clarendon Press, pp. 511–40.

Davy, J.E. and Patten, N.L. (2007). Morphological diversity of virus-like particles within the surface microlayer of scleractinian corals. *Aquatic Microbial Ecology* **47**:37–44.

Davy, S.K., Allemand, D. and Weis, V.M. (2012). Cell biology of cnidarian–dinoflagellate symbiosis. *Microbiology and Molecular Biology Reviews* **76**:229–61.

Davy, S.K. and Cook, C.B. (2001a). The influence of 'host release factor' on carbon release by zooxanthellae isolated from fed and starved *Aiptasia pallida* (Verrill). *Comparative Biochemistry Physiology A* **129**:487–94.

Davy, S.K. and Cook, C.B. (2001b). The relationship between nutritional status and carbon flux in the zooxanthellate sea anemone *Aiptasia pallida*. *Marine Biology* **139**:999–1005.

Davy, S.K., Trautman, D.A., Borowitzka, M.A. and Hinde, R. (2002). Ammonium excretion by a symbiotic sponge supplies the nitrogen requirements of its rhodophyte partner. *Journal of Experimental Biology* **205**:3505–11.

Davy, S.K., Withers, K.J.T. and Hinde, R. (2006). Effects of host nutritional status and seasonality on the nitrogen status of zooxanthellae in the temperate coral *Plesiastrea versipora* (Lamarck). *Journal of Experimental Marine Biology Ecology* **335**:256–65.

de Bakker, D.M., Meesters, E.H., Bak, R.P.M., Nieuwland, G. and Van Duyl, F.C. (2016). Long-term shifts in coral communities on shallow to deep reef slopes of Curaçao and Bonaire: Are there any winners? *Frontiers in Marine Science* **3**:247.

de Bakker, D.M., van Duyl, F.C., Bak, R.P.M., Nugues, M.M., Nieuwland, G. and Meesters, E.H. (2017). 40 Years of benthic community change on the Caribbean reefs of Curaçao and Bonaire: The rise of slimy cyanobacterial mats. *Coral Reefs* **36**:355–67.

De'ath, G., Fabricius, K.E., Sweatman, H. and Puotinen, M. (2012). The 27-year decline of coral cover on the Great Barrier Reef and its causes. *Proceedings of the National Academy of Sciences of the United States of America* **109**:17995–9.

Deane, E.M. and O'Brien, R.W. (1981). Uptake of phosphate by symbiotic and free-living dinoflagellates. *Archives of Microbiology.* **128**:307–10.

Delesalle, B., Pichon, M., Frankignoulle, M and Gattuso, J.-P. (1993). Effects of a cyclone on coral reef phytoplankton biomass, primary production and composition (Moorea island, French Polynesia). *Journal of Plankton Research* **15**:1413–23.

den Haan, J., Visser, P.M., Ganase, A.E., Gooren, E.E., Stal, L.J., van Duyl, F.C., Vermeij, M.J.A. and Huisman, J. (2014). Nitrogen fixation rates in algal turf communities of a degraded versus less degraded coral reef. *Coral Reefs* **33**:1003–15.

Depczynski, M. and Bellwood, D.R. (2005). Shortest recorded vertebrate lifespan found in a coral reef fish. *Current Biology* **15**: R288–9.

Dikou, A. and van Woesik, R. (2006). Survival under chronic stress from sediment load: Spatial patterns of hard coral communities in the southern islands of Singapore. *Marine Pollution Bulletin* **52**:1340–54.

Dinsdale, E.A., Pantos, O., Smriga, S., Edwards, R.A., Angly, F., Wegley, L., et al. (2008). Microbial ecology of four coral atolls in the northern Line Islands. *PLOS ONE* **3**:e1584.

Done, T.J. (1992). Effects of tropical cyclone waves on ecological and geomorphological structures on the Great Barrier Reef. *Continental Shelf Research* **12**:859–72.

Done, T.J. (1999). Coral community adaptability to environmental change at the scale of reefs, regions and reef zones. *American Zoologist* **39**:66–79.

Done, T.J., Ogden, J.C. and Wiebe, W.J. (1996). Biodiversity and ecosystem function of coral reefs. In: H.A. Mooney, J.H. Cushman, E. Medina, O.E. Sala and E.-D. Schulze (eds), *Functional Roles of Biodiversity: A Global Perspective.* Wiley, pp. 393–429.

Donner, S.D., Knutson, T.R. and Openheimer, M. (2007). Model-based assessment of the role of human-induced climate change in the 2005 Caribbean coral bleaching event. *Proceedings of the National Academy of Sciences of the United States of America* **104**:5483–8.

Donner, S.D., Skirving, W.J., Little, C.M., Oppenheimer, M. and Hoegh-Guldberg, O. (2005). Global assessment of coral bleaching and required rates of adaptation under climate change. *Global Change Biology* **11**:1–15.

Doty, M.S. (1974). Coral reef roles played by free-living algae. *Proceedings of the 2nd International Coral Reef Symposium* **1**:27–33.

Douglas, A.E. (1983). Uric acid utilization in Platymonas convolutae and symbiotic *Convoluta roscoffensis*. *Journal of the Marine Biological Association of the United Kingdom* **63**:435–47.

Douglas, A.E. (1994). *Symbiotic Interactions*. Oxford University Press.

Douglas, A.E. (2003). Coral bleaching: How and why? *Marine Pollution Bulletin* **46**:385–92.

Downing, N., Buckley, R., Stobart, B., LeClair, L. and Teleki, K. (2005). Reef fish diversity at Aldabra atoll, Seychelles, during the five years following the 1998 coral bleaching event. *Philosophical Transactions of the Royal Society A* **363**:257–61.

Dubinsky, Z., Stambler, N., Ben-Zion, M., McCloskey, L.R., Muscatine, L. and Falkowski, P.G. (1990). The effect of external nutrient resources on the optical properties and photosynthetic efficiency of *Stylophora pistillata*. *Proceedings of the Royal Society of London. Series B, Biological Sciences* **239**:231–46.

Ducklow, H.W. (1990). The biomass, production and fate of bacteria in coral reefs. In: Z. Dubinsky (ed.), *Coral Reefs*. Elsevier, pp. 265–90.

Ducklow, H.W. and Mitchell, R. (1979). Bacterial populations and adaptations in the mucus layers on living corals. *Limnology and Oceanography* **24**:715–25.

Dufresne, A., Ostrowski, M., Scanlan, D.J., Garczarek, L., Mazard, S., Palenik, B.P. (2008). Unravelling the genomic mosaic of a ubiquitous genus of marine cyano-bacteria. *Genome Biology* **9**:R90.

Dulvy, N.K., Freckleton, R.P. and Polunin, N.V.C. (2004). Coral reef cascades and the indirect effects of predator removal by exploitation. *Ecology Letters* **7**:410–16.

Dulvy, N.K., Polunin, N.V.C., Mill, A.C. and Graham, N.A.J. (2004). Size structural change in lightly exploited coral reef fish communities: Evidence for weak indirect effects. *Canadian Journal of Fisheries and Aquatic Sciences* **61**:466–75.

Dulvy, N.K., Sadovy, Y. and Reynolds, J.D. (2003). Extinction vulnerability in marine populations. *Fish and Fisheries* **4**:25–64.

Dunlap, W.C. and Shick, J.M. (1998). Ultraviolet radiation absorbing mycosporine-like amino acids in coral reef organisms: A biochemical and environmental perspective. *Journal of Phycology* **34**:418–30.

Dunn, S.R., Thomason, J.C., Le Tissier, M.D.A. and Bythell, J.C. (2004). Heat stress induces different forms of cell death in sea anemones and their endosymbiotic algae depending on temperature and duration. *Cell Death and Differentiation* **11**:1213–22.

Dustan, P. (1975). Growth and form in the reef-building coral *Montastrea annularis*. *Marine Biology* **33**:101–7.

Dustan, P. (2000). Florida keys. In: C.R.C Sheppard (ed.), *Seas at the Millennium*, Vol. 1. Elsevier, pp. 405–14.

Edinger, E.N., Limmon, G.V., Jompa, J., Widjatmoko, W., Heikoop, J. M. and Risk, M.J. (2000). Normal coral growth rates on dying reefs: Are coral growth rates good indicators of reef health? *Marine Pollution Bulletin* **40**:404–25.

Edmunds, P.J. (1994). Evidence that reef-wide patterns of coral bleaching may be the result of the distribution of bleaching susceptible clones. *Marine Biology* **121**:137–42.

Edmunds, P.J. and Davies, P.S. (1986). An energy budget for *Porites porites* (Scleractinia). *Marine Biology* **92**:339–47.

Edwards, E. and Gomez, E. (2007). *Reef Restoration Concepts and Guidelines: Making Sensible Management Choices in the Face of Uncertainty*. Coral Reef Targeted Research & Capacity Building for Management Programme.

Ehrenberg, G.C. (1834). Über die Natur und Bildung der Corallenbänke des rothen Meeres und über einen neuen Fortschritt in der Kenntniss der Organisation im kleinsten Räume durch Verbesserung des Mikroskops von Pistor und Schick. *Physikalische Mathematische Abhandlungen der Königlichen Akademie der Wissenschaften zu Berlin* **1832**:381–438.

Ekebom, J., Patterson, D.J. and Vors, N. (1996). Heterotrophic flagellates from coral reef sediments (Great Barrier Reef, Australia*). Archiv fur Protistenkunde* **146**:251–72.

Elliot, J.K. and Mariscal, R.N. (1996). Ontogenetic and interspecific variation in the protection of anemonefishes from sea anemones. *Journal of Experimental Marine Biology Ecology* **208**:57–72.

Emanuel, K. (2005a). *Divine Wind: The History and Science of Hurricanes.* Oxford University Press.

Emanuel, K. (2005b). Increasing destructiveness of tropical cyclones over the past 30 years. *Nature* **436**:686–8.

Endean, R. (1977). *Acanthaster planci* infestations of reefs of the Great Barrier Reefs. *Proceedings of the 3rd International Coral Reef Symposium* **1**:185–91.

Entsch, B., Boto, K.G., Sim, R.G. and Wellington, J.T. (1983). Phosphorus and nitrogen in coral reef sediments. *Limnology and Oceanography* **28**:465–76.

Entsch, B., Sim, R.G. and Hatcher, B.G. (1983). Indications from photosynthetic components that iron is a limiting nutrient in primary producers on coral reefs. *Marine Biology* **73**:17–30.

Fabricius, K.E. and Alderslade, P. (2001). *Soft Corals and Sea Fans: A Comprehensive Guide to the Tropical Shallow Water Genera of the Central-West Pacific, the Indian Ocean and the Red Sea.* Australian Institute of Marine Science.

Fabricius, K.E., De'Ath, G., McCook, L., Turak, E. and Williams, D.M. (2005). Changes in algal, coral and fish assemblages along water quality gradients on the inshore Great Barrier Reef. *Marine Pollution Bulletin* **51**:384–98.

Fabricius, K.E., Mieog, J.C., Colin, P.L., Idip, D. and van Oppen, M.J.H. (2004). Identity and diversity of coral endosymbionts (zooxanthellae) from three Palauan reefs with contrasting bleaching, temperature and shading histories. *Molecular Ecology* **13**:2445–58.

Fabricius, K.E. and Wolanski, E. (2000). Rapid smothering of coral reef organisms by muddy marine snow. *Estuarine, Coastal and Shelf Science* **50**:115–20.

Fagoonee, I., Wilson, H.B., Hassell, M.P. and Turner, J.R. (1999). The dynamics of zooxanthellae populations: A long-term study in the field. *Science* **283**:843–5.

Falkowski, P.G. and Dubinsky, Z. (1981). Light-shade adaptation of *Stylophora pistillata*, a hermatypic coral from the Gulf of Eilat. *Nature* **289**:172–4.

Falkowski, P.G., Dubinsky, Z., Muscatine, L. and McCloskey, L. (1993). Population control in symbiotic corals. *Bioscience* **43**:606–11.

Falkowski, P.G., Dubinsky, Z., Muscatine, L. and Porter, J.W. (1984). Light and the bioenergetics of a symbiotic coral. *Bioscience* **34**:705–9.

FAO. (2003). *The State of Food Insecurity in the World 2003.* Available at <ftp://ftp.fao.org/docrep/fao/006/j0083e/j0083e00.pdf>.

Fautin, D.G. (1991). The anemonefish symbiosis: What is known and what is not. *Symbiosis* **10**:23–46.

Fautin, D.G. and Allen, G.R. (1997). *Anemone Fishes and Their Host Sea Anemones.* Western Australian Museum.

Fautin, D.G. and Buddemeier, R.W. (2004). Adaptive bleaching: A general phenomenon. *Hydrobiologia* **1**:459–67.

Ferrier, M.D. (1991). Net uptake of dissolved free amino acids by four scleractinian corals. *Coral Reefs* **10**:183–7.

Ferrier-Pagès, C. and Gattuso, J.-P. (1998). Biomass, production and grazing rates of pico- and nanoplankton in coral reef waters (Miyako Island, Japan). *Microbial Ecology* **35**:46–57.

Ferrier-Pagès, C.,Gattuso, J.-P. and Jaubert, J. (1999). Effect of small variations in salinity on the rates of photosynthesis and respiration of the zooxanthellate coral *Stylophora pistillata*. *Marine Ecology Progress Series* **181**:309–14.

Ferrier-Pagès, C., Godinot, C., D'Angelo, C., Wiedenmann, J. and Grover, R. (2016). Phosphorus metabolism of reef organisms with algal symbionts. *Ecological Monographs* **86**:262–77.

Ferrier-Pagès, C., Peirano, A., Abbate, M., Cocito, S., Negri, A., Rottier, C., et al. (2011). Summer autotrophy and winter heterotrophy in the temperate symbiotic coral *Cladocora caespitosa*. *Limnology and Oceanography* **56**:1429–38.

Fiore, C.L., Jarett, J.K., Olson, N.D. and Lesser, M.P. (2010). Nitrogen fixation and nitrogen transformations in marine symbioses. *Trends in Microbiology* **18**:455–63.

Fisher, R., Bellwood, D.R. and Job, S.D. (2000). Development of swimming abilities in reef fish larvae. *Marine Ecology Progress Series* **202**:163–73.

Fitt, W.K. (1984). The role of chemosensory behavior of *Symbiodinium microadriaticum*, intermediate hosts, and host behavior in the infection of coelenterates and mollusks with zooxanthellae. *Marine Biology* **81**:9–17.

Fitt, W.K., Gates, R.D., Hoegh-Guldberg, O., Bythell, J.C., Jatkar, A., Grottoli, A.G., et al. (2009). Response of two species of Indo-Pacific corals, *Porites cylindrica* and *Stylophora pistillata*, to short-term thermal stress: The host does matter in determining the tolerance of corals to bleaching. *Journal of Experimental Marine Biology Ecology* **373**:102–10.

Fitt, W.K., McFarland, F.K., Warner, M.E. and Chilcoat, G.C. (2000). Seasonal patterns of tissue biomass and densities of symbiotic dinoflagellates in reef corals and relation to coral bleaching. *Limnology and Oceanography* **45**:677–85.

Fitt, W.K. and Trench, R.K. (1983). Endocytosis of the symbiotic dinoflagellate *Symbiodinium microadriaticum* Freudenthal by endodermal cells of the scyphistomae of *Cassiopeia xamachana* and resistance of the algae to host digestion. *Journal of Cell Science* **64**:195–212.

Fitt, W.K. and Warner, M.E. (1995). Bleaching patterns of four species of Caribbean reef corals. *Biological Bulletin* **189**:298–307.

Fosså, J.H., Mortensen, P.B. and Furevik, D.M. (2002). The deep-water coral *Lophelia pertusa* in Norwegian waters: Distribution and fishery impacts. *Hydrobiologia* **471**:1–12.

Foster, T. and Gilmour, J.P. (2016). Seeing red: Coral larvae are attracted to healthy-looking reefs. *Marine Ecology Progress Series* **559**:65–71.

Fox, R.J. and Bellwood, D.R. (2013). Niche partitioning of feeding microhabitats produces a unique function for herbivorous rabbitfishes (Perciformes, Siganidae) on coral reefs. *Coral Reefs* **32**:13–23.

Frade, P.R., Roll, K., Bergauer, K. and Herndl, G.J. (2016). Archaeal and bacterial communities associated with the surface mucus of Caribbean corals differ in their degree of host specificity and community turnover over reefs. *PLOS ONE* **11**:e0144702.

Fransolet, D., Roberty, S. and Plumier, J.C. (2012). Establishment of endosymbiosis: The case of cnidarians and *Symbiodinium. Journal of Experimental Marine Biology Ecology* **420**:1–7.

Frieler, K., Meinsausen, M., Golly, A., Mengel, M., Lebek, K., Donner, S.D., et al. (2013). Limiting global warming to 2 degrees C is unlikely to save most coral reefs. *Nature Climate Change* **3**:165–70.

Friedlander, A.M. and DeMartini, E.E. (2002). Contrasts in density, size, and biomass of reef fishes between the northwestern and the main Hawaiian Islands: The effects of fishing down apex predators. *Marine Ecology Progress Series* **230**:253–64.

Fuhrman, J.A. (1999). Marine viruses and their biogeochemical and ecological effects. *Nature* **399**:541–8.

Fulton, C.J., Bellwood, D.R. and Wainwright, P.C. (2005). Wave energy and swimming performance shape coral reef fish assemblages. *Proceedings of the Royal Society of London. Series B, Biological Sciences* **272**:827–32.

Furla, P., Galgani, I., Durand, I. and Allemand, D. (2000). Sources and mechanisms of inorganic carbon transport for coral calcification and photosynthesis. *Journal of Experimental Biology* **203**:3445–57.

Furnas, M.J. (2003). *Catchments and Corals: Terrestrial Runoff to the Great Barrier Reef.* Australian Institute of Marine Science and CRC Reef Research Centre.

Furnas, M.J., Alongi, D.M., McKinnon, D.A., Trott, L.A. and Skuza, M.S. (2011). Regional-scale nitrogen and phosphorus budgets for the northern (14°S) and central (17°S) Great Barrier Reef shelf ecosystem. *Continental Shelf Research* **31**:1967–90.

Furnas, M.J. and Mitchell, A.W. (1996). Nutrient inputs into the central Great Barrier Reef (Australia) from subsurface intrusions of Coral Sea waters: A two-dimensional displacement model. *Continental Shelf Research* **16**:1127–48.

Furnas, M.J., Mitchell, A.W. and Skuza, M.S. (1997a). River inputs of nutrients and sediment to the Great Barrier Reef. In: D. Wachenfeld, J.K. Oliver and K. Davis (eds), *State of the Great Barrier Reef World Heritage Area Workshop.* GBRMPA Workshop Series 23. Great Barrier Reef Marine Park Authority, pp. 46–68.

Furnas, M.J., Mitchell, A.W. and Skuza, M.S. (1997b). Shelf-scale nitrogen and phosphorus budgets for the Central Great Barrier Reef (16–19°S). *Proceedings of the 8th International Coral Reef Symposium* **1**:809–14.

Furnas, M.J., Mitchell, A.W., Skuza, M.S. and Brodie, J.E. (2005). In the other 90%: Phytoplankton responses to enhanced nutrient availability in the Great Barrier Reef Lagoon. *Marine Pollution Bulletin* **51**:253–65.

Gall, J.P. and Pardue, M.L. (1969). Formation and detection of RNA–DNA hybrid molecules in cytological preparations. *Proceedings of the National Academy of Sciences of the United States of America* **63**:378–83.

Gallop, S.L., Young, I.R., Ranasinghe, R., Durrant, T.H. and Haigh, I.D. (2014). The large-scale influence of the Great Barrier Reef matrix on wave attenuation. *Coral Reefs* **33**:1167–78.

Ganase, A., Bongaerts, P., Visser, P.M. and Dove, S.G. (2016). The effect of seasonal temperature extremes on sediment rejection in three scleractinian coral species. *Coral Reefs* **35**:187–91.

Gantner, S., Andersson, A.F., Alonso-Saez, L. and Bertilsson, S. (2011). Novel primers for 16S rRNA-based archaeal community analyses in environmental samples. *Journal of Microbiological Methods* **84**:12–18.

Gardner, T.A., Cote, I.M., Gill, J.A., Gerant, A. and Watkinson, A.R. (2003). Long-term region-wide declines in Caribbean corals. *Science* **301**:958–60.

Gardner, T.A., Cote, I.M., Gill, J.A., Grant, A. and Watkinson, A.R. (2005). Hurricanes and Caribbean coral reefs: Impacts recovery patterns, and role in long-term decline. *Ecology* **86**:174–84.

Garrison, V.H., Shinn, E.A., Foreman, W.T., Griffin, D.W., Holmes, C.W., Kellogg, C.A., et al. (2003). African and Asian dust: From desert soils to coral reefs. *BioScience* **53**:469–80.

Gast, G.J., Wiegman, S., Wieringa, E., van Duyl, F.C. and Bak, R.P.M. (1998). Bacteria in coral reef water types: Removal of cells, stimulation of growth and mineralization. *Marine Ecology Progress Series* **167**:37–45.

Gaston, K.J. (2003). Ecology: The how and why of biodiversity. *Nature* **421**:900–1.

Gates, R.D., Hoegh-Guldberg, O., McFall-Ngai, M.J., Bil, K.Y. and Muscatine, L. (1995). Free amino acids exhibit anthozoan host factor activity: They induce the release of photosynthate from symbiotic dinoflagellates in vitro. *Proceedings of the National Academy of Sciences of the United States of America* **92**:7430–4.

Gattuso, J.P., Allemand, D. and Frankignoulle, M. (1999). Photosynthesis and calcification at cellular, organismal and community levels in coral reefs: A review on interactions and control by carbonate chemistry. *American Zoologist* **39**:160–83.

Gattuso, J.P., Gentili, B., Duarte, C.M., Kleypas, J.A., Middelburg, J.J. and Antoine, D. (2006). Light availability in the coastal ocean: Impact on the distribution of benthic photosynthetic organisms and their contribution to primary production. *Biogeosciences* **3**:489–513.

Gattuso, J.P., Reynaud-Vaganay, S., Furla, P., Romaine-Lioud, S., Jaubert, J., Bourge, I., et al. (2000). Calcification does not stimulate photosynthesis in the zooxanthellate scleractinian coral *Stylophora pistillata*. *Limnology and Oceanography* **45**:246–50.

Gell, F.R. and Roberts, C.M. (2003). Benefits beyond boundaries: The fishery effects of marine reserves. *Trends in Ecology and Evolution* **18**:448–55.

Ginsburg, R.N. (1983). Geological and biological roles of cavities in coral reefs. In: D.J. Barnes (ed.), *Perspectives on Coral Reefs*. Australian Institute of Marine Science, pp. 148–53.

Glynn, P.W. (1993). Coral reef bleaching: Ecological perspectives. *Coral Reefs* **12**:1–17.

Glynn, P.W. (1996). Coral bleaching: Facts, hypotheses and implications. *Global Change Biology* **2**:495–509.

Glynn, P.W. (1997).Bioerosion and coral reef growth: A dynamic balance. In: C. Birkeland (ed.), *Life and Death of Coral Reefs*. Chapman & Hall, pp. 68–95.

Glynn, P.W. and Ault, J.S. (2000). A biogeographic analysis and review of the far eastern Pacific coral reef region. *Coral Reefs* **19**:1–23.

Glynn, P.W. and Stewart, R.H. (1973). Distribution of coral reefs in the Pearl Islands (Gulf of Panama) in relation to thermal conditions. *Limnology and Oceanography* **18**:367–79.

Godinot, C., Ferrier-Pagès, C. and Grover, R. (2009). Control of phosphate uptake by zooxanthellae and host cells in the scleractinian coral *Stylophora pistillata*. *Limnology and Oceanography* **54**:1627–33.

Godinot, C., Gaysinski, M., Thomas, O.P., Ferrier-Pagès, C. and Grover, R. (2016). On the use of P-31 NMR for the quantification of hydrosoluble phosphorus-containing compounds in coral host tissues and cultured zooxanthellae. *Scientific Reports* **6**:21760.

Godinot, C., Grover, R., Allemand, D. and Ferrier-Pagès, C. (2011). High phosphate uptake requirements of the scleractinian coral *Stylophora pistillata*. *Journal of Experimental Biology* **214**:2749–54.

Golden, A.S., Naisilsisili, W., Ligairi, I. and Drew, J.A. (2014). Combining natural history collections with fisher knowledge for community-based conservation in Fiji. *PLOS ONE* **9**:e98036.

Gonzalez, J.M., Torreton, J.P., Dufour, P. and Charpy, L. (1998). Temporal and spatial dynamics of the pelagic microbial food web in an atoll lagoon. *Aquatic Microbial Ecology* **16**:53–64.

Goreau, T.F. (1959). The physiology of skeleton formation in corals. I. A method for measuring the rate of calcium deposition by corals under different conditions. *Biological Bulletin* **116**:59–75.

Goreau, T.F. (1961). Problems of growth and calcium deposition in reef corals. *Endeavour* **20**:32–9.

Goreau, T.F. (1964). Mass expulsion of zooxanthellae from Jamaican reef communities after Hurricane Flora. *Science* **145**:383–6.

Gottschalk, S., Uthicke, S. and Heimann, K. (2007). Benthic diatom community composition in three regions of the Great Barrier Reef, Australia. *Coral Reefs* **26**:345–57.

Gou, W., Sun, J., Li, X., Zhen, Y., Xin, Z., Yu, Z., et al. (2003). Phylogenetic analysis of a free-living strain of *Symbiodinium* isolated from Jiaozhou Bay, P.R. China. *Journal of Experimental Marine Biology Ecology* **296**:135–44.

Graham, N.A.J. (2007). Ecological versatility and the decline of coral feeding fishes following climate driven coral mortality. *Marine Biology* **153**:119–27.

Graham, N.A.J., Chabanet, P., Evans, R.D., Jennings, S., Letourneur, Y., MacNeil, M.A., et al. (2011). Extinction vulnerability of coral reef fishes. *Ecology Letters* **14**:341–8.

Graham, N.A.J., Jennings, S., MacNeil, M.A., Mouillot, D. and Wilson, S.K. (2015). Predicting climate-driven regime shifts versus rebound potential in coral reefs. Nature **518**:94–7.

Graham, N.A.J., McClanahan, T.R., MacNeil, M.A., Wilson, S.K., Cinner, J.E., Huchery, C., et al. (2017). Human disruption of coral reef trophic structure. *Current Biology* **27**:231–6.

Graham, N.A.J., McClanahan, T.R., MacNeil, M.A., Wilson, S.K., Polunin, N.V., Jennings, S., et al. (2008). Climate warming, marine protected areas and the ocean-scale integrity of coral reef ecosystems. *PLOS ONE* **3**:e3039.

Graham, N.A.J. and Nash, K.L. (2013). The importance of structural complexity in coral reef ecosystems. *Coral Reefs* **32**:315–26.

Graham, N.A.J., Wilson, S.K., Jennings, S., Polunin, N.V.C., Bijoux, J.P. and Robinson, J. (2006). Dynamic fragility of oceanic coral reef ecosystems. *Proceedings of the National Academy of Sciences of the United States of America* **103**:8424–9.

Graham, N.A.J., Wilson, S.K., Jennings, S., Polunin, N.V.C., Robinson, J., Bijoux, J.P., et al. (2007). Lag effects in the impacts of mass coral bleaching on coral reef fish, fisheries and ecosystems. *Conservation Biology* **21**:1291–300.

Grant, A.J., Remond, M. and Hinde, R. (1998). Low molecular-weight factor from *Plesiastrea versipora* (Scleractinia) that modifies release and glycerol metabolism of isolated symbiotic algae. *Marine Biology* **130**:553–7.

Grant, A.J., Remond, M., Starke-Peterkovic, T. and Hinde, R. (2006). A cell signal from the coral *Plesiastrea versipora* reduces starch synthesis in its symbiotic alga, *Symbiodinium* sp. *Comparative Biochemistry and Physiology Part A* **144**:458–63.

Grant, A.J., Trautman, D.A., Menz, I. and Hinde, R. (2006). Separation of two cell signalling molecules from a symbiotic sponge that modify algal carbon metabolism. *Biochemical and Biophysical Research Communications* **348**:92–8.

Grigg, R.W. (1981). Coral reef development at high latitudes in Hawaii. *Proceedings of the 4ᵗʰ International Coral Reef Symposium* **1**:687–93.

Grigg, R.W. (2006). Depth limit for reef building corals in the Au'au Channel, SE Hawaii. *Coral Reefs* **25**:77–84.

Grigg, R.W. and Epp, D. (1989). Critical depth for the survival of coral islands: Effects on the Hawaiian Archipelago. *Science* **243**:638–41.

Grover, R., Maguer, J.F., Allemand, D. and Ferrier-Pagès, C. (2003). Nitrate uptake in the scleractinian coral *Stylophora pistillata*. *Limnology and Oceanography* **48**:2266–74.

Grover, R., Maguer, J.F., Allemand, D. and Ferrier-Pagès, C. (2008). Uptake of dissolved free amino acids by the scleractinian coral *Stylophora pistillata*. *Journal of Experimental Biology* **211**:860–5.

Grover, R., Maguer, J.F., Reynaud-Vaganay, S. and Ferrier-Pagès, C. (2002). Uptake of ammonium by the scleractinian coral *Stylophora pistillata*: Effect of feeding, light, and ammonium concentrations. *Limnology and Oceanography* **47**:782–90.

Grutter, A.S. (1997). Effect of the removal of cleaner fish on the abundance and species composition of reef fish. *Oecologia* **111**:137–43.

Guilcher, A. (1988). A heretofore neglected type of coral reef: The ridge reef. Morphology and origin. *Proceedings of the 6ᵗʰ International Coral Reef Symposium* **3**:399–402.

Guinotte, J.M., Buddemeier, R.W. and Kleypas, J.A. (2003). Future coral reef habitat marginality: Temporal and spatial effects of climate change in the Pacific basin. *Coral Reefs* **22**:551–8.

Guinotte, J.M., Orr, J., Cairns, S., Freiwald, A., Morgan, L. and George, R. (2006). Will human-induced changes in seawater chemistry alter the distribution of deep-sea scleractinian corals? *Frontiers in Ecology and the Environment* **4**:141–6.

Guzman, H.M., Jackson, J.B.C. and Weil, E. (1991). Short-term ecological consequences of a major oil spill on Panamanian subtidal reef corals. *Coral Reefs* **10**:1–12.

Hadas, E., Marie, D., Shpigel, M. and Ilan, M. (2006). Virus predation by sponges is a new nutrient-flow pathway in coral reef food webs. *Limnology and Oceanography* **51**:1548–50.

Hallock, P. (2005). Global change and modern coral reefs new opportunities to understand shallow-water carbonate depositional processes. *Sedimentary Geology* **175**:19–33.

Hallock, P. and Schlager, W. (1986). Nutrient excess and the demise of coral reefs and carbonate platforms. *Palaios* **1**:389–98.

Hancock, G. (1993). *Lords of Poverty*. Mandarin.

Harborne, A.R., Mumby, P.J., Zychaluk, K., Hedley, J.D. and Blackwell, P.G. (2006). Modelling the beta diversity of coral reefs. *Ecology* **87**:2871–81.

Hardin, G.G. (1999). *The Ostrich Factor: Our Population Myopia*. Oxford University Press.

Harland, A.D. and Davies, P.S. (1995). Symbiont photosynthesis increases both respiration and photosynthesis in the symbiotic sea anemone *Anemonia viridis*. *Marine Biology* **123**:715–22.

Harriott, V.J. (1999). Coral growth in subtropical eastern Australia. *Coral Reefs* **18**:281–91.

Harriott, V.J. and Banks, S.A. (2002). Latitudinal variation in coral communities in eastern Australia: A qualitative biophysical model of factors regulating coral reefs. *Coral Reefs* **21**:83–94.

Harriott, V.J., Harrison, P.L. and Banks, S.A. (1995). The coral communities of Lord Howe Island. *Marine Freshwater Research* **46**:457–65.

Haszprunar, G. and Spies, M. (2014). An integrative approach to the taxonomy of the crown-of-thorns starfish species group (Asteroidea: *Acanthaster*): A review of names and comparison to recent molecular data. *Zootaxa* **3841**:271–84.

Hatcher, A.I. (1985). The relationship between coral reef structure and nitrogen dynamics. *Proceedings of the 5th International Coral Reef Symposium* 3:407–13.

Hatcher, B.G. (1997). Organic production and decomposition. In: C. Birkeland (ed.), *Life and Death of Coral Reefs*. Springer, pp. 140–74.

Hawkins, T.D., Bradley, B.J. and Davy, S.K. (2013). Nitric oxide mediates coral bleaching through an apoptotic-like cell death pathway: Evidence from a model sea anemone–dinoflagellate symbiosis. *FASEB Journal* **27**:4790–8.

Hawkins, T.D. and Davy, S.K. (2012). Nitric oxide production and tolerance differ among *Symbiodinium* types exposed to heat stress. *Plant and Cell Physiology* **53**:1889–98.

Hawkins, T.D., Krueger, T., Becker, S., Fisher, P.L. and Davy, S.K. (2014). Differential nitric oxide synthesis and host apoptotic events correlate with bleaching susceptibility in reef corals. *Coral Reefs* **33**:141–53.

Heil, C.A., Chaston, K., Jones, A., Bird, P., Longstaff, B., Costanzo, S., et al. (2004). Benthic microalgae in coral reef sediments of the southern Great Barrier Reef, Australia. *Coral Reefs* **23**:336–43.

Heiss, G.A. (1995). Carbonate production by scleractinian corals in Aqaba, Gulf of Aqaba, Red Sea. *Facies* **33**:19–34.

Helfman, G.S., Collette, B.B. and Facey, D.E. (1997). *The Diversity of Fishes*. Blackwell Science.

Hennige, S.J., Wicks, L.C., Kamenos, N.A., Perna, G., Findlay, H.S. and Roberts, J.M. (2015). Hidden impacts of ocean acidification to live and dead coral framework. *Proceedings of the Royal Society of London. Series B, Biological Sciences* **282**:20150990.

Henry, L.-A., Moreno Navas, J., Hennige, S.J., Wicks, L., Vad, J. and Roberts, J.M. (2013). Cold-water coral reef habitats benefit recreationally valuable sharks. *Biological Conservation* **161**:67–70.

Henry, L.-A. and Roberts, J.M. (2007). Biodiversity and ecological composition of macrobenthos on cold-water coral mounds and adjacent off-mound habitat in the bathyal Porcupine Seabight, NE Atlantic. *Deep Sea Research Part I* **54**:654–72.

Hewson, I. and Fuhrman, J.A. (2006). Spatial and vertical biogeography of coral reef sediment bacterial and diazotroph communities. *Marine Ecology Progress Series* **306**:79–86.

Hewson, I., Moisander, P.H., Morrison, A.E. and Zehr, J.P. (2007). Diazotrophic bacterioplankton in a coral reef lagoon: Phylogeny, diel nitrogenase expression and response to phosphate enrichment. *The ISME Journal* **1**:78–91.

Heyward, A.J. and Negri, A.P. (1999). Natural inducers for coral larval metamorphosis. *Coral Reefs* **18**:273–9.

Hilborn, R., Stokes, K. and Maguire, J.J. (2004). When can marine reserves improve fisheries management? *Ocean and Coastal Management* **47**:197–205.

Hill, M.S. (1996). Symbiotic zooxanthellae enhance boring and growth rates of the tropical sponge *Anthosigmella varians* forma *varians*. *Marine Biology* **125**:649–54.

Hinde, R. (1988). Factors produced by symbiotic marine invertebrates which affect translocation between symbionts. In: S. Scannerini, D.C. Smith, P. Bonfante-Fasolo and V. Gianinazzi-Pearson (eds), *Cell to Cell Signals in Plant, Animal and Microbial Symbiosis*. NATO ASI Subseries H, Book 17. Springer, pp. 311–24.

Hirose, M., Kinzie, R.A., III and Hidaka, M. (2001). Timing and process of entry of zooxanthellae into oocytes of hermatypic corals. *Coral Reefs* **20**:273–80.

Hodgson, G. (1990). Sediment and the settlement of larvae of the reef coral *Pocillopora damicornis*. *Coral Reefs* **9**:41–3.

Hoegh-Guldberg, O. (1999). Climate change, coral bleaching and the future of the world's coral reefs. *Marine and Freshwater Research* **50**:839–66.

Hoegh-Guldberg, O. (2004). Coral reefs in a century of rapid environmental change. *Symbiosis* **37**:1–31.

Hoegh-Guldberg, O. and Fine, M. (2004). Low temperatures cause coral bleaching. *Coral Reefs* **23**:444.

Hoegh-Guldberg, O., Jones, R.J., Ward, S. and Loh, W.K. (2002). Ecology: Is coral bleaching really adaptive? *Nature* **415**:601–2.

Hoegh-Guldberg, O., McCloskey, L.R. and Muscatine, L. (1987). Expulsion of zooxanthellae by symbiotic cnidarians from the Red Sea. *Coral Reefs* **5**:201–4.

Hoegh-Guldberg, O., Mumby, P.J., Hooten, A.J., Steneck, R.S., Greenfield, P., Gomez, E., et al. (2007). Coral reefs under rapid climate change and ocean acidification. *Science* **318**:1737–42.

Hoegh-Guldberg, O. and Smith, G.J. (1989a). Influence of the population density of zooxanthellae and supply of ammonium on the biomass and metabolic characteristics of the reef corals *Seriatopora hystrix* and *Stylophora pistillata*. *Marine Ecology Progress Series* **57**:173–86.

Hoegh-Guldberg, O. and Smith, G.J. (1989b). The effect of sudden changes in temperature, light and salinity on the population density and export of zooxanthellae from the reef corals *Stylophora pistillata* Esper and *Seriatopora hystrix* Dana. *Journal of Experimental Marine Biology Ecology* **129**:279–303.

Hoey, A.S. and Bellwood, D.R. (2008). Cross-shelf variation in the role of parrotfishes on the Great Barrier Reef. *Coral Reefs* **27**:37–47.

Hoey, A.S. and Bellwood, D.R. (2009). Limited functional redundancy in a high diversity system: Single species dominates key ecological process on coral reefs. Ecosystems **12**:1316–28.

Hoitink, A.J.F. (2004). Tidally-induced clouds of suspended sediment connected to shallow-water coral reefs. *Marine Geology* **208**:13–31.

Hoitink, A.J.F. and Hoekstra, P. (2003). Hydrodynamic control of the supply of reworked terrigenous sediment to coral reefs in the Bay of Banten (NW Java, Indonesia). *Estuarine, Coastal and Shelf Science* **58**:743–55.

Holcomb, M., Tambutté, É., Allemand, D. and Tambutté, S. (2014). Light enhanced calcification in *Stylophora pistillata*: Effects of glucose, glycerol and oxygen. *Peerj* **2**.

Hooper, J.N.A. and van Soest, R.W.M. (2002). *Systema Porifera: A Guide to the Classification of Sponges*. Kluwer Academic Press.

Hopley, D. (1982). *The Geomorphology of the Great Barrier Reef*. Wiley.

Hopley, D., Smithers, S.G. and Parnell, K. (2008). *The Geomorphology of the Great Barrier Reef: Development, Diversity and Change*. Cambridge University Press.

Houlbreque, F., Tambutte, E., Allemand, D. and Ferrier-Pagès, C. (2004). Interactions between zooplankton feeding, photosynthesis and skeletal growth in the scleractinian coral *Stylophora pistillata*. *Journal of Experimental Biology* **207**:1461–9.

Houlbrèque, F., Tambutté, E., Richard, C. and Ferrier-Pagès, C. (2004). Importance of a micro-diet for scleractinian corals. *Marine Ecology Progress Series* **282**:151–60.

Howells, E.J., Berkelmans, R., van Oppen, M.J.H., Willis, B.L. and Bay, L.K. (2013). Historical thermal regimes define limits to coral acclimatization. *Ecology* **94**:1078–88.

Hubbard, D., Gischler, E., Davies, P., Montaggioni, L., Camoin, G., Dullo, W.C., et al. (2014). Island outlook: Warm and swampy. *Science* **345**:1461.

Hughes, T.P. (1994). Catastrophes, phase shifts, and large-scale degradation of a Caribbean coral reef. *Science* **265**:1547–51.

Hughes, T.P. (2016). The 2016 coral bleaching event in Australia. In: *Abstract Book, 13th International Coral Reef Symposium*. International Society for Reef Studies, p. 153.

Hughes, T.P., Baird, A.H., Bellwood, D.R., Card, M., Connolly, S.R. and Folke, C. (2003). Climate change, human impacts, and the resilience of coral reefs. *Science* **301**:929–33.

Hughes, T.P., Bellwood, D.R., Folke, C., Steneck, R.S. and Wilson, J. (2005). New paradigms for supporting the resilience of marine ecosystems. *Trends in Ecology and Evolution* **20**:380–6.

Hughes, T.P., Rodrigues, M.J., Bellwood, D.R., Ceccarelli, D., Hoegh-Guldberg, O., McCook, L., et al. (2007). Phase shifts, herbivory, and the resilience of coral reefs to climate change. *Current Biology* **17**:360–5.

Hume, B.C.C., D'Angelo, C., Smith, E.G., Stevens, J.R., Burt, J. and Wiedenmann, J. (2015). *Symbiodinium thermophilum* sp. nov., a thermotolerant symbiotic alga prevalent in corals of the world's hottest sea, the Persian/Arabian Gulf. *Scientific Reports* **5**:8652.

Hutchings, P. and Haynes, D. (eds) (2000). Sources, fates and consequences of pollutants in the Great Barrier Reef. *Marine Pollution Bulletin* **41**:265–434.

Hutchings, P. and Haynes, D. (eds) (2005). Catchment to reef: Water quality issues in the Great Barrier Reef region. *Marine Pollution Bulletin* **51**:1–480.

Hutchings, P., Peyrot-Clausade, M. and Osnorno, A. (2005). Influence of land run-off on rates and agents of bioerosion of coral substrates. *Marine Pollution Bulletin* **51**:438–47.

Hviding, E. (1998). Contextual flexibility: Present status and future of customary marine tenure in Solomon Islands. *Ocean and Coastal Management* **40**:253–69.

Idjadi, J.A., Lee, S.C., Bruno, J.F., Precht, W.F., Allen-Requa, L. and Edmunds, P.J. (2006). Rapid phase shift reversal on a Jamaican coral reef. *Coral Reefs* **25**:209–11.

Iglesias-Prieto, R., Beltran, V.H., LaJeunesse, T.C., Reyes-Bonilla, H. and Thome, P.E. (2004). Different algal symbionts explain the vertical distribution of dominant reef corals in the eastern Pacific. *Proceedings of the Royal Society of London. Series B, Biological Sciences* **271**:1757–63.

IPCC. (2007). *Climate Change 2007: Synthesis Report. Contribution of Working Groups I, II and III to the Fourth Assessment Report of the Intergovernmental Panel on Climate Change*. Available at <http://www.ipcc.ch/publications_and_data/publications_ipcc_fourth_assessment_report_synthesis_report.htm>.

ISRS. (2007). *Coral Reefs And Ocean Acidification*. Briefing Paper Number 5. International Society for Reef Studies. Available at <http://coralreefs.org/

wp-content/uploads/2014/05/ISRS-Briefing-Paper-5-Coral-Reefs-and-Ocean-Acidification.pdf >.

Jackson, A.E. and Yellowlees, D. (1990). Phosphate uptake by zooxanthellae isolated from corals. *Proceedings of the Royal Society B Biological Sciences* **242**:201–4.

Jackson, J.B.C (2001). What was natural in the oceans? *Proceedings of the National Academy of Sciences of the United States of America* **98**:5411–18.

Jackson, J.B.C., Kirby, M.X., Berger, W.H., Bjorndal, K.A., Botsford, L.W., Bourque, B.J., et al. (2001). Historical fishing and the recent collapse of coastal ecosystems. *Science* **293**:629–37.

James, N.P. and Wood,R.A.(2010). Reefs. In: R. Dalrymple and N.P. James (eds), *Facies Models: Response to Sea Level Change*. Geological Association of Canada, p. 421–47.

Jameson, S. (2008). Reefs in trouble: The real root cause. *Marine Pollution Bulletin* **56**:1513–14.

Jankowski, M.W., Graham, N.A.J. and Jones, G.P. (2015). Depth gradients in diversity, distribution and habitat specialisation in coral reef fishes: Implications for the depth–refuge hypothesis. *Marine Ecology Progress Series* **540**:203–15.

Jarrett, B.D., Hine, A.C., Halley, R.B., Naar, D.F., Locker, S.D., Neumann, A.C., et al. (2005). Strange bedfellows: A deep-water hermatypic coral reef superimposed on a drowned barrier island; southern Pulley Ridge, SW Florida platform margin. *Marine Geology* **214**:295–307.

Jennings, S. and Polunin, N.V.C. (1997). Impact of predator depletion by fishing on the biomass and diversity of non-target reef fish communities. *Coral Reefs* **16**:71–82.

Jennings, S., Reynolds, J.D. and Polunin, N.V.C. (1999). Predicting the vulnerability of tropical reef fishes to exploitation with phylogenies and life histories. *Conservation Biology* **13**:1466–75.

Jeong, H.J., Lee, S.Y., Kang, N.S.,Yoo, Y.D., Lim, A.S., Lee, M.J., et al. (2014). Genetics and morphology characterize the dinoflagellate *Symbiodinium voratum*, n. sp., (Dinophyceae) as the sole representative of *Symbiodinium* Clade E. *Journal of Eukaryotic Microbiology* **61**:75–94.

Johannes, R.E. (1978). Traditional marine conservation methods in oceania and their demise. *Annual Review of Ecological Systems* **9**:349–64.

Johannes, R.E. (1998). Government-supported, village-based management of marine resources in Vanuatu. *Ocean and Coastal Management* **40**:165–86.

Johannes, R.E. and Gerber, R. (1974). Import and export of net plankton by an Eniwetok coral reef community. *Great Barrier Reef Committee (Brisbane, Australia)* **1**:97–104.

Johnston, I.S. and Rohwer, F. (2007). Microbial landscapes on the outer tissue surfaces of the reef-building coral *Porites compressa*. *Coral Reefs* **26**:375–83.

Jokiel, P.L. and Coles, S.L. (1977). Effects of temperature on the mortality and growth of Hawaiian reef corals. *Marine Biology* **43**:201–8.

Jokiel, P.L. and Maragos, J.E. (1978). Reef corals of Canton Island. *Atoll Research Bulletin* **221**:71–97.

Jokiel, P.L., Rodgers, K.S., Kuffner, I.B., Andersson, A.J., Cox, E.F. and Mackenzie, F.T. (2008). Ocean acidification and calcifying reef organisms: A mesocosm investigation. *Coral Reefs* **27**:473–83.

Jones, G.P., McCormick, M.I., Srinivasan, M. and Eagle, J. V. (2004). Coral decline threatens fish biodiversity in marine reserves. *Proceedings of the National Academy of Sciences of the United States of America* **101**:8251–3.

Jones, G.P., Milicich, M.J., Emslie, M.J. and Lunow, C. (1999). Self-recruitment in a coral reef fish population. Nature **402**:802–4.

Jones, R.J. (1997). Zooxanthellae loss as a bioassay for assessing stress in corals. *Marine Ecology Progress Series* **149**:163–71.

Jones, R.J. (2005). The ecotoxicological effects of Photosystem II herbicides on corals. *Marine Pollution Bulletin* **51**:495–506.

Jones, R.J., Hoegh-Guldberg, O., Larkum, A.W.D. and Schreiber, U. (1998). Temperature-induced bleaching of corals begins with impairment of the CO_2 fixation mechanism in zooxanthellae. *Plant, Cell and Environment* **21**:1219–30.

Jones, R.J., Ricardo, G.F. and Negri, A.P. (2015). Effects of sediments on the reproductive cycle of corals. *Marine Pollution Bulletin* **100**:13–33.

Kan, H., Hori, N., Nakashima, Y. and Ichikawa, K. (1995). The evolution of narrow reef flats at high latitude in the Ryukyu Islands. *Coral Reefs* **14**:123–30.

Karlson, R.H. and Cornell, H.V. (1998). Scale-dependent variation in local vs. regional effects on coral species richness. *Ecological Monographs* **68**:259–74.

Karner, M.B., DeLong, E.F. and Karl, D.M. (2001). Archaeal dominance in the mesopelagic zone of the Pacific Ocean. *Nature* **409**:507–10.

Karplus, I. (1979). The tactile communication between *Cryptocentrus steinitzi* (Pisces, Gobiidae) and *Alpheus purpurilenticularis* (Crustacea, Alpheidae). *Zeitschrift fur Tierpsychologie* **49**:173–96.

Kayal, M., Vercelloni, J., De Loma, T.L., Bosserelle, P., Chancerelle, Y., Geoffroy, S., et al. (2012). Predator crown-of-thorns starfish (*Acanthaster planci*) outbreak, mass mortality of corals, and cascading effects on reef fish and benthic communities. *PLOS ONE* **7**:e47363.

Kemp, D.W., Oakley, C.A., Thornhill, D.J., Newcomb, L.A., Schmidt, G.W. and Fitt, A.K. (2011). Catastrophic mortality on inshore coral reefs of the Florida Keys due to severe low-temperature stress. *Global Change Biology* **17**:3468–77.

Kench, P.S., Parnell, K.E. and Brander, R.W. (2009). Monsoonally influenced circulation around coral reef islands and seasonal dynamics of reef island shorelines. *Marine Geology* **266**:91–108.

Kerswell, A.P. and Jones, R.J. (2003). Effects of hypo-osmosis on the coral *Stylophora pistillata*: Nature and cause of 'low-salinity bleaching'. *Marine Ecology Progress Series* **253**:145–54.

Kettler, G.C., Martiny, A.C., Huang, K., Zucker, J., Coleman, M.L., Rodrigue, S., et al. (2007). Patterns and implications of gene gain and loss in the evolution of *Prochlorococcus*. *PLOS Genetics* **3**:e231.

Kiessling, W., Aberhan, M. and Villier, L. (2008). Phanerozoic trends in skeletal mineralogy driven by mass extinctions. *Nature Geoscience* **1**:527–30.

Kim, H.J., Ryu, J.O., Lee, S.Y., Kim, E.S. and Kim, H.Y. (2015). Multiplex PCR for detection of the *Vibrio* genus and five pathogenic *Vibrio* species with primer sets designed using comparative genomics. *BMC Microbiology* **15**:239.

Kinsman, D.J.J. (1964). Reef coral tolerance of high temperatures and salinities. *Nature* **202**:1280–2.

Kirk, J.T.O. (1994). *Light and Photosynthesis in Aquatic Ecosystems*. Cambridge University Press.

Kleypas, J.A. (1994). A diagnostic model for predicting global coral reef distribution. In: *Proceedings of PACON 1994: Recent Advances in Marine Science and Technology*. PACON International and James Cook University of North Queensland, pp. 211–20.

Kleypas, J.A. (1996). Coral reef development under naturally turbid conditions: Fringing reefs near Broad Sound, Australia. *Coral Reefs* **15**:153–67.

Kleypas, J.A. (1997). Modeled estimates of global reef habitat and carbonate production since the last glacial maximum. *Paleoceanography* **12**:533–45.

Kleypas, J.A., Buddemeier, R.W., Archer, D., Gattuso, J.-P., Langdon, C. and Opdyke, B.N. (1999). Geochemical consequences of increased atmospheric carbon dioxide on coral reefs. *Science* **284**:118–20.

Kleypas, J.A., Danabasoglu, G. and Lough, J.M. (2008). Potential role of the ocean thermostat in determining regional differences in coral reef bleaching events. *Geophysical Research Letters* **35**:L03613.

Kleypas, J.A., Feely, R.A., Fabry, V.J., Langdon, C.L., Sabine, C.L. and Robbins, L.L. (2006). *Impacts of Increasing Ocean Acidification on Coral Reefs and Other Marine Calcifiers: A Guide for Future Research*. Available at <https://www.researchgate.net/profile/Joan_Kleypas/publication/248700866_Impacts_of_Ocean_Acidification_on_Coral_Reefs_and_Other_Marine_Calcifiers_A_Guide_for_Future_Research/links/54b577eb0cf2318f0f998b54/Impacts-of-Ocean-Acidification-on-Coral-Reefs-and-Other-Marine-Calcifiers-A-Guide-for-Future-Research.pdf>.

Kleypas, J.A., McManus, J.W. and Meñez, L.A.B. (1999). Environmental limits to coral reef development: Where do we draw the line? *American Zoologist* **39**:146–59.

Klunzinger, C.B. (1878). *Upper Egypt: Its People and Its Products*. Blackie and Sons.

Knowlton, N. (2001). The future of coral reefs. *Proceedings of the National Academy of Sciences of the United States of America* **98**:5419–25.

Knowlton, N. and Rohwer, F. (2003). Multispecies microbial mutualisms on coral reefs: The host as a habitat. *American Naturalist* **162**: S51–62.

Kolber, Z.S., Barber, R.T., Coale, K.H., Fitzwater, S.E., Greene, R.M., Johnson, K.S., et al. (1994). Iron limitation of phytoplankton photosynthesis in the equatorial Pacific Ocean. *Nature* **371**:145–9.

Kopp, C., Pernice, M., Domart-Coulon, I., Djediat, C., Spangenberg, J.E., Alexander, D.T.L., et al. (2013). Highly dynamic cellular-level response of symbiotic coral to a sudden increase in environmental nitrogen. *mBio* **4**: e00052–13.

Koslow, J.A., Hanley, F. and Wicklund, R. (1988). Effects of fishing on reef fish communities at Pedro bank and Port Royal Cays, Jamaica. *Marine Ecology Progress Series* **43**:201–12.

Kramarsky-Winter, E., Harel, M., Siboni, N., Ben Dov, E., Brickner, I., Loya, Y., et al. (2006). Identification of a protist–coral association and its possible ecological role. *Marine Ecology Progress Series* **317**:67–73.

Krediet, C.J., Ritchie, K.B., Paul, V.J. and Teplitski, M. (2013). Coral-associated micro-organisms and their roles in promoting coral health and thwarting diseases. *Proceedings of the Royal Society of London. Series B, Biological Sciences* **280**:20122328.

Krueger, T., Hawkins, T.D., Becker, S., Pontasch, S., Dove, S., Hoegh-Guldberg, O., et al. (2015). Differential coral bleaching: Contrasting the activity and response of enzymatic antioxidants in symbiotic partners under thermal stress. *Comparative Biochemistry and Physiology Part A* **190**:15–25.

Kumara, P.B.T.P., Cumaranathunga, P.R.T. and Linden, O. (2005). Present status of the sea cucumber fishery in southern Sri Lanka: A resource deleted industry. *SPC Beche-de-mer Information Bulletin.* **22**:24–29.

Kurten, B., Khomayis, H.S., Devassy, R., Audritz, S., Sommer, U., Struck, U., et al. (2015). Ecohydrographic constraints on biodiversity and distribution of phytoplankton and zooplankton in coral reefs of the Red Sea, Saudi Arabia. *Marine Ecology: An Evolutionary Perspective* **36**:1195–1214.

Kvennefors, E.C.E., Leggat, W., Hoegh-Guldberg, O., Degnan, B.M. and Barnes, A.C. (2008). An ancient and variable mannose-binding lectin from the coral *Acropora millepora* binds both pathogens and symbionts. *Developmental and Comparative Immunology* **32**:1582–92.

Kvennefors, E.C.E. and Roff, G. (2009). Evidence of cyanobacteria-like endosymbionts in Acroporid corals from the Great Barrier Reef. *Coral Reefs* **28**:547–547.

LaJeunesse, T.C. (2001). Investigating the biodiversity, ecology, and phylogeny of endosymbiotic dinoflagellates in the genus *Symbiodinium* using the ITS region: In search of a 'species' level marker. *Journal of Phycology* **37**:866–80.

LaJeunesse, T.C. (2002). Diversity and community structure of symbiotic dinoflagellates from Caribbean reef corals. *Marine Biology* **141**:387–400.

LaJeunesse, T.C., Bhagooli, R., Hidaka, M., DeVantier, L., Done, T., Schmidt, G.W., et al. (2004). Closely related *Symbiodinium* spp. differ in relative dominance in coral reef host communities across environmental, latitudinal and biogeographic gradients. *Marine Ecology Progress Series* **284**:147–61.

LaJeunesse, T.C., Forsman, Z.H. and Wham, D.C. (2016). An Indo-West Pacific 'zooxanthella' invasive to the western Atlantic finds its way to the Eastern Pacific via an introduced Caribbean coral. *Coral Reefs* **35**:577–82.

LaJeunesse, T.C., Loh, W.K.W., van Woesik, R., Hoegh-Guldberg, O., Schmidt, G.W. and Fitt, W.K. (2003). Low symbiont diversity in southern Great Barrier Reef corals, relative to those of the Caribbean. *Limnology and Oceanography* **48**:2046–54.

LaJeunesse, T.C., Thornhill, D.J., Cox, E.F., Stanton, F.G., Fitt, W.K. and Schmidt, G.W. (2004). High diversity and host specificity observed among symbiotic dinoflagellates in reef coral communities from Hawaii. *Coral Reefs* **23**:596–603.

LaJeunesse, T.C., Wham, D.C., Pettay, D.T., Parkinson, J.E., Keshavmurthy, S. and Chen, C.A. (2014). Ecologically differentiated stress-tolerant endosymbionts in the dinoflagellate genus *Symbiodinium* (Dinophyceae) Clade D are different species. *Phycologia* **53**:305–19.

Lapointe, B.E., Littler, M.M. and Littler, D.S. (1993). Modification of benthic community structure by natural eutrophication: The Belize Barrier Reef. *Proceedings of the 7th International Coral Reef Symposium* **1**:323–34.

Lapointe, B.E., Littler, M.M. and Littler, D.S. (1997). Macroalgal overgrowth of fringing coral reefs at Discovery Bay, Jamaica: Bottom-up versus top-down control. *Proceedings of the 8th International Coral Reef Symposium* **1**:927–32.

Larcombe, P., Ridd, P.V., Prytz, A. and Wilson, B. (1995). Factors controlling suspended sediment on inner-shelf coral reefs, Townsville, Australia. *Coral Reefs* **14**:163–71.

Larcombe, P. and Woolfe, K.J. (1999). Increased sediment supply to the Great Barrier Reef will not increase sediment accumulation at most coral reefs. *Coral Reefs* **18**:163–9.

Larkum, A.W.D., Kennedy, I.R. and Muller, W.J. (1988). Nitrogen fixation on a coral reef. *Marine Biology* **98**:143–55.

Lawrence, S.A., Davy, J.E., Wilson, W.H., Hoegh-Guldberg, O. and Davy, S.K. (2015). *Porites* white patch syndrome: Associated viruses and disease physiology. *Coral Reefs* **34**:249–57.

Lawrence, S.A., Wilkinson, S.P., Davy, J.E., Arlidge, W.N.S., Williams, G.J., Wilson, W.H., et al. (2015). Influence of local environmental variables on the viral consortia associated with the coral *Montipora capitata* from Kane'ohe Bay, Hawaii, USA. *Aquatic Microbial Ecology* **74**:251–62.

Le Campion-Alsumard, T., Goubic, S. and Hutchings, P. (1995). Microbial endoliths in skeletons of live and dead corals: *Porites lobata* (Moorea, French Polynesia). *Marine Ecology Progress Series* **117**:149–57.

Leclercq, N., Gattuso, J.-P. and Jaubert, J. (2000). CO2 partial pressure controls the calcification rate of a coral community. *Global Change Biology* **6**:329–34.

Lee Long, W.J., Mellors, J.E. and Coles, R.G. (1993). Seagrasses between Cape York and Hervey Bay, Queensland, Australia. *Australian Journal Marine Freshwater Research* **44**:19–31.

Lee, M.J., Jeong, H.J., Jang, S.H., Lee, S.Y., Kang, N.S., Lee, K.H., et al. (2016). Most low-abundance 'background' *Symbiodinium* spp. are transitory and have minimal functional significance for symbiotic corals. *Microbial Ecology* **71**:771–83.

Lee, S.Y., Jeong, H.J., Kang, N.S., Jang, T.Y., Jang, S.H. and Lim, A.S. (2014). Morphological characterization of *Symbiodinium minutum* and *S. psygmophilum* belonging to Clade B. *Algae* **29**:299–310.

Leggat, W., Hoegh-Guldberg, O., Dove, S. and Yellowlees, D. (2007). Analysis of an EST library from the dinoflagellate (*Symbiodinium* sp.) symbiont of reef-building corals. *Journal of Phycology* **43**:1010–21.

Lesser, M.P. (2006). Benthic–pelagic coupling on coral reefs: Feeding and growth of Caribbean sponges. *Journal of Experimental Marine Biology and Ecology* **328**:277–88.

Lesser, M.P., Falcon, L.I., Rodriguez-Roman, A., Enriquez, S., Hoegh-Guldberg, O. and Iglesias-Prieto, R. (2007). Nitrogen fixation by symbiotic bacteria provides a source of nitrogen for the scleractinian coral *Montastraea cavernosa*. *Marine Ecology Progress Series* **346**:143–52.

Lesser, M.P., Mazel, C.H., Gorbunov, M.Y. and Falkowski, P.G. (2004). Discovery of symbiotic nitrogen-fixing cyanobacteria in corals. *Science* **305**:997–1000.

Levy, O., Mizrahi, L., Chadwick-Furman, N.E. and Achituv, Y. (2001). Factors controlling the expansion behavior of *Favia favus* (Cnidaria: Scleractinia). Effects of light, flow and planktonic prey. *Biological Bulletin* **200**:118–26.

Lewis, D.H. and Smith, D.C. (1971). The autotrophic nutrition of symbiotic marine coelenterates with special reference to hermatypic corals. I. Movement of photosynthetic products between the symbionts. *Proceedings of the Royal Society of London. Series B, Biological Sciences* **178**:111–29.

Lewis, J.B. (1989). The ecology of *Millepora*: A review. *Coral Reefs* **8**:99–107.

Lewis, S.M. (1986). The role of herbivorous fishes in the organization of a Caribbean reef community. *Ecological Monographs* **56**:183–200.

Lillis, A., Bohnenstiehl, D., Peters, J.W. and Eggleston, D. (2016). Variation in habitat soundscape characteristics influences settlement of a reef-building coral. *PeerJ* **4**:e2557.

Lin, K.L., Wang, J.T. and Fang, L.S. (2000). Participation of glycoproteins on zooxanthellal cell walls in the establishment of a symbiotic relationship with the sea anemone *Aiptasia pulchella*. *Zoological Studies* **39**:172–8.

Lirman, D., Schopmeyer, S., Manzello, D., Gramer, L.J., Precht, W.F., Muller-Karger, F., et al. (2011). Severe 2010 cold-water event caused unprecedented mortality to corals of the Florida Reef Tract and reversed previous survivorship patterns. *PLOS ONE* **6**:e23047.

Littler, M.M. and Doty, M.S. (1975). Ecological components structuring the seaward edges of tropical Pacific Reefs: The distribution, communities and productivity of *Porolithon*. *Journal of Ecology* **63**:117–29.

Lobban, C.S. and Harrison, P.J. (1994). *Seaweed Ecology and Physiology*. Cambridge University Press.

Lobban, C.S., Modeo, L., Verni, F. and Rosati, G. (2005). *Euplotes uncinatus* (Ciliophora, Hypotrichia), a new species with zooxanthellae. *Marine Biology* **147**:1055–61.

Lobban, C.S., Schefter, M., Simpson, A.G.B., Pochon, X., Pawlowski, J. and Foissner, W. (2002). *Maristentor dinoferus* n. gen., n. sp., a giant heterotrich ciliate (Spirotrichea: Heterotrichida) with zooxanthellae, from coral reefs on Guam, Mariana Islands. *Marine Biology* **140**:411–23.

Logan, D.D.K., LaFlamme, A.C., Weis, V.M. and Davy, S.K. (2010). Flow-cytometric characterization of the cell-surface glycans of symbiotic dinoflagellates (*Symbiodinium* spp.). *Journal of Phycolpgy* **46**:525–33.

Lokrantz, J., Nyström, M., Thyresson, M. and Johansson, C. (2008). The nonlinear relationship between body size and function in parrotfishes. *Coral Reefs* **27**:967–74.

Long R. and Rodríguez Chaves, M. (2015). Anatomy of a new international instrument for marine biodiversity beyond national jurisdiction. *Environmental Liability* **23**:213–29.

Lopez, G.R. and Levinton, J.S. (1987). Ecology of deposit-feeding animals in marine sediments. *Quarterly Review of Biology* **62**:235–60.

Loya, Y. and Rinkevich, B. (1980). Effects of oil pollution on coral reef communities. *Marine Ecology Progress Series* **3**:167–80.

Loya, Y. and Sakai, K. (2008). Bidirectional sex change in mushroom stony corals. *Proceedings of the Royal Society of London. Series B, Biological Sciences* **275**: 2335–43.

Lugo-Fernández, A., Roberts, H.H. and Suhayda, J.N. (1998). Wave transformations across a Caribbean fringing-barrier coral reef. *Continental Shelf Research* **18**:1099–124.

Lugo-Fernández, A., Roberts, H.H. and Wiseman, W.J. (1998). Tide effects on wave attenuation and wave set-up on a Caribbean coral reef. *Estuarine, Coastal and Shelf Science* **47**:385–93.

Lugo-Fernández, A., Roberts, H.H. and Wiseman, W.J. (2004). Currents, water levels, and mass transport over a modern Caribbean coral reef: Tague Reef, St. Croix, USVI. *Continental Shelf Research* **24**:1989–2009.

Madhupratap, M., Achuthankutty, C.T., Sreekumaran Nair, S.R. (1991). Estimates of high absolute densities and emergence rates ofdemersal zooplankton from the Agatti Atoll, Laccadives. *Limnology and Oceanography* **36**:585–8.

Madin, J.S. and Connolly, S.R. (2006). Ecological consequences of major hydrodynamic disturbances on coral reefs. *Nature* **444**:477–80.

Mahon, R. and Hunte, W. (2001). Trap mesh selectivity and the management of reef fisheries. *Fish and Fisheries* **2**:356–75.

Mantyka, C.S. and Bellwood, D.R. (2007). Macroalgal grazing selectivity among herbivorous coral reef fishes. *Marine Ecology Progress Series* **352**:177–85.

Manzello, D.P., Brandt, M., Smith, T.B., Lirman, D., Hendee, J. and Nemeth, R.S. (2007). Hurricanes benefit bleached corals. *Proceedings of the National Academy of Science of the United States of America* **104**:12035–9.

Manzello, D.P., Kleypas, J.A., Budd, D.A., Eakin, C.M., Glynn, P.W. and Langdon, C. (2008). Poorly cemented coral reefs of the eastern tropical Pacific: Possible insights into reef development in a high-CO2 world. *Proceedings of the National Academy of Science of the United States of America* **105**:10450–5.

Manzello, D. and Lirman, D. (2003). The photosynthetic resilience of *Porites furcata* to salinity disturbance. *Coral Reefs* **22**:537–40.

Marcelino, V.R. and Verbruggen, H. (2016). Multi-marker metabarcoding of coral skeletons reveals a rich microbiome and diverse evolutionary origins of endolithic algae. *Scientific Reports* **6**:31508.

Mariscal, R.N. (1970). The nature of the symbiosis between Indo-Pacific anemone fishes and sea anemones. *Marine Biology* **6**:58–65.

Markell, D.A. and Trench, R.K. (1993). Macromolecules exuded by symbiotic dino-flagellates in culture: Amino acid and sugar composition. *Journal of Phycology* **29**:64–8.

Markell, D.A., Trench, R.K. and Iglesias-Prieto, R. (1992). Macromolecules associated with the cell walls of symbiotic dinoflagellates. *Symbiosis* **12**:19–31.

Markell, D.A. and Wood-Charlson, E. (2010). Immunocytochemical evidence that symbiotic algae secrete potential recognition signal molecules *in hospite*. *Marine Biology* **157**:1105–11.

Marubini, F., Ferrier-Pagès, C. and Cuif, J.P. (2003). Suppression of skeletal growth in scleractinian corals by decreasing ambient carbonate-ion concentration: A cross-family comparison. *Proceedings of the Royal Society of London. Series B, Biological Sciences* **270**:179–84.

Massel, S.R. and Done, T.J. (1993). Effects of cyclone waves on massive coral assemblages on the Great Barrier Reef: Meteorology, hydrodynamics and demography. *Coral Reefs* **12**:153–66.

Mathews, E., Veitayaki, J. and Bidesi, V.R. (1998). Fijian villagers adapt to changes in local fisheries. *Ocean and Coastal Management* **38**:207–24.

Maxwell, K. and Johnson, G.N. (2000). Chlorophyll fluorescence: A practical guide. *Journal of Experimental Botany* **51**:659–68.

Mayer, A.G. (1915). The lower temperature at which reef corals lose their ability to capture food. *Carnegie Institute of Washington Year Book* **14**:212.

Mayfield, A.B. and Gates, R.D. (2007). Osmoregulation in anthozoan–dinoflagellate symbiosis. *Comparitive Biochemistry Physiology A* **147**:1–10.

McClanahan, T.R. (1994). Coral-eating snail *Drupella cornus* population increases in Kenyan coral reef lagoons. *Marine Ecology Progress Series* **115**:131–7.

McClanahan, T.R. (2000). Recovery of a coral reef keystone predator, *Balistapus undulatus*, in East African marine parks. *Biological Conservation* **94**:191–8.

McClanahan, T.R., Ateweberhan, M., Muhando, C., Maina, J. and Mohammed, S.M. (2007). Effects of climate and seawater temperature variation on coral bleaching and mortality. *Ecological Monographs* **74**:503–25.

McClanahan, T.R., Ateweberhan, M. and Omukoto, J. (2008). Long-term changes in coral colony size distributions on Kenyan reefs under different management regimes and across the 1998 bleaching event. *Marine Biology* **152**:755–68.

McClanahan, T.R., Hicks, C.C. and Darling, E.S. (2008). Malthusian overfishing and efforts to overcome it on Kenyan coral reefs. *Ecological Applications* **18**:1516–29.

McClanahan, T.R. and Obura, D. (1997). Sedimentation effects on shallow coral communities in Kenya. *Journal of Experimental Marine Biology Ecology* **209**:103–22.

McConnaughey, T. (1991). Calcification in *Chara corallina*: CO_2 hydroxylation generates protons for bicarbonate assimilation. *Limnology and Oceanography* **36**:619–28.

McConnaughey, T. and Whelan, J.F. (1997). Calcification generates protons for nutrient and bicarbonate uptake. *Earth Science Reviews* **42**:95–117.

McCook, L.J. (1997). Effects of herbivory on zonation of *Sargassum* spp. within fringing reefs of the central Great Barrier Reef. *Marine Biology* **129**:713–22.

McKenney, B. (2005). *Environmental Offset Policies, Principles, and Methods: A Review of Selected Legislative Frameworks*. Available at <http://www.issuelab.org/resource/environmental_offset_policies_principles_and_methods_a_review_of_selected_legislative_frameworks>.

McLaughlin, C.J., Smith, C.A., Buddemeier, R.W., Bartley, J.D. and Maxwell, B.A. (2003). Rivers, runoff, and reefs. *Global and Planetary Change* **39**:191–9.

McManus, J.W., Reyes, R.B. and Nanola, C.L. (1997). Effects of some destructive fishing methods on coral cover and potential rates of recovery. *Environmental Management* **21**:69–78.

McMurray, S.E., Johnson, Z.I., Hunt, D.E., Pawlik, J.R. and Finelli, C.M. (2016). Selective feeding by the giant barrel sponge enhances foraging efficiency. *Limnology and Oceanography* **61**:1271–86.

Mees, C.C., Pilling, G.M. and Barry, C.J. (1999). Commercial inshore fishing activity in the British Indian Ocean Territory. In: C.R.C Sheppard and M.R.D. Seward (eds), *Ecology of the Chagos Archipelago*. Westbury Academic and Scientific Publishing, pp. 327–46.

Mellin, C., MacNeil, M.A., Cheal, A.J., Emslie, M.J. and Caley, M.J. (2016). Marine protected areas increase resilience among coral reef communities. Ecology Letters **19**:629–37.

Miller, D.J. and Yellowlees, D. (1989). Inorganic nitrogen uptake by symbiotic marine cnidarians: A critical review. *Proceedings of the Royal Society of London. Series B, Biological Sciences* **237**:109–25.

Miller, W.I., Montgomery, R.T. and Collier, A.W. (1977). A taxonomic survey of the diatoms associated with Florida Keys coral reefs. *Proceedings of the 3rd International Coral Reef Symposium* **1**:349–55.

Milligan, R.J., Spence, G., Roberts, J.M. and Bailey, D.M. (2016). Fish communities associated with cold-water corals vary with depth and habitat type. *Deep Sea Research Part I* **114**:43–54.

Mills, M.M. and Sebens, K.P. (2004). Ingestion and assimilation of nitrogen from benthic sediments by three species of coral. *Marine Biology* **145**:1097–106.

Mitchell, A.W. and Furnas, M.J. (1997). Terrestrial inputs of nutrients and suspended sediments to the GBR Lagoon. In: Great Barrier Reef Marine Park Authority, *The Great Barrier Reef: Science, Use and Management: A National Conference*, Vol. 1. Great Barrier Reef Marine Park Authority and CRC Reef Research, pp. 59–71.

Mitchell, A.W., Reghenzani, J., Hunter, H.M. and Bramley, R.G.V. (1996). Water quality and nutrient fluxes from river systems draining to the Great Barrier Reef. In: Hunter, H.M., Eyles, A.G. and Rayment, G.E. (eds), *Downstream Effects of Land Use*. Queensland Department of Natural Resources, pp. 23–34.

Moberg, F. and Folke, C. (1999). Ecological goods and services of coral reef ecosystems. *Ecological Economics* **29**:215–33.

Moberg, F., Nyström, M., Kautsky, N., Tedengren, M. and Jarayabhand, P. (1997). Effects of reduced salinity on the rates of photosynthesis and respiration in the hermatypic corals *Porites lutea* and *Pocillopora damicornis*. *Marine Ecology Progress Series* **157**:53–9.

Moland, E., Eagle, J.V. and Jones, G.P. (2005). Ecology and evolution of mimicry in coral reef fishes. *Oceanography and Marine Biology* **43**:457–84.

Monismith, S.G., Rogers, J.S., Koweek, D. and Dunbar, R.B. (2015). Frictional wave dissipation on a remarkably rough reef. *Geophysical Research Letters* **42**:4063–71.

Moore, R.B., Ferguson, K.M., Loh, W.K.W., Hoegh-Guldberg, O., and Carter, D.A. (2003). Highly organized structure in the non-coding region of the psbA minicircle from clade C *Symbiodinium*. *International Journal of Systematic and Evolutionary Microbiology* **53**: 1725–34.

Mora, C. (2015). *Ecology of Fishes on Coral Reefs*. Cambridge University Press.

Mora, C., Adrefouet, S., Costello, M.J., Kranenburg, C., Rollo, A., Veron, J., et al. (2006). Coral reefs and the global network of marine protected areas. *Science* **312**:1750–1.

Mora, C., Graham, N.A.J. and Nyström, M. (2016). Ecological limitations to the resilience of coral reefs. *Coral Reefs*. **35**:1271–80.

Moran, P.J. (1986). The *Acanthaster* phenomenon. *Oceanography And Marine Biology* **24**:379–480.

Moriarty, D.J.W. and Hansen, J.A. (1990). Productivity and growth rates of coral reef bacteria on hard calcareous substrates and in sandy sediments in summer. *Australian Journal Marine Freshwater Research* **41**:785–94.

Moriarty, D.J.W., Pollard, P.C., Alongi, D.M., Wilkinson, C.R. and Gray, J.S. (1985). Bacterial productivity and trophic relationships with consumers on a coral reef (Mecor I). *Proceedings of the 5th International Coral Reef Symposium* **3**:457–62.

Moriarty, D.J.W., Pollard, P.C., Hunt, W.G., Moriarty, C.M. and Wassenberg, T.J. (1985). Productivity of bacteria and microalgae and the effect of grazing by holothurians in sediments on a coral reef flat. *Marine Biology* **85**:293–300.

Morrison-Gardiner, S. (2002). Dominant fungi from Australian coral reefs. *Fungal Diversity* **9**:105–21.

Morse, A.N.C., Iwao, K., Baba, M., Shimoike, K., Hayashibara, T., Omori, M. (1996). An ancient chemosensory mechanism brings new life to coral reefs. *Biological Bulletin* **191**:149–54.

Morse, A.N.C and Morse, D.E. (1996). Flypapers for coral and other planktonic larvae. *BioScience* **46**:254–62.

Mouillot, D., Villéger, S., Parravicini, V., Kulbicki, M., Ernesto Arias-González, J., Bender, M., et al. (2014). Functional over-redundancy and high functional vulnerability in global fish faunas on tropical reefs. *Proceedings of the National Academy of Sciences of the United States of America* **111**:13757–62.

Muir, P.R., Wallace, C.C., Done, T. and Aguirre, J.D. (2015). Limited scope for latitudinal extension of reef corals. *Science* **348**:1135–8.

Muller-Parker, G. (1984). Dispersal of zooxanthellae on coral reefs by predators on cnidarians. *Biological Bulletin* **167**:159–67.

Muller-Parker, G. and D'Elia, C.F. (1997). Interactions between corals and their symbiotic algae. In: C. Birkeland (ed.), *Life and Death of Coral Reefs*. Chapman & Hall, pp. 96–105.

Muller-Parker, G., Cook, C.B. and D'Elia, C.F. (1990). Feeding affects phosphate fluxes in the symbiotic sea anemone *Aiptasia pallida*. *Marine Ecology Progress Series* **60**:283–90.

Muller-Parker, G. and Davy, S.K. (2001). Temperate and tropical algal–sea anemone symbioses. *Invertebrate Biology* **120**:104–23.

Muller-Parker, G., Lee, K.W. and Cook, C.B. (1996). Changes in the ultrastructure of symbiotic zooxanthellae (*Symbiodinium* sp, Dinophyceae) in fed and starved sea anemones maintained under high and low light. *Journal of Phycology* 32:987–94.

Mumby, P.J., Dahlgren, C.P., Harborne, A.R., Kappel, C.V., Micheli, F., Brumbaugh, D.R., et al. (2006). Fishing, trophic cascades, and the process of grazing on coral reefs. *Science* 311:98–101.

Mumby, P.J., Green, E.P., Edwards, A.J. and Clark, C.D. (1997). Coral reef habitat mapping: How much detail can remote sensing provide? *Marine Biology* 130:193–202.

Mumby, P.J., Harborne, A.R., Hedley, J.D., Zychaluk, K. and Blackwell, P.G. (2006). Revisiting the catastrophic die-off of the urchin *Diadema antillarum* on Caribbean coral reefs: Fresh insights on resilience from a simulation model. *Ecological Modelling* 196:131–48.

Mumby, P.J., Hastings, A. and Edwards, H.J. (2007). Thresholds and the resilience of Caribbean coral reefs. *Nature* 450:98–101.

Mumby, P.J. and Steneck, R.S. (2008). Coral reef management and conservation in light of rapidly-evolving ecological paradigms. *Trends in Ecology and Evolution* 23:555–63.

Mumby, P.J., Wolff, N.H., Bozec, Y.M., Chollett, I. and Halloran, P.R. (2014). Operationalizing the resilience of coral reefs in an era of climate change. *Conservation Letters* 7:176–87.

Munday, P.L. (2004). Habitat loss, resource specialisation, and extinction on coral reefs. *Global Change Biology* 10:1642–7.

Munday, P.L., Jones, G.P. and Caley, M.J. (1997). Habitat specialisation and the distribution and abundance of coral-dwelling gobies. *Marine Ecology Progress Series* 152:227–39.

Munday, P.L., Jones, G.P. and Caley, M.J. (2001). Interspecific competition and coexistence in a guild of coral-dwelling fishes. *Ecology* 82:2177–89.

Munday, P.L., Jones, G.P., Pratchett, M.S. and Williams, A.J. (2008). Climate change and the future for coral reef fishes. *Fish and Fisheries* 9:1–25.

Munday, P.L., Jones, G.P., Sheaves, M., Williams, A.J. and Hoby, G. (2007). Vulnerability of fishes on the Great Barrier Reef to climate change. In: J. Johnson and P. Marshall (eds), *Climate Change and the Great Barrier Reef*. Great Barrier Reef Marine Park Authority, pp. 357–92.

Munk, W.H. and Sargent, M.C. (1954). Adjustment of Bikini Atoll to ocean waves. *U.S. Geological Survey Professional Paper* 260:275–80.

Murray, S.P., Roberts, H.H., Conlon, D.M. and Rudder, G.M. (1977). Nearshore current fields around coral islands: Control on sediment accumulation and reef growth. *Proceedings of the 3rd International Coral Reef Symposium* 2:53–9.

Muscatine, L. (1967). Glycerol excretion by symbiotic algae from corals and *Tridacna* and its control by the host. *Science* 156:516–19.

Muscatine, L. and Cernichiari, E. (1969). Assimilation of photosynthetic products of zooxanthellae by a reef coral. *Biological Bulletin* 137:506–23.

Muscatine, L. and D'Elia, C.F. (1978). The uptake, retention, and release of ammonium by reef corals. *Limnology and Oceanography* 23:725–34.

Muscatine, L., Falkowski, P.G., Dubinsky, Z., Cook, P.A. and McCloskey, L.R. (1989). The effect of external nutrient resources on the population dynamics of zooxanthellae in a reef coral. *Proceedings of the Royal Society of London. Series B, Biological Sciences* 236:311–24.

Muscatine, L., Falkowski, P.G., Porter, J.W. and Dubinsky, Z. (1984). Fate of photosynthetically fixed carbon in light- and shade-adapted colonies of the symbiotic coral *Stylophora pistillata*. *Proceedings of the Royal Society of London. Series B, Biological Sciences* **222**:181–202.

Muscatine, L., Gates, R.D. and Lafontaine, I. (1994). Do symbiotic dinoflagellates secrete lipid droplets? *Limnology and Oceanography* **39**:925–9.

Muscatine, L., Goiran, C., Land, L., Jaubert, J., Cuif, J.P. and Allemand, D. (2005). Stable isotopes ($\delta^{13}C$ and $\delta^{15}N$) of organic matrix from coral skeleton. *Proceedings of the National Academy of Sciences of the United States of America* **102**:1525–30.

Muscatine, L. and Hand, C. (1958). Direct evidence for the transfer of materials from symbiotic algae to the tissues of a coelenterate. *Proceedings of the National Academy of Sciences of the United States of America* **44**:1259–63.

Muscatine, L., McCloskey, L.R. and Marian, R.E. (1981). Estimating the daily contribution of carbon from zooxanthellae to coral animal respiration. *Limnology and Oceanography* **26**:601–11.

Muthiga, N.A. and Szmant, A.M. (1987). The effects of salinity stress on the rates of aerobic respiration and photosynthesis in the hermatypic coral *Siderastrea siderea*. *Biological Bulletin* **173**:539–51.

Muyzer, G., Brinkhoff, T., Nübel, U., Santegoeds, C., Schäfer, H. and Wawer, C. (1987). Denaturing gradient gel electrophoresis (DGGE) in microbial ecology. In: A.D.L. Akkermans, J.D. van Elsas and F.J. de Bruijn (eds), *Molecular Microbial Ecology Manual*, Section 3.4.4. Kluwer Academic Publishers, pp. 1–27.

Nadon, M.O., Baum, J.K., Williams, I.D., McPherson, J.M., Zgliczynski, B.J., Richards, B.L., et al. (2012). Re-creating missing population baselines for Pacific reef sharks. *Conservation Biology* **26**:493–503.

Nahon, S., Richoux, N.B., Kolasinski, J., Desmalades, M., Ferrier-Pagès, C.F., Lecellier, G., et al. (2013). Spatial and temporal variations in stable carbon ($\delta^{13}C$) and nitrogen ($\delta^{15}N$) isotopic composition of symbiotic scleractinian corals. *PLOS ONE* **8**:e81247.

Nash, K.L., Graham, N.A.J., Jennings, S., Wilson, S.K. and Bellwood, D.R. (2016). Herbivore cross-scale redundancy supports response diversity and promotes coral reef resilience. *Journal of Applied Ecology* **53**:646–55.

Nash, K.L., Welsh, J.Q., Graham, N.A.J. and Bellwood, D.R. (2015). Home-range allometry in coral reef fishes: Comparison to other vertebrates, methodological issues and management implications. *Oecologia* **177**:73–83.

Negri, A.P. and Heyward, A.J. (2000). Inhibition of fertilization and larval metamorphosis of the coral *Acropora millepora* (Ehrenberg, 1834) by petroleum products. *Marine Pollution Bulletin* **41**:420–7.

Neil, D.T. (1996). Sediment concentrations in streams and coastal waters in the North Queensland humid tropics: Land use, rainfall and wave resuspension contributions. In: H.M. Hunter, A.G. Eyles and G.E. Rayment (eds), *Downstream Effects of Land Use*. Queensland Department of Natural Resources, pp. 97–101.

Neil, D.T., Orpin, A.R., Ridd, P.V. and Yu, B. (2002). Sediment yield and impacts from river catchments to the Great Barrier Reef lagoon. *Marine Freshwater Research* **53**:733–52.

Newton, K., Cote, I.M., Pilling, G.M., Jennings, S. and Dulvy, N.K. (2007). Current and future sustainability of island coral reef fisheries. *Current Biology* **17**:1–4.

Nguyen-Kim, H., Bouvier, T., Bouvier, C., Doan-Nhu, H., Nguyen-Ngoc, L., Rochelle-Newall, E., et al. (2014). High occurrence of viruses in the mucus layer of scleractinian corals. *Environmental Microbiology Reports* **6**:675–82.

Niebuhr, M. (1792). *Travels through Arabia and Other Countries in the Far East, Performed by M. Niebuhr, Now a Captain of Engineers in the Service of the King of Denmark*, Vol. 1. Libraire du Liban.

Nitschke, M.R., Davy, S.K. and Ward, S. (2016). Horizontal transmission of *Symbiodinium* cells between adult and juvenile corals is aided by benthic sediment. *Coral Reefs* **35**:335–44.

NOAA. (2001). *Oil Spills in Coral Reefs: Planning and Response Considerations*. National Oceanic and Atmospheric Administration.

Nugues, M.M. and Bak, R.P.M (2006). Differential competitive abilities between Caribbean coral species and a brown alga: A year of experiments and a long-term perspective. *Marine Ecology Progress Series* **315**:75–86.

Odum, H.T. and Odum, E.P. (1955). Trophic structure and productivity of a windward coral reef community on Eniwetok Atoll. *Ecological Monographs* **25**:291–320.

Orpin, A.R. and Ridd, P.V. (2012). Exposure of inshore corals to suspended sediments due to wave-resuspension and river plumes in the central Great Barrier Reef: A reappraisal. *Continental Shelf Research* **47**:55–67.

Palumbi, S.R. (1997). Molecular biogeography of the Pacific. *Coral Reefs* **16**:47–52.

Palumbi, S.R., Barshis, D.J., Traylor-Knowles, N. and Bay, R.A. (2014). Mechanisms of reef coral resistance to future climate change. *Science*, **344**:895–8.

Pandolfi, J.M., Jackson, J.B.X.C., Baron, N., Bradbury, R.H., Guzman, H.M., Hughes, T.P., Kappel, C.V., et al. (2005). Are US coral reefs on the slippery slope to slime? *Science* **307**:1725–6.

Paracer, S. and Ahmadjian, V. (2000). *Symbiosis: An Introduction to Biological Associations*. Oxford University Press.

Pari, N., Peyrot-Clausade, M. and Hutchings, P.A. (2002). Bioerosion of experimental substrates on high islands and atoll lagoons (French Polynesia) five years of exposure. *Journal of Experimental Marine Biology and Ecology* **276**:109–27.

Parkinson, J.E. and Coffroth, M.A. (2015). New species of Clade B *Symbiodinium* (Dinophyceae) from the Greater Caribbean belong to different functional guilds: *S. aenigmaticum* sp. nov., *S. antillogorgium* sp. nov., *S. endomadracis* sp. nov., and *S. pseudominutum* sp. nov. *Journal of Phycology* **51**:850–8.

Parmentier, E., Berten, L., Rigo, P., Aubrun, F., Nedelec, S.L., Simpson, S.D., et al. (2015). The influence of various reef sounds on coral-fish larvae behavior. *Journal of Fish Biology* **86**:1507–18.

Parmentier, E. and Das, K. (2004). Commensal vs. parasitic relationship between Carapini fish and their hosts: Some further insight through δ^{13} C and δ^{15} N measurements. *Journal of Experimental Marine Biology Ecology* **310**:47–58.

Parmentier, E. and Vandewalle, P. (2005). Further insight on carapid–holothuroid relationships. *Marine Biology* **146**:455–65.

Patten, N.L., Harrison, P.L. and Mitchell, J.G. (2008). Prevalence of virus-like particles within the staghorn coral (*Acropora muricata*) from the Great Barrier Reef. *Coral Reefs*: **27**:569–80.

Patten, N.L., Mitchell, J.G., Middelboe, M., Eyre, B.D., Seuront, L., Harrison, P.L., et al. (2008). Bacterial and viral dynamics during a mass coral spawning period on the Great Barrier Reef. *Aquatic Microbial Ecology* **50**:209–20.

Patten, N.L., Seymour, J.R. and Mitchell, J.G. (2006). Flow cytometric analysis of virus-like particles and heterotrophic bacteria within coral-associated reef water. *Journal of the Marine Biology Association of the United Kingdom* **86**:563–6.

Paul, J.H., Rose, J.B., Jiang, S.C., Kellogg, C.A. and Dickson, L. (1993). Distribution of viral abundance in the reef environment of Key Largo, Florida. *Applied Environmental Microbiology* **59**:718–24.

Paulay, G and Meyer, C. (2006). Dispersal and divergence across the greatest ocean region: Do larvae matter? *Integrative and Comparative Biology* **46**:269–81.

Pauly, D. (1995). Anecdotes and the shifting baseline syndrome. *Trends in Ecology and Evolution* **10**:430.

Pearse, V.B. (1971). Sources of carbon in the skeleton of the coral *Fungia scutaria*. In: Lenhoff, H.M. and Muscatine, L. (eds), *Experimental Coelenterate Biology*. University of Hawaii Press, pp. 239–45.

Peng, S.-E., Chen, W.-N.U., Chen, H.-K., Lu, C.-Y., Mayfield, A.B., Fang, L.-S., et al. (2011). Lipid bodies in coral-dinoflagellate endosymbiosis: Proteomic and ultrastructural studies. *Proteomics* **11**:3540–55.

Perry, C.T. and Morgan, K.M. (2016). Bleaching drives collapse in reef carbonate budgets and reef growth potential on southern Maldives reefs. *Scientific Reports* **7**:40581.

Perry, C.T., Murphy, G.N., Graham, N.A.J., Wilson, S.K., Januchowski-Hartley, F.A. and East, H.K. (2015). Remote coral reefs can sustain high growth potential and may match future sea-level trends. *Scientific Reports* **5**:18289.

Perry, C.T., Smithers, S.G., Kench, P.S. and Pears, B. (2014). Impacts of Cyclone Yasi on nearshore, terrigenous sediment-dominated reefs of the central Great Barrier Reef, Australia. *Geomorphology* **222**:92–105.

Perry, C.T., Steneck, R.S., Murphy, G.N., Kench, P.S., Edinger, E.N., Smithers, S.G., et al. (2015). Regional-scale dominance of non-framework building corals on Caribbean reefs affects carbonate production and future reef growth. *Global Change Biology* **21**:1153–64.

Peters, E.C. (1997). Diseases of coral reef organisms. In: C. Birkeland (ed.), *Life and Death of Coral Reefs*. Chapman & Hall, pp. 114–39.

Pettay, D.T., Wham, D.C., Smith, R.T., Iglesias-Prieto, R. and LaJeunesse, T.C. (2015). Microbial invasion of the Caribbean by an Indo-Pacific coral zooxanthella. *Proceedings of the National Academy of Sciences of the United States of America* **112**:7513–18.

Pichon, M. (1978). Recherches sur les peuplements a dominance d'anthozoaires dans les recifs coralliens de Tulear (Madagascar). *Atoll Research Bulletin* **222**:1–447.

Pickard, G.L. (1986). Effects of wind and tide on upper-layer currents at Davies Reef, Great Barrier Reef, during Mecor (July–August 1984). *Australian Journal Marine Freshwater Research* **37**:545–65.

Pierrehumbert, R.T. (1995). Thermostats, radiator fins and the local runaway greenhouse. *Journal Atmospheric Sciences* **52**:1784–806.

Pile, A.J. (1997). Finding Reiswig's missing carbon: Quantification of sponge feeding using dual-beam flow cytometry. *Proceedings of the 8th International Coral Reef Symposium* **2**:1403–10.

Pile, A.J., Grant, A., Hinde, R. and Borowitzka, M.A. (2003). Heterotrophy on ultraplankton communities is an important source of nitrogen for a sponge–rhodophyte symbiosis. *Journal Experimental Biology* **206**:4533–8.

Pile, A.J., Patterson, M.R. and Witman, J.D. (1996). In situ grazing on plankton <10 μm by the boreal sponge *Mycale lingua*. *Marine Ecology Progress Series* **141**:95–102.

Piller, W.E. and Kleemann, K. (1992). Distribution and composition of coral reefs in and outside the northern bay of Safaga, Red Sea, Egypt. *Proceedings of the 7th International Coral Reef Symposium* **1**:582.

Pilling, G.M., Harley, S.J., Nicol, S., Williams, P. and Hampton, J. (2015). Can the tropical Western and Central Pacific tuna purse seine fishery contribute to Pacific Island population food security? *Food Security* **7**:67–81.

Piniak, G.A. and Lipschultz, F. (2004). Effects of nutritional history on nitrogen assimilation in congeneric temperate and tropical scleractinian corals. *Marine Biology* **145**:1085–96.

Pochon, X. and Gates, R.D. (2010). A new *Symbiodinium* clade (Dinophyceae) from soritid foraminifera in Hawai'i. *Molecular Phylogenetics and Evolution* **56**:492–7.

Pomeroy, L.R. (1974). The ocean's food web, a changing paradigm. *BioScience* **24**:499–504.

Pomeroy, L.R. and Kuenzler, E.J. (1969). Phosphorus turnover by coral reef animals. In: D.J. Nelson and F.C. Evans (eds), *Proceedings of the Second Annual Symposium on Radioecology*. Clearinghouse for Federal Scientific and Technical Information, National Bureau of Standards, U.S. Department of Commerce, pp. 474–82.

Pomeroy, L.R., Williams, P.B., Azam, F. and Hobbie, J.E. (2007). The microbial loop. *Oceanography* **20**:28–33.

Pomeroy, R.S. and Berkes, F. (1997). Two to tango: The role of government in fisheries co-management. *Marine Policy* **21**:465–80.

Porat, D. and Chadwick-Furman, N.E. (2004). Effects of anemonefish on giant sea anemones: Expansion behavior, growth, and survival. *Hydrobiologia* **530**:513–20.

Porat, D. and Chadwick-Furman, N.E. (2005). Effects of anemonefish on giant sea anemones: Ammonium uptake, zooxanthella content and tissue regeneration. *Marine and Freshwater Behaviour and Physiology* **38**:43–51.

Porter, J.W. (1974). Zooplankton feeding by the Caribbean reef-building coral *Montastrea cavernosa*. *Proceedings of the 2nd International Coral Reef Symposium* **2**:111–25.

Porter, J.W., Battey, J. and Smith, G.J. (1982). Perturbation and change in coral reef communities. *Proceedings of the National Academy of Science of the United States* **79**:1678–81.

Porter, J.W., Muscatine, L., Dubinsky, Z. and Falkowski, P.G. (1984). Primary production and photoadaptation in light-adapted and shade-adapted colonies of the symbiotic coral, *Stylophora pistillata*. *Proceedings of the Royal Society of London. Series B, Biological Sciences* **222**:161–80.

Porter, J.W. and Porter, K.G. (1977). Quantitative sampling of demersal plankton migrating from different substrates. *Limnology and Oceanography* **22**:553–6.

Porter, J.W. and Tougas, J.I. (2001). Reef ecosystems: Threats to their biodiversity. In: S.A. Levin (ed.), *Encyclopedia of Biodiversity*, Vol. 5. Academic Press, pp. 73–95.

Pratchett, M.S., Caballes, C., Rivera-Posada, J.A. and Sweatman, H.P.A. (2014). Limits to understanding and managing outbreaks of crown-of-thorns starfish (*Acanthaster* spp.). *Oceanography and Marine Biology* **52**:133–200.

Pratchett, M.S., Munda, P.L., Wilson, S.K., Graham, N.A.J., Cinner, J.E., Bellwood, D.R., et al. (2008). Effects of climate-induced coral bleaching on coral-reef fishes: Ecological and economic consequences. *Oceanography and Marine Biology* **46**:251–96.

Pratchett, M.S., Wilson, S.K. and Baird, A.H. (2006). Declines in the abundance of Chaetodon butterflyfishes following extensive coral depletion. *Journal of Fish Biology* **69**:1269–80.

Pratchett, M.S., Wilson, S.K., Berumen, M.L. and McCormick, M.I. (2004). Sublethal effects of coral bleaching on an obligate coral feeding butterflyfish. *Coral Reefs* **23**:352–6.

Precht, W.F. (2006). *Coral Reef Restoration Handbook.* Taylor and Francis.

Precht, W.F. and Aronson, R.B. (2004). Climate flickers and range shifts of reef corals. *Frontiers in Ecology and the Environment* **2**:307–14.

Precht, W.F. and Aronson, R.B. (2006). Death and resurrection of Caribbean coral reefs: A palaeoecological perspective. In: I.M. Cote and J.D. Reynolds (eds), *Coral Reef Conservation.* Cambridge University Press, pp. 40–77.

Preston, G.L. (1997). Exploitation, ecology and management of fisheries for sea cucumbers (bêche-de-mer). Paper presented at the Sea Cucumber (Bêche-de-Mer) Fishery Management Workshop, Brisbane, 8–9 December 1997.

Price, A.R.G., Evans, L.E., Rowlands, N. and Hawkins, J.P. (2013). Negligible recovery in Chagos holothurians (sea cucumbers). *Aquatic Conservation: Marine and Freshwater Ecosystems* **23**: 811–19.

Price, A.R.G., Harris, A., McGowan, A., Venkatachalam, A. and Sheppard, C.R.C. (2010). Chagos feels the pinch: Assessment of holothurian (sea cucumber) abundance, illegal harvesting and conservation prospects in British Indian Ocean Territory. *Aquatic Conservation: Marine and Freshwater Ecosystems* **20**:117–26.

Purcell, S.W., Eriksson, H. and Byrne, M. (2016). Rotational zoning systems in multispecies sea cucumber fisheries. *SPC Bêche-de-Mer Information Bulletin* **36**:3–8.

Purdy, E.G. (1974). Reef configurations: Cause and effect. In: L.F. Laporte (ed.), *Reefs in Time and Space.* Special Publication 18. Society of Economic Paleontologists and Mineralogists, pp. 9–76.

Purkis, S.J. (2005). A 'reef up' approach to classifying coral habitats from IKONOS imagery. *IEEE Transactions on Geoscience and Remote Sensing* **43**:1375–90.

Purkis, S.J., Graham, N.A.J. and Riegl, B.M. (2007). Predictability of reef fish diversity and abundance using remote sensing data in Diego Garcia (Chagos Archipelago). *Coral Reefs* **27**:167–78.

Purkis, S.J. and Riegl, B. (2005). Spatial and temporal dynamics of Arabian Gulf coral assemblages quantified from remote-sensing and in-situ monitoring data. *Marine Ecology Progress Series* **287**:99–113.

Putnam, H.M., Stat, M., Pochon, X. and Gates, R.D. (2012). Endosymbiotic flexibility associates with environmental sensitivity in scleractinian corals. *Proceedings of the Royal Society of London. Series B, Biological Sciences* **279**:4352–61.

Radford, C.A., Stanley, J.A. and Jeffs, A.G. (2014). Adjacent coral reef habitats produce different underwater sound signatures. *Marine Ecology Progress Series* **505**:19–28.

Raikar, V. and Wafar, M. (2006). Surge ammonium uptake in macroalgae from a coral atoll. *Journal of Experimental Marine Biology and Ecology* **339**:236–40.

Raina, J.B., Tapiolas, D., Motti, C.A., Foret, S., Seemann, T., Tebben, J., et al. (2016). Isolation of an antimicrobial compound produced by bacteria associated with reef-building corals. *PeerJ* **4**:e2275.

Ralph, P.J., Larkum, A.W.D. and Kuhl, M. (2007). Photobiology of endolithic microorganisms in living coral skeletons: 1. Pigmentation, spectral reflectance and variable chlorophyll fluorescence analysis of endoliths in the massive corals *Cyphastrea serailia*, *Porites lutea* and *Goniastrea australensis*. *Marine Biology* **152**:395–404.

Randall, J.E. and Randall, H.A. (1960). Examples of mimicry and protective resemblance in tropical marine fishes. *Bulletin of Marine Science* **10**:444–80.

Rasher, D.B., Hoey, A.S. and Hay, M.E. (2013). Consumer diversity interacts with prey defenses to drive ecosystem function. *Ecology* **94**:1347–58.

Reading, J.E., Myers-Miller, R.L., Baker, D.M., Fogel, M., Raymundo, L.J. and Kim, K. (2013). Link between sewage-derived nitrogen pollution and coral disease severity in Guam. *Marine Pollution Bulletin* **73**:57–63.

Reis, J.B., Stanley, S.M. and Hardie, L.A. (2006). Scleractinian corals produce calcite, and grow more slowly, in artificial Cretaceous seawater. *Geology* **34**:525–8.

Reiswig, H.M. (1971). Particle feeding in natural populations of three marine demosponges. *Biological Bulletin* **141**:568–91.

Renema, W., Bellwood, D.R., Braga, J.C., Bromfield, K., Hall, R., Johnson, K.G., et al. (2008). Hopping hotspots: Global shifts in marine biodiversity. *Science* **321**:654–7.

Richards, Z.T., Beger, M., Pinca, S. and Wallace, C.C. (2008). Bikini Atoll coral biodiversity five decades after nuclear testing. *Marine Pollution Bulletin* **56**:503–15.

Richardson, L.L. (2004). Black band disease. In: E. Rosenberg and Y. Loya (eds), *Coral Health and Disease*. Springer, pp.323–36.

Riegl, B. (2002). Effects of the 1996 and 1998 positive sea-surface temperature anomalies on corals, coral diseases and fish in the Arabian Gulf (Dubai, UAE). *Marine Biology* **140**:29–40.

Riegl, B. and Branch, G.M. (1995). Effects of sediment on the energy budgets of four scleractinian (Bourne 1900) and five alcyonacean (Lamouroux 1816) corals. *Journal of Experimental Marine Biology Ecology* **186**:259–75.

Riegl, B. and Piller, W.E. (2003). Possible refugia for reefs in times of environmental stress. *International Journal Earth Science* **92**:520–31.

Rinkevich, B. (1989). The contribution of photosynthetic products to coral reproduction. *Marine Biology* **101**:259–63.

Rinkevich, B. (2008). Management of coral reefs: We have gone wrong when neglecting active reef restoration. *Marine Pollution Bulletin* **56**:1821–4.

Rioja Nieto, R. and Sheppard, C.R.C. (2008). Effects of management strategies on the landscape ecology of a Marine Protected Area. *Ocean and Coastal Management* **51**:397–404.

Risk, M.J. and Sluka, R. (2000). The Maldives: A nation of atolls. In: T.R. McClanaham, C.R.C. Sheppard, and D.O. Obura (eds), *Coral Reefs of the Indian Ocean: Their Ecology and Conservation*. Oxford University Press, pp. 325–51.

Ritchie, K.B. and Smith, G.W. (1997). Physiological comparisons of bacterial communities from various species of scleractinian corals. *Proceedings of the 8th International Coral Reef Symposium* **1**:521–6.

Ritchie, K.B. and Smith, G.W. (2004). Microbial communities of coral surface mucopolysaccharide layers. In: E. Rosenberg and Y. Loya (eds), *Coral Health and Disease*. Springer, pp. 259–64.

Rivera-Posada, J., Pratchett, M.S., Aguilar, C., Grand, A. and Caballes, C.F. (2014). Bile salts and the single-shot lethal injection method for killing crown-of-thorns sea stars (*Acanthaster planci*). *Ocean and Coastal Management* **102**:383–90.

Roberts, C.M. (1995). Rapid build-up of fish biomass in a Caribbean marine reserve. *Conservation Biology* **9**:816–26.

Roberts, C.M. (1997). Connectivity and management of Caribbean coral reefs. *Science* **278**:1454–7.

Roberts, C.M. (2003). Our shifting perspective on the oceans. *Oryx* **37**:166–77.

Roberts, C.M. (2007). *The Unnatural History of the Seas*. Island Press.

Roberts, C.M., McClean, C.J., Veron, J.E., Hawkins, J.P., Allen, G.R., McAllister, D.E., et al. (2002). Marine biodiversity hotspots and conservation priorities for tropical reefs. *Science* **295**:1280–4.

Roberts, H.H., Murray, S.P. and Suhayda, J.N. (1977). Physical processes on a fore-reef shelf environment. *Proceedings of the 3rd International Coral Reef Symposium* **2**:507–15.

Roberts, H.H. and Suhayda, J.N. (1983). Wave–current interactions on a shallow reef (Nicaragua, Central America). *Coral Reefs* **1**:209–14.

Roberts, H.H., Wilson, P.A. and Lugo-Fernández, A. (1992). Biologic and geologic responses to physical processes: Examples from modern reef systems of the Caribbean-Atlantic region. *Continental Shelf Research* **12**:809–34.

Roberts, J.M., Fixter, L.M. and Davies, P.S. (2001). Ammonium metabolism in the symbiotic sea anemone *Anemonia viridis*. *Hydrobiologia* **461**:25–35.

Roberts, J.M., Wheeler, A.J., Freiwald, A. and Cairns, S.D. (2009). *Cold-Water Corals: The Biology and Geology of Deep-Sea Coral Habitats*. Cambridge University Press.

Rocha, L.A. and Bowen, B.W. (2008). Speciation in coral reef fishes. *Journal of Fish Biology* **72**:1101–21.

Rodriguez-Lanetty, M., Loh, W., Carter, D. and Hoegh-Guldberg, O. (2001). Latitudinal variability in symbiont specificity within the widespread scleractinian coral *Plesiastrea versipora*. *Marine Biology* **138**:1175–81.

Rodríguez-Troncoso, A.P., Carpizo-Ituarte, E., Pettay, D.T., Warner, M.E. and Cupul-Magana, A.L. (2014). The effects of an abnormal decrease in temperature on the Eastern Pacific reef-building coral *Pocillopora verrucosa*. *Marine Biology* **161**:131–9.

Rogers, A., Harbourne, A.R., Brown, C.J., Bozec, Y.-M., Castro, C., Chollett, I., et al. (2014). Anticipative management for coral reef ecosystem services in the 21st century. *Global Change Biology* **21**:504–14.

Rogers, C.S. (1990). Responses of reef corals and organisms to sedimentation. *Marine Ecology Progress Series* **62**:185–202.

Rogers, C.S. and Miller, J. (2006). Permanent 'phase shifts' or reversible declines in coral cover? Lack of recovery of two coral reefs in St John, US Virgin Islands. *Marine Ecology Progress Series* **306**:103–14.

Rohwer, F. (2010). *Coral Reefs in the Microbial Seas*. Plaid Press.

Rohwer, F. and Kelley, S. (2004). Culture-independent analyses of coral-associated microbes. In: E. Rosenberg and Y. Loya (eds), *Coral Health and Disease*. Springer, pp. 265–78.

Rohwer, F., Seguritan, V., Azam, F. and Knowlton, N. (2002). Diversity and distribution of coral-associated bacteria. *Marine Ecology Progress Series* **243**:1–10.

Rosenberg, E. and Loya, Y. (eds). (2004). *Coral Health and Disease*. Springer.

Rosic, N.N., Pernice, M., Dove, S., Dunn, S. and Hoegh-Guldberg, O. (2011). Gene expression profiles of cytosolic heat shock proteins Hsp70 and Hsp90 from symbiotic dinoflagellates in response to thermal stress: Possible implications for coral bleaching. *Cell Stress and Chaperones* **16**:69–80.

Ross, S.W. and Quattrini, A.M. (2007). The fish associated with deep coral banks off the southeastern United States. *Deep Sea Research Part I* **54**:975–1007.

Rougerie, F., Fagerstrom, J.A. and Andrie, C. (1992). Geothermal endo-upwelling: A solution to the reef nutrient paradox? *Continental Shelf Research* **12**:785–98.

Rovelli, L., Attard, K., Bryant, L.D., Flögel, S., Stahl, H.J., Roberts, J.M., et al. (2015). Benthic O_2 uptake of two cold-water coral communities estimated with the non-invasive eddy-correlation technique. *Marine Ecology Progress Series* **525**:97–104.

Rowan, R. (1991). Molecular systematics of symbiotic algae. *Journal of Phycology* **27**:661–6.

Rowan, R. (2004). Coral bleaching: Thermal adaptation in reef coral symbionts. *Nature* **430**:742.

Rowan, R. and Knowlton, N. (1995). Intraspecific diversity and ecological zonation in coral algal symbiosis. *Proceedings of the National Academy of Science of the United States of America* **92**:2850–3.

Rowan, R., Knowlton, N., Baker, A. and Jara, J. (1997). Landscape ecology of algal symbionts creates variation in episodes of coral bleaching. *Nature* **388**:265–9.

Rowan, R. and Powers, D.A. (1991a). A molecular genetic classification of zooxanthellae and the evolution of animal–algal symbioses. *Science* **251**:1348–51.

Rowan, R. and Powers, D.A. (1991b). Molecular genetic identification of symbiotic dinoflagellates (zooxanthellae). *Marine Ecology Progress Series* **71**:65–73.

Rowan, R. and Powers, D.A. (1992). Ribosomal-RNA sequences and the diversity of symbiotic dinoflagellates (zooxanthellae). *Proceedings of the National Academy of Science of the United States of America* **89**:3639–43.

Ruddle, K. (1996). Traditional management of reef fishing. In: N.V.C. Polunin and C.M. Roberts (eds), *Reef Fisheries*. Chapman & Hall, pp. 137–60.

Ruppert, E.E., Fox, R.S. and Barnes, R.D. (2004). *Invertebrate Zoology: A Functional Evolutionary Approach* (7th edition). Thomson, Brooks/Cole.

Ruttenberg, B.J. and Lester, S.E. (2015). Patterns and processes in geographic range size in coral reef fishes. In: C. Mora (ed.), *Ecology of Fishes on Coral Reefs*. Cambridge University Press, pp. 97–103.

Rützler, K. (1978). Sponges in coral reefs. In: D.E. Stoddart and J.E. Johannes (eds), *Coral Reefs: Research Methods*. Monographs on Oceanographic Methodology 5. UNESCO, pp. 299–313.

Sadally, S.B., Taleb-Hossenkhan, N. and Bhagooli, R. (2014). Spatio-temporal variation in density of microphytoplankton genera in two tropical coral reefs of Mauritius. *African Journal of Marine Science* **36**:423–38.

Sadovy, Y.J. (2005). Trouble on the reef: The imperative for managing vulnerable and valuable fisheries. *Fish and Fisheries* **6**:167–85.

Sadovy, Y.J. and Cheung, W.L. (2003). Near extinction of a highly fecund fish: The one that nearly got away. *Fish and Fisheries* **4**:86–99.

Sadovy, Y.J. and Domeier, M.L. (2005). Are aggregation fisheries sustainable? Reef fish fisheries as a case study. *Coral Reefs* **24**:254–62.

Sadovy, Y.J., Donaldson, T.J., Graham, T.R., McGilvray, F., Muldoon, G.J., Phillips, M.J., et al. (2003). *While Stocks Last: The Live Reef Food Fish Trade*. Asian Development Bank.

Sadovy, Y.J. and Vincent, A.C.J. (2002). Ecological issues and the trades in live reef fishes. In: P.F. Sale (ed.), *Coral Reef Fishes: Dynamics and Diversity in a Complex Ecosystem*. Academic Press, pp. 391–420.

Sakka, A., Legendre, L., Gosselin, M., Niquil, N. and Delesalle, B. (2002). Carbon budget of the planktonic food web in an atoll lagoon (Takapoto, French Polynesia). *Journal of Plankton Research* **24**:301–20.

Sale, P.F. (1977). Maintenance of high diversity in coral reef fish communities. *The American Naturalist* **111**:337–59.

Sale, P.F. (2002). *Coral Reef Fishes: Dynamics and Diversity in a Complex Ecosystem.* Academic Press.

Sale, P.F. (2008). Management of coral reefs: Where we have gone wrong and what we can do about it. *Marine Pollution Bulletin* **56**:805–9.

Salih, A., Larkum, A., Cox, G., Kuhl, M. and Hoegh-Guldberg, O. (2000). Fluorescent pigments in corals are photoprotective. *Nature* **408**:850–3.

Sandberg, P.A. (1983). An oscillating trend in Phanerozoic non-skeletal carbonate mineralogy. *Nature* **305**:19–22.

Sansone, F.J., Tribble, G.W., Andrews, C.C. and Chanton, J.P. (1990). Anaerobic diagenesis within Recent, Pleistocene and Eocene marine carbonate frameworks. *Sedimentology* **37**:997–1009.

Santos, S.R., Taylor, D.J., Kinzie, R.A., III, Hidaka, M., Sakai, K. and Coffroth, M.A. (2002). Molecular phylogeny of symbiotic dinoflagellates inferred from partial chloroplast large subunit (23S)-rDNA sequences. *Molecular Phylogenetics and Evolution* **23**:97–111.

Sara, M., Bavestrello, G., Cattaneo-Vietti, R. and Cerrano, C. (1998). Endosymbiosis in sponges: Relevance for epigenesis and evolution. *Symbiosis* **25**:57–70.

Saxby, T., Dennison, W.C. and Hoegh-Guldberg, O. (2003). Photosynthetic responses of the coral *Montipora digitata* to cold temperature stress. *Marine Ecology Progress Series* **248**:85–97.

Scheffers, S.R., Nieuwland, G., Bak, R.P.M. and van Duyl, F.C. (2004). Removal of bacteria and nutrient dynamics within the coral reef framework of Curaçao (Netherlands Antilles). *Coral Reefs* **23**:413–22.

Schiel, D.R., Kingsford, M.J. and Choat, J.H. (1986). Depth distribution and abundance of benthic organisms and fishes at the subtropical Kermadec Islands. *New Zealand Journal Marine Freshwater Research* **20**:521–35.

Schill, S.R., Raber, G.T., Roberts, J.J., Treml, E.A., Brenner, J. and Halpin, P.N. (2015). No reef is an island: Integrating coral reef connectivity data into the design of regional-scale marine protected area networks. *PLOS ONE* **10**:e0144199.

Schleyer, M.H. (2000). South African coral communities. In: T.R. McClanahan, C.R.C. Sheppard and D.O. Obura (eds), *Coral Reefs of the Indian Ocean: Their Ecology and Conservation.* Oxford University Press, pp. 83–105.

Schleyer, M.H. and Celliers, L. (2005). Modelling reef zonation in the Greater St Lucia Wetland Park, South Africa. *Estuarine and Coastal Shelf Science* **63**:373–84.

Schleyer, M.H., Kruger, A. and Celliers, L. (2008). Long-term community changes on high-latitude coral reefs in the Greater St Lucia Wetland Park, South Africa. *Marine Pollution Bulletin* **56**:493–502.

Schleyer, M.H. and Tomalin, B.J. (2000). Ecotourism and damage on South African coral reefs with an assessment of their carrying capacity. *Bulletin Marine Science* **67**:1025–42.

Schlichter, D., Kampmann, H. and Conrady, S. (1997). Trophic potential and photoecology of endolithic algae living within coral skeletons. *Marine Ecology* **18**:299–317.

Schmidt, H.E. (1973). The vertical distribution and diurnal migration of some zooplankton in the Bay of Eilat (Red Sea). *Helgoland Marine Research* **24**:333–40.

Schoenberg, C.H.L and Wilkinson, C.R. (2001). Induced colonization of corals by a clionid bioeroding sponge. *Coral Reefs* **20**:69–76.

Schoenberg, D.A. and Trench, R.K. (1980a). Genetic variation in *Symbiodinium* (= *Gymnodinium*) *microadriaticum* Freudenthal, and specificity in its symbiosis with marine invertebrates. I. Isoenzyme and soluble protein patterns of axenic cultures of *S. microadriaticum*. *Proceedings of the Royal Society of London. Series B, Biological Sciences* 207:405–27.

Schoenberg, D.A. and Trench, R.K. (1980b). Genetic variation in *Symbiodinium* (= *Gymnodinium*) *microadriaticum* Freudenthal, and specificity in its symbiosis with marine invertebrates. II. Morphological variation in *S. microadriaticum*. *Proceedings of the Royal Society of London. Series B, Biological Sciences* 207:429–44.

Schoenberg, D.A. and Trench, R.K. (1980c). Genetic variation in *Symbiodinium* (= *Gymnodinium*) *microadriaticum* Freudenthal, and specificity in its symbiosis with marine invertebrates. III. Specificity and infectivity of *Symbiodinium microadriaticum*. *Proceedings of the Royal Society of London. Series B, Biological Sciences* 207:445–60.

Schofield, P.J. (2009). Geographic extent and chronology of the invasion of nonnative lionfish (*Pterois volitans* [Linnaeus 1758] and *P. miles* [Bennett 1828]) in the Western North Atlantic and Caribbean Sea. *Aquatic Invasions* 4:473–9.

Schonberg, C.H.L. and Loh, W.K. (2005). Molecular identity of the unique symbiotic dinoflagellates found in the bioeroding demosponge *Cliona orientalis*. *Marine Ecology Progress Series* 299:157–66.

Schreiber, U (2004). Pulse-amplitude-modulation (PAM) fluorometry and saturation pulse method: An overview. In: G.C. Papageorgiou and Govindjee (eds), *Chlorophyll Fluorescence: A Signature of Photosynthesis*. Kluwer Academic Publishers, pp. 279–319.

Schubert, R., Schellnhuber, H.-J., Buchmann, N., Epiney, A., Grießhammer, R., Kulessa, M., et al. (2007). *The Future Oceans: Warming up, Rising High, Turning Sour: Special Report*. German Advisory Council on Global Change. Available at <http://www.wbgu.de/fileadmin/user_upload/wbgu.de/templates/dateien/veroeffentlichungen/sondergutachten/sn2006/wbgu_sn2006_en.pdf>.

Schuhmacher, H. (1992). Impact of some corallivorous snails on stony corals in the Red Sea. *Proceedings of the 7th International Coral Reef Symposium* 2:840–6.

Schuhmacher, H. (1997). Soft corals as reef builders. *Proceedings of the 8th International Coral Reef Symposium* 1:499–502.

Schuhmacher, H. (2002). Use of artificial reefs with special reference to the rehabilitation of coral reefs. *Bonner Zoologische Monographien* 50:81–108.

Scoffin, T.P. (1993). The geological effects of hurricanes on coral reefs and the interpretation of storm deposits. *Coral Reefs* 12:203–21.

Scott, F.J., Wetherbee, R. and Kraft, G.T. (1984). The morphology and development of some prominently stalked southern Australian Halymeniaceae (Cryptonemiales, Rhodophyta). II. The sponge-associated genera *Thamnoclonium* Kuetzing and *Codiophyllum* Gray. *Journal of Phycology* 20:286–95.

Seveso, D., Montano, S., Strona, G., Orlandi, I., Galli, P. and Vai, M. (2013). Exploring the effect of salinity changes on the levels of Hsp60 in the tropical coral *Seriatopora caliendrum*. *Marine Environmental Research* 90:96–103.

Seymour, J.R., Patten, N., Bourne, D.G. and Mitchell, J.G. (2005). Spatial dynamics of virus-like particles and heterotrophic bacteria within a shallow coral reef system. *Marine Ecology Progress Series* 288:1–8.

Shashar, N., Banaszak, A.T., Lesser, M.P. and Amrami, D. (1997). Coral endolithic algae: Life in a protected environment. *Pacific Science* 51:167–73.

Sheppard, C.R.C. (1979). Interspecific aggression between reef corals with reference to their distribution. *Marine Ecology Progress Series* 1:237–47.

Sheppard, C.R.C. (1981). The groove and spur structures of Chagos atolls and their coral zonation. *Estuarine Coastal Shelf Science* **12**:549–60.

Sheppard, C.R.C. (1985). Unoccupied substrate in the central Great Barrier Reef: Role of coral interactions. *Marine Ecology Progress Series* **25**:259–68.

Sheppard, C.R.C. (1988). Similar trends, different causes: Responses of corals to stressed environments in Arabian seas. *Proceedings of the 6th International Coral Reef Symposium* **3**:297–302.

Sheppard, C.R.C. (1995). The shifting baseline syndrome. *Marine Pollution Bulletin* **30**:766–7.

Sheppard, C.R.C. (1996). Making a mark in the scientific aid business. *Marine Pollution Bulletin.* **32**:692–3.

Sheppard, C.R.C. (2000). Coral reefs of the Western Indian Ocean: An overview. In: T.R. McClanahan, C.R.C. Sheppard and D.O. Obura (eds), *Coral Reefs of the Western Indian Ocean: Their Ecology and Conservation.* Oxford University Press, pp. 3–38.

Sheppard, C.R.C. (2003a). Environmental carpetbaggers. *Marine Pollution Bulletin* **46**:1–2.

Sheppard, C.R.C. (2003b). Predicted recurrences of mass coral mortality in the Indian Ocean. *Nature* **425**:294–7.

Sheppard, C.R.C. (2003c). Rates and totals: Population pressures on habitats. *Marine Pollution Bulletin* **46**:1517–18.

Sheppard, C.R.C. (2007). Extinction muddles and swindles. *Marine Pollution Bulletin* **54**:1309–10.

Sheppard, C.R.C. (2008). Coral reefs. In: *Bahrain Marine Habitat Atlas.* Geomatec.

Sheppard, C.R.C. (2016). Coral reefs in the Gulf are mostly dead now, but can we do anything about it? *Marine Pollution Bulletin* **105**:593–8.

Sheppard, C.R.C.,Dixon, D.J.,Gourlay, M., Sheppard, A.L.S. and Payet, R. (2005). Coral mortality increases wave energy reaching shores protected by reef flats: Examples from the Seychelles. *Estuarine, Coastal and Shelf Science* **64**:223–34.

Sheppard, C.R.C.,Harris, A. and Sheppard, A.L.S. (2008). Archipelago-wide coral recovery patterns since 1998 in Chagos, central Indian Ocean. *Marine Ecology Progress Series* **362**:109–17.

Sheppard, C.R.C. and Loughland, R. (2002). Coral mortality, recovery and temperature patterns in the extreme tropical conditions of the Arabian Gulf. *Aquatic Ecosystem Health and Management* **5**:395–402.

Sheppard, C.R.C., Matheson, K., Bythell, J.C., Murphy, P., Blair-Myers, C. and Blake, B. (1995). Habitat mapping in the Caribbean for management and conservation: Use and assessment of aerial photography. *Aquatic Conservation: Marine and Freshwater Ecosystems* **5**:277–98.

Sheppard, C.R.C. and Obura, D. (2005). Corals and reefs of Cosmoledo and Aldabra atolls: Extent of damage, assemblage shifts and recovery following the severe mortality of 1998. *Journal Natural History* **39**:103–21.

Sheppard, C.R.C., Price, A.R.G. and Roberts, C.J. (1992). *Marine Ecology of the Arabian Area: Patterns and Processes in Extreme Tropical Environments.* Academic Press.

Sheppard, C.R.C. and Rioja-Nieto, R. (2005). Sea surface temperature 1871–2099 in 38 cells in the Caribbean region. *Marine Environmental Research* **60**:389–96.

Sheppard, C.R.C. and Sheppard, A.L.S. (1985). Reefs and coral assemblages of Saudi Arabia. 1. The central Red Sea at Yanbu al Sanaiyah. *Fauna of Saudi Arabia* **7**:17–36.

Sheppard, C.R.C., Spalding, M., Bradshaw, C. and Wilson, S. (2002). Erosion vs. recovery of coral reefs after 1998 El Niño: Chagos reefs, Indian Ocean. *Ambio* **31**:40–8.

Shinn, E.A., Smith, G.W., Prospero, J.M., Betzer, P., Hayes, M.L., Garrison, V., et al. (2000). African dust and the demise of Caribbean coral reefs. *Geophysical Research Letters* **27**:3029–32.

Shoguchi, E., Shinzato, C., Kawashima, T., Gyoja, F., Mungpakdee, S., Koyanagi, R., et al. (2013). Draft assembly of the *Symbiodinium minutum* nuclear genome reveals dinoflagellate gene structure. *Current Biology* **23**:1399–408.

Siebeck, U.E. (2004). Communication in coral reef fishes: The role of ultraviolet colour patterns for the territorial behaviour of *Pomacentrus amboinensis*. *Animal Behaviour* **68**:273–82.

Siebeck, U.E. and Marshall, N.J. (2001). Ocular media transmission of coral reef fish: Can coral reef fish see ultraviolet light? *Vision Research* **41**:133–49.

Siebeck, U.E., Wallis, G.M. and Litherland, L. (2008). Colour vision in coral reef fish. *Journal of Experimental Biology* **211**:354–60.

Silverstein, R.N., Cunning, R. and Baker, A.C. (2015). Change in algal symbiont communities after bleaching, not prior heat exposure, increases heat tolerance of reef corals. *Global Change Biology* **21**:236–49.

Simpson, S.D., Meekan, M., Montgomery, J., McCauley, R. and Jeffs, A. (2005). Homeward sound. *Science* **308**:221.

Smith, D.C. and Douglas, A.E. (1987). *The Biology of Symbiosis*. Edward Arnold.

Smith, D.J., Suggett, D.J. and Baker, N.R. (2005). Is photoinhibition of zooxanthellae photosynthesis the primary cause of thermal bleaching in corals? *Global Change Biology* **11**:1–11.

Smith, G.J. and Muscatine, L. (1999). Cell cycle of symbiotic dinoflagellates: Variation in G(1) phase-duration with anemone nutritional status and macronutrient supply in the *Aiptasia pulchella–Symbiodinium pulchrorum* symbiosis. *Marine Biology* **134**:405–18.

Smith, G.W., Ives, L.D., Nagelkerken, I.A. and Ritchie, K.B. (1996). Caribbean sea fan mortalities. *Nature* **383**:487.

Smith, G.W. and Weil, E. (2004). Aspergillosis of gorgonians. In: E. Rosenberg and Y. Loya (eds), *Coral Health and Disease*. Springer, pp. 279–87.

Smith, S.H. (1988). Cruise ships: A serious threat to coral reefs and associated organisms. *Ocean and Shoreline Management* **11**:231–48.

Smithers, S.G. and Woodroffe, C.D. (2000). Microatolls as sea-level indicators on a mid-ocean atoll. *Marine Geology* **168**:61–78.

Smithers, S.G. and Woodroffe, C.D. (2001). Coral microatolls and 20th century sea level in the eastern Indian Ocean. *Earth and Planetary Science Letters* **191**:173–84.

Sneed, J.M., Ritson-Williams, R. and Paul, V.J. (2015). Crustose coralline algal species host distinct bacterial assemblages on their surfaces. *ISME Journal* **9**:2527–36.

Soetart, K., Mohn, C., Rengstorf, A., Grehan, A. and van Oevelen, D. (2016). Ecosystem engineering creates a direct nutritional link between 600-m deep cold-water coral mounds and surface productivity. *Scientific Reports* **6**:35057.

Sorokin, Y.I. (1993). *Coral Reef Ecology*. Springer.

Souter, D.W. and Linden, O. (2000). The health and future of coral reef systems. *Ocean and Coastal Management* **43**:657–88.

Spalding, M.D., Ravilious, C. and Green, E.P. (2001). *World Atlas of Coral Reefs*. University of California Press.

Stafford-Smith, M.G. (1992). Mortality of the hard coral *Leptoria phrygia* under persistent sediment influx. *Proceedings of the 7th International Coral Reef Symposium* 1:289–99.

Stafford-Smith, M.G. (1993). Sediment-rejection efficiency of 22 species of Australian scleractinian corals. *Marine Biology* 115:229–43.

Stafford-Smith, M.G. and Ormond, R.F.G. (1992). Sediment-rejection mechanisms of 42 species of Australian scleractinian corals. *Australian Journal Marine Freshwater Research* 43:683–705.

Starzak, D.E., Quinnell, R.G., Nitschke, M.R. and Davy, S.K. (2014). The influence of symbiont type on photosynthetic carbon flux in a model cnidarian–dinoflagellate symbiosis. *Marine Biology* 161:711–24.

Stat, M. Carter, D. and Hoegh-Guldberg, O. (2006). The evolutionary history of *Symbiodinium* and scleractinian hosts: Symbiosis, diversity, and the effect of climate change. *Perspectives in Plant Ecology, Evolution and Systematics* 8:23–43.

Stehli, F.G. and Wells, J.W. (1971). Diversity and age patterns in hermatypic corals. *Systematic Zoology* 20:115–18.

Steneck, R.S. (1998). Human influences on coastal ecosystems: Does overfishing create trophic cascades? *Trends in Ecology and Evolution* 13:429–30.

Steneck, R.S. and Dethier, M.N. (1994). A functional group approach to the structure of algal-dominated communities. *Oikos* 69:476–98.

Steuber, T. (2002). Plate tectonic control on the evolution of Cretaceous platform-carbonate production. *Geology* 30:259–62.

Stevenson, C., Katz, L.S., Micheli, F., Block, B., Heiman, K.W., Perle, C., et al. (2007). High apex predator biomass on remote Pacific islands. *Coral Reefs* 26:47–51.

Stoddart, D.R. (ed.). (2007). Tsunamis and coral reefs. *Atoll Research Bulletin* 544:1–163.

Streamer, M., McNeil, Y.R. and Yellowlees, D. (1993). Photosynthetic carbon dioxide fixation in zooxanthellae. *Marine Biology* 115:195–8.

Streit, R.P., Hoey, A.S. and Bellwood, D.R. (2015). Feeding characteristics reveal functional distinctions among browsing herbivorous fishes on coral reefs. *Coral Reefs* 34:1037–47.

Suhayda, J.N. and Roberts, H.H. (1977). Wave action and sediment transport on fringing reefs. *Proceedings of the 3rd International Coral Reef Symposium* 2:65–70.

Sutherland, K.P. and Ritchie, K.B. (2004). White pox disease of the Caribbean Elkhorn coral, *Acropora palmata*. In: E. Rosenberg and Y. Loya (eds), *Coral Health and Disease*. Springer, pp. 289–300.

Sutton, D.C. and Hoegh-Guldberg, O. (1990). Host–zooxanthella interactions in four temperate marine invertebrate symbioses: Assessment of effect of host extracts on symbionts. *Biological Bulletin* 178:175–86.

Sweet, M.J. and Sere, M.G. (2016). Ciliate communities consistently associated with coral diseases. *Journal of Sea Research* 113:119–31.

Sweetman, A.K., Thurber, A.R., Smith, C.R., Levin, L.A., Mora, C., Wei, C.-L., et al. (2017). Global climate change effects on deep seafloor ecosystems. *Elementa: Science of the Anthropocene* 5:4.

Tambutté, S., Holcomb, M., Ferrier-Pagès, C., Reynaud, S., Tambutté, É., Zoccola, D., et al. (2011). Coral biomineralization: From the gene to the environment. *Journal of Experimental Marine Biology Ecology* 408:58–78.

Tchernov, D., Gorbunov, M.Y., de Vargas, C., Yadav, S.N., Milligan, A.J., Haggblom, M., et al. (2004). Membrane lipids of symbiotic algae are diagnostic of sensitivity to thermal bleaching in corals. *Proceedings of the National Academy of Science of the United States of America* **101**:13531–5.

Teh, L.S.L., Teh, L.C.L. and Sumaila, U.R. (2013). A global estimate of the number of coral reef fishers. *PLOS ONE* **8**:e65397.

Thomas, T., Moitinho-Silva, L., Lurgi, M., Björk, J.R., Easson, C., Astudillo-García, C., et al. (2016). Diversity, structure and convergent evolution of the global sponge microbiome. *Nature Communications* **7**:11870.

Thompson, A.R. (2004). Habitat and mutualism affect the distribution and abundance of a shrimp-associated goby. *Marine Freshwater Research* **55**:105–13.

Thompson, J.R., Rivera, H.E., Closek, C.J. and Medina, M. (2015). Microbes in the coral holobiont: Partners through evolution, development, and ecological interactions. *Frontiers in Cellular and Infection Microbiology* **4**:176.

Thurber, R.L.V. and Correa, A.M.S. (2011). Viruses of reef-building scleractinian corals. *Journal of Experimental Marine Biology Ecology* **408**:102–13.

Titlyanov, E.A., Titlyanova, T.V., Leletkin, V.A., Tsukahara, J., vanWoesik, R. and Yamazato, K. (1996). Degradation of zooxanthellae and regulation of their density in hermatypic corals. *Marine Ecology Progress Series* **139**:167–78.

Tout, J., Jeffries, T.C., Webster, N.S., Stocker, R., Ralph, P.J. and Seymour, J.R. (2014). Variability in microbial community composition and function between different niches within a coral reef. *Microbial Ecology* **67**:540–52.

Trapido-Rosenthal, H., Zielke, S., Owen, R., Buxton, L., Boeing, B., Bhagooli, R., et al. (2005). Increased zooxanthellae nitric oxide synthase activity is associated with coral bleaching. *Biological Bulletin* **208**:3–6.

Trautman, D.A. and Hinde, R. (2001). Sponge/algal symbioses: A diversity of associations. In: J. Seckbach (ed.), *Symbiosis*. Kluwer Academic Publishers, pp. 521–37.

Tremblay, P., Maguer, J.F., Grover, R. and Ferrier-Pagès, C. (2015). Trophic dynamics of scleractinian corals: Stable isotope evidence. *Journal of Experimental Biology* **218**:1223–34.

Trenberth, K. (2005). Uncertainty in hurricanes and global warming. *Science* **308**:1753–4.

Trench, R.K. (1971). The physiology and biochemistry of zooxanthellae symbiotic with marine coelenterates. II. Liberation of fixed ^{14}C by zooxanthellae in vitro. *Proceedings of the Royal Society of London. Series B, Biological Sciences* **177**:237–50.

Trench, R.K. (1993). Microalgal–invertebrate symbioses: A review. *Endocytobiosis Cell Research* **9**:135–75.

Tribollet, A., Langdon, C., Golubic, S. and Atkinson, M. (2006). Endolithic microflora are major primary producers in dead carbonate substrates of Hawaiian coral reefs. *Journal of Phycology* **42**:292–303.

Tudhope, A.W. and Risk, M.J. (1985). Rate of dissolution of carbonate sediments by microboring organisms, Davies Reef, Australia. *Journal of Sedimentary Petrology* **55**:440–7.

Turon, X., Galera, J. and Uriz, M.J. (1997). Clearance rates and aquiferous systems in two sponges with contrasting life-history strategies. *Journal of Experimental Zoology* **278**:22–36.

Ullman, W.J. and Sandstrom, M.W. (1987). Dissolved nutrient fluxes from the nearshore sediments of Bowling Green Bay, Central Great Barrier Reef Lagoon (Australia). *Estuarine, Coastal and Shelf Science* **24**:289–303.

Unson, M.D. and Faulkner, D.J. (1993). Cyanobacterial symbiont biosynthesis of chlorinated metabolites from *Dysidea herbacea* (Porifera). *Experientia* **49**:349–53.

Unson, M.D., Holland, N.D. and Faulkner, D.J. (1994). A brominated secondary metabolite synthesized by the cyanobacterial symbiont of a marine sponge and accumulation of the crystalline metabolite in the sponge tissue. *Marine Biology* **119**:1–11.

Usher, K.M., Fromont, J., Sutton, D.C. and Toze, S. (2004). The biogeography and phylogeny of unicellular cyanobacterial symbionts in sponges from Australia and the Mediterranean. *Microbial Ecology* **48**:167–77.

Uthicke, S., Furnas, M.J. and Lønborg, C. (2014). Coral reefs on the edge? Carbon chemistry on inshore reefs of the Great Barrier Reef. *PLOS ONE* **9**:e109092.

Uthicke, S. and Klumpp, D.W. (1998). Microphytobenthos community production at a near-shore coral reef: Seasonal variation and response to ammonium recycled by holothurians. *Marine Ecology Progress Series* **169**:1–11.

Uthicke, S. and McGuire, K. (2007). Bacterial communities in Great Barrier Reef calcareous sediments: Contrasting 16S rDNA libraries from nearshore and outer shelf reefs. *Estuarine Coastal Shelf Science* **72**:188–200.

Uthicke, S., Schaffelke, B. and Byrne, M. (2009). A boom–bust phylum? Ecological and evolutionary consequences of density variations in echinoderms. *Ecological Monographs* **79**:3–24.

van Hooidonk, R., Maynard, J., Tamelander, J., Gove, J., Ahmadia, G., Raymundo, L., et al. (2016). Local-scale projections of coral reef futures and implications of the Paris Agreement *Scientific Reports* **6**:39666.

Van Woesik, R., De Vantier, L.M. and Glazebrook, J.S. (1995). Effects of Cyclone 'Joy' on nearshore coral communities of the Great Barrier Reef. *Marine Ecology Progress Series* **128**:261–70.

Van Woesik, R., Tomascik, T. and Blake, S. (1999). Coral assemblages and physico-chemical characteristics of the Whitsunday Islands: Evidence of recent community changes. *Marine Freshwater Research* **50**:427–40.

Vaughan, G.O. and Burt, J.A. (2016). The changing dynamics of coral reef science in Arabia. *Marine Pollution Bulletin* **105**:441–58.

Vega Thurber, R., Payet, J.P., Thurber, A.R. and Correa, A.M.S. (2017). Virus-host interactions and their roles in coral reef health and disease. *Nature Reviews Microbiology* **15**:205–16.

Venn, A.A., Wilson, M.A., Trapido-Rosenthal, H.G., Keely, B.J. and Douglas, A.E. (2006). The impact of coral bleaching on the pigment profile of the symbiotic alga, *Symbiodinium. Plant, Cell and Environment* **29**:2133–42.

Venn, A.A., Loram, J.E. and Douglas, A.E. (2008). Photosynthetic symbioses in animals. *Journal of Experimental Botany* **59**:1069–80.

Veron, J.E.N. (1993). *A Biogeographic Database of Hermatypic Corals*. Australian Institute of Marine Science Monograph Series 10. Australian Institute of Marine Science.

Veron, J.E.N. (1995). *Corals in Space and Time: The Biogeography and Evolution of the Scleractinia*. University of New South Wales Press.

Veron, J.E.N. (2000). *Corals of the World* (3 vols). Australian Institute Marine Sciences.

Veron, J.E.N. (2007). *A Reef in Time: The Great Barrier Reef From Beginning to End*. Harvard University Press.

Veron, J.E.N. (2008). Mass extinctions and ocean acidification: Biological constraints on geological dilemmas. *Coral Reefs* **27**:459–72.

Veron, J.E.N., Stafford-Smith, M., DeVantier, L. and Turak, E. (2015). Overview of distribution patterns of zooxanthellate Scleractinia. *Frontiers in Marine Science* **1**:81.

Vine, P.J. (1973). Crown of thorns (*Acanthaster planci*) plagues: The natural causes theory. *Atoll Research Bulletin* **166**:1–10.

Vogel, S. and LaBarbera, M. (1978). Simple flow tanks for research and teaching. *Bioscience* **28**:638–43.

Vogler, C., Benzie, J., Lessios, H., Barber, P. and Worhelde, G. (2008). A threat to coral reefs multiplied? Four species of crown-of-thorns starfish. *Biology Letters* **4**:696–9.

Wainwright, S.A. (1965). Reef communities visited by the South Red Sea Expedition, 1962. *Bulletin of Sea Fisheries Research Station, Israel* **38**:40–53.

Walker, N.D., Roberts, H.H., Rouse, L.J. and Huh, O.K. (1982). Thermal history of reef-associated environments during a record cold-air outbreak event. *Coral Reefs* **1**:83–7.

Wang, J.T. and Douglas, A.E. (1998). Nitrogen recycling or nitrogen conservation in an alga–invertebrate symbiosis? *Journal of Experimental Biology* **201**:2445–53.

Ware, J.R., Smith, S.V. and Reaka-Kudla, M.L. (1991). Coral reefs: Sources or sinks of atmospheric CO^2? *Coral Reefs* **11**:127–30.

Warner, M.E., Fitt, W.K. and Schmidt, G.W. (1999). Damage to photosystem II in symbiotic dinoflagellates: A determinant of coral bleaching. *Proceedings of the National Academy of Science of the United States of America* **96**:8007–12.

Watanabe, T., Fukuda, I., China, K. and Isa, Y. (2003). Molecular analyses of protein components of the organic matrix in the exoskeleton of two scleractinian coral species. *Comparative Biochemistry and Physiology Part B* **136**:767–74.

Webster, N.S., Luter, H.M., Soo, R.M., Botté, E.S., Simister, R.L., Abdo, D., et al. (2013). Same, same but different: Symbiotic bacterial associations in GBR sponges. *Frontiers in Microbiology* **3**:444.

Webster, N.S. and Taylor, M.W. (2012). Marine sponges and their microbial symbionts: Love and other relationships. *Environmental Microbiology* **14**:335–46.

Webster, P.J., Holland, G.J., Curry, J.A. and Chang, H.R. (2005). Changes in tropical cyclone number, duration and intensity in a warming environment. *Science* **309**:1844–6.

Wegley, L., Yu, Y., Breitbart, M., Casas, V., Kline, D.I. and Rohwer, F. (2004). Coral-associated archaea. *Marine Ecology Progress Series* **273**:89–96.

Weil, E. (2004). Coral reef diseases in the wider Caribbean. In: E. Rosenberg and Y. Loya (eds), *Coral Health and Disease*. Springer, pp. 35–68.

Weis, V.M. (1991). The induction of carbonic anhydrase in the symbiotic sea anemone *Aiptasia pulchella*. *Biological Bulletin* **180**:496–504.

Weis, V.M. (2008). Cellular mechanisms of cnidarian bleaching: Stress causes the collapse of symbiosis. *Journal of Experimental Biology* **211**:3059–66.

Weis, V.M., Davy, S.K., Hoegh-Guldberg, O., Rodriguez-Lanetty, M. and Pringle, J. (2008). Cell biology in model systems as the key to understanding corals. *Trends in Ecology and Evolution* **23**:369–76.

Wellington, G.M. and Victor, B.C. (1989). Planktonic larval duration of one hundred species of Pacific and Atlantic damselfishes (Pomacentridae). *Marine Biology* **101**:557–67.

Wellstead, J.R. (1840). *Travels to the City of the Caliphs, along the Shores of the Persian Gulf and Mediterranean. Including a tour of the Island of Socotra*, Vols 1 and 2. Henry Colburn.

Wiebe, W.J. (1985). Nitrogen dynamics on coral reefs. *Proceedings of the 5th International Coral Reef Symposium* **3**:401–6.

Wilcox, T.P. (1998). Large-subunit ribosomal RNA systematics of symbiotic dinoflagellates: Morphology does not recapitulate phylogeny. *Molecular Phylogenetics and Evolution*, **10**:436–48.

Wilkerson, F.P. and Kremer, P. (1992). DIN, DON and PO4 flux by a medusa with algal symbionts. *Marine Ecology Progress Series* **90**:237–50.

Wilkerson, F.P. and Trench, R.K. (1986). Uptake of dissolved inorganic nitrogen by the symbiotic clam *Tridacna gigas* and the coral *Acropora* sp. *Marine Biology* **93**:237–46.

Wilkinson, C.R. (1980). Cyanobacteria symbiotic in marine sponges. In: W. Schwemmler and H.E.A. Schenck (eds), *Endocytobiology, Endosymbiosis and Cell Biology*. De Gruyter, pp. 993–1002.

Wilkinson, C.R. (1983). Net primary productivity in coral reef sponges. *Science* **219**:410–12.

Wilkinson, C.R. (1984). Immunological evidence for the Precambrian origin of bacterial symbioses in marine sponges. *Proceedings of the Royal Society of London. Series B, Biological Sciences* **220**:509–17.

Wilkinson, C.R. (1987). Interocean differences in size and nutrition of coral reef sponge populations. *Science* **236**:1654–7.

Wilkinson, C.R. (1996). Global change and coral reefs: Impacts on reefs, economies and human cultures. *Global Change Biology* **2**:547–58.

Wilkinson, C.R. (1998). The role of sponges in coral reefs. In: C. Lévi (ed.), *Sponges of the New Caledonian Lagoon*. Orstom, pp. 55–60.

Wilkinson, C.R. (ed.). (2004). *Status of Coral Reefs of the World* (2 vols). Global Coral Reef Monitoring Network.

Wilkinson, C.R. and Fay, P. (1979). Nitrogen fixation in coral reef sponges with symbiotic cyanobacteria. *Nature* **279**:527–9.

Wilkinson, C.R., Lindén, O., Cesar, H.S.J., Hodgeson, G., Rubens, J. and Strong, A.E. (1999). Ecological and socioeconomic impacts of the 1998 coral mortality in the Indian Ocean: An ENSO impact and a warning of future change? *Ambio* **28**:188–96.

Wilkinson, C.R. and Sammarco, P.W. (1983). Effects of fish grazing and damselfish territoriality on coral reef algae. II. Nitrogen fixation. *Marine Ecology Progress Series* **13**:15–19.

Wilkinson, C.R. and Vacelet, J. (1979). Transplantation of marine sponges to different conditions of light and current. *Journal of Experimental Marine Biology Ecology* **37**:91–104.

Wilkinson, C.R., Williams, D.M., Sammarco, P.W., Hogg, R.W. and Trott, L.A. (1984). Rates of nitrogen-fixation on coral reefs across the continental-shelf of the Central Great Barrier Reef. *Marine Biology* **80**:255–62.

Wilkinson, S.P., Fisher, P.L., van Oppen, M.J. H. and Davy, S.K. (2015). Intra-genomic variation in symbiotic dinoflagellates: Recent divergence or recombination between lineages? *BMC Evolutionary Biology* **15**:46.

Williams, A.J., Loeun, J., Nicol, S.J., Chavance, P., Ducrocq, M., Harley, S.J., et al. (2013). Population biology and vulnerability to fishing of deep-water Eteline snappers. *Journal of Applied Ichthyology* **29**: 395–403.

Williams, G.J., Knapp, I.S., Maragos, J.E. and Davy, S.K. (2010). Modeling patterns of coral bleaching at a remote Central Pacific atoll. *Marine Pollution Bulletin* **60**:1467–76.

Williams, G.J., Price, N.N., Ushijima, B., Aeby, G.S., Callahan, S., Davy, S.K., et al. (2014). Ocean warming and acidification have complex interactive effects on the dynamics of a marine fungal disease. *Proceedings of the Royal Society B: Biological Sciences* **281**:20133068.

Williamson, D.H., Harrison, H.B., Almany, G.R., Berumen, M.L., Bode, M., Bonin, M.C., et al. (2016). Large-scale, multidirectional larval connectivity among coral reef fish populations in the Great Barrier Reef Marine Park. *Molecular Ecology* **25**:6039–54.

Willis, B.L., Page, C.A. and Dinsdale, E. (2004). Coral disease on the Great Barrier Reef. In: E. Rosenberg and Y. Loya (eds), *Coral Health and Disease*. Springer, pp. 69–104.

Wilson, S.K., Bellwood, D.R., Choat, J.H. and Furnas, M.J. (2003). Detritus in the epilithic algal matrix and its use by coral reef fishes. *Oceanography and Marine Biology* **41**:279–309.

Wilson, S.K., Fisher, R., Pratchett, M.S., Graham, N.A.J., Dulvy, N.K., Turner, R.A., et al. (2008). Exploitation and habitat degradation as agents of change within coral reef fish communities. *Global Change Biology* **14**:2796–809.

Wilson, S.K., Graham, N.A.J., Pratchett, M.S., Jones, G.P. and Polunin, N.V.C. (2006). Multiple disturbances and the global degradation of coral reefs: Are reef fishes at risk or resilient? *Global Change Biology* **12**:2220–34.

Wolanski, E. (2001). *Oceanographic Processes of Coral Reefs: Physical and Biological Links in the Great Barrier Reef*. CRC Press.

Wolanski, E., Drew, E., Abel, K.M. and O'Brien, J. (1988). Tidal jets, nutrient upwelling and their influence on the productivity of the alga *Halimeda* in the Ribbon Reefs, Great Barrier Reef. *Estuarine Coastal and Shelf Science* **26**:169–201.

Wolanski, E. and Gibbs, R.J. (1995). Flocculation of suspended sediment in the Fly River Estuary, Papua New Guinea. *Journal Coastal Research* **11**:754–62.

Wolanski, E. and Hamner, W.M. (1988). Topographically controlled fronts in the ocean and their biological influence. *Science* **241**:177–81.

Wolanski, E. and Pickard, G.L. (1983). Upwelling by internal tides and Kelvin waves at the continental shelf break on the Great Barrier Reef. *Australian Journal of Marine Freshwater Research* **34**:65–80.

Wolanski, E. and Spagnol, S. (2000). Pollution by mud of Great Barrier Reef coastal waters. *Journal of Coastal Research* **16**:1151–6.

Wolfe, K., Graba-Landry, A., Dworjanyn, S.A. and Byrne, M. (2015). Larval starvation to satiation: Influence of nutrient regime on the success of *Acanthaster planci*. *PLOS ONE* **10**:e0122010.

Wommack, K.E. and Colwell, R.R. (2000). Virioplankton: Viruses in aquatic ecosystems. *Microbiology and Molecular Biology Reviews* **64**:69–114.

Wood-Charlson, E.M., Weynberg, K.D., Suttle, C.A., Roux, S. and van Oppen, M.J.H. (2015). Metagenomic characterization of viral communities in corals: Mining biological signal from methodological noise. *Environmental microbiology* **17**:3440–9.

Wooldridge, S.A. and Brodie, J.E. (2015). Environmental triggers for primary outbreaks of crown-of-thorns starfish on the Great Barrier Reef, Australia. *Marine Pollution Bulletin* **101**:805–15.

Wright, V.P. and Burgess, P.M. (2005). The carbonate factory continuum, facies mosaics and microfacies: An appraisal of some of the key concepts underpinning carbonate sedimentology. *Facies* **51**:17–23.

Wulff, J.L. (2006). Ecological interactions of marine sponges. *Canadian Journal of Zoology* **84**:146–66.

Wulff, J.L. and Buss, L.W. (1979). Do sponges help hold coral reefs together? *Nature* **281**:474–5.

Yahel, G., Post, A.F., Fabricius, K.E., Marie, D., Vaulot, D. and Genin, A. (1998). Phytoplankton distribution and grazing near coral reefs. *Limnology and Oceanography* **43**:551–63.

Yahel, G., Sharp, J.H., Marie, D., Häse, C. and Genin, A. (2003). In situ feeding and element removal in the symbiont-bearing sponge *Theonella swinhoei*: Bulk DOC is the major source for carbon. *Limnology and Oceanography* **48**:141–9.

Yahel, R., Yahel, G., Berman, T., Jaffe, J.S. and Genin, A. (2005). Diel pattern with abrupt crepuscular changes of zooplankton over a coral reef. *Limnology and Oceanography* **50**:930–44.

Yahel, R., Yahel, G. and Genin, A. (2005). Near-bottom depletion of zooplankton over coral reefs: I: Diurnal dynamics and size distribution. *Coral Reefs* **24**:75–85.

Yamano, H., Kayanne, H., Yonekura, N. and Nakamura, H. (1998). Water circulation in a fringing reef located in a monsoon area: Kabira Reef, Ishigaki Island, southwest Japan. *Coral Reefs* **17**:89–99.

Yancey, P.H., Heppenstall, M., Ly, S., Andrell, R.M., Gates, R.D., Carter, V.L., et al. (2010). Betaines and dimethylsulfoniopropionate as major osmolytes in cnidaria with endosymbiotic dinoflagellates. *Physiological and Biochemical Zoology* **83**:167–73.

Yoshioka, R.M., Kim, C.J.S., Tracy, A.M., Most, R. and Harvell, C.D. (2016). Linking sewage pollution and water quality to spatial patterns of *Porites lobata* growth anomalies in Puako, Hawaii. *Marine Pollution Bulletin* **104**:313–21.

Zahn, L.P. and Bolton, L. (1985). The distribution, abundance and ecology of the blue coral *Heliopora coerulea* (Pallas) in the Pacific. *Coral Reefs* **4**:125–34.

Index